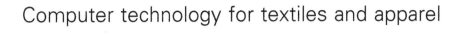

Computer technology for textiles and apparel

The Textile Institute and Woodhead Publishing

The Textile Institute is a unique organisation in textiles, clothing and footwear. Incorporated in England by a Royal Charter granted in 1925, the Institute has individual and corporate members in over 90 countries. The aim of the Institute is to facilitate learning, recognise achievement, reward excellence and disseminate information within the global textiles, clothing and footwear industries.

Historically, The Textile Institute has published books of interest to its members and the textile industry. To maintain this policy, the Institute has entered into partnership with Woodhead Publishing Limited to ensure that Institute members and the textile industry continue to have access to high calibre titles on textile science and technology.

Most Woodhead titles on textiles are now published in collaboration with The Textile Institute. Through this arrangement, the Institute provides an Editorial Board which advises Woodhead on appropriate titles for future publication and suggests possible editors and authors for these books. Each book published under this arrangement carries the Institute's logo.

Woodhead books published in collaboration with The Textile Institute are offered to Textile Institute members at a substantial discount. These books, together with those published by The Textile Institute that are still in print, are offered on the Woodhead website at www.woodheadpublishing.com. Textile Institute books still in print are also available directly from the Institute's website at www.textileinstitute-books.com.

A list of Woodhead books on textile science and technology, most of which have been published in collaboration with The Textile Institute, can be found towards the end of the contents pages.

Woodhead Publishing Series in Textiles: Number 121

Computer technology for textiles and apparel

Edited by
Jinlian Hu

The Textile Institute

WOODHEAD
PUBLISHING

Oxford   Cambridge   Philadelphia   New Delhi

Published by Woodhead Publishing Limited in association with The Textile Institute
Woodhead Publishing Limited,
80 High Street, Sawston, Cambridge CB22 3HJ, UK
www.woodheadpublishing.com

Woodhead Publishing, 1518 Walnut Street, Suite 1100, Philadelphia,
PA 19102-3406, USA

Woodhead Publishing India Private Limited, G-2, Vardaan House,
7/28 Ansari Road, Daryaganj, New Delhi – 110002, India
www.woodheadpublishingindia.com

First published 2011, Woodhead Publishing Limited
© Woodhead Publishing Limited, 2011
The authors have asserted their moral rights.

British Library Cataloguing in Publication Data
A catalogue record for this book is available from the British Library.

Library of Congress Control Number: 2011934552

ISBN 978-0-08-101703-6 (print)
ISBN 978-0-85709-360-8 (online)
ISSN 2042-0803 Woodhead Publishing Series in Textiles (print)
ISSN 2042-0811 Woodhead Publishing Series in Textiles (online)

The publisher's policy is to use permanent paper from mills that operate a sustainable forestry policy, and which has been manufactured from pulp which is processed using acid-free and elemental chlorine-free practices. Furthermore, the publisher ensures that the text paper and cover board used have met acceptable environmental accreditation standards.

Typeset by Toppan Best-set Premedia Limited, Hong Kong

Contents

Contributor contact details		*xi*
Woodhead Publishing Series in Textiles		*xv*
Introduction		*xxiii*

Part I Computer-based technology for textile materials   1

1 Digital technology for yarn structure and appearance analysis   3
B. G. Xu, The Hong Kong Polytechnic University, Hong Kong

1.1	Introduction	3
1.2	Measurement of yarn evenness	4
1.3	Analysis of yarn hairiness	5
1.4	Measurement of yarn twist	9
1.5	Recognition of yarn snarl	11
1.6	Analysis of yarn blend	14
1.7	Grading of yarn appearance	16
1.8	Future trends	19
1.9	Conclusions	20
1.10	Acknowledgement	20
1.11	References	20

2 Digital-based technology for fabric structure analysis   23
B. Xin, Shanghai University of Engineering Science, China and J. Hu, G. Baciu and X. Yu, The Hong Kong Polytechnic University, Hong Kong

2.1	Introduction	23
2.2	Background and literature review	24
2.3	The digital system for weave pattern recognition	27

2.4	Theoretical background for weave pattern analysis	29
2.5	Methodology for active grid model (AGM) construction and weave pattern extraction	35
2.6	Conclusions	42
2.7	References	42
3	Computer vision-based fabric defect analysis and measurement	45
	A. KUMAR, The Hong Kong Polytechnic University, Hong Kong	
3.1	Introduction	45
3.2	Fabric inspection for quality assurance	46
3.3	Fabric defect detection methods	49
3.4	Fabric defect classification	52
3.5	Fabric properties and color measurement using image analysis	59
3.6	Conclusions and future trends	60
3.7	Acknowledgments	61
3.8	References	62
Part II	Modelling and simulation of textiles and garments	67
4	Key techniques for 3D garment design	69
	Y. ZHONG, Donghua University, China	
4.1	Introduction	69
4.2	Sketch-based garment design	70
4.3	Surface flattening for virtual garments	73
4.4	Online garment-shopping system: problems and solutions	80
4.5	Challenges and future trends	90
4.6	Sources of further information	91
4.7	References	91
5	Modelling and simulation of fibrous yarn materials	93
	X. CHEN, University of Manchester, UK	
5.1	Introduction	93
5.2	Modelling and simulation of yarns	94
5.3	Weave modelling	103
5.4	Geometrical modelling of woven fabrics	105
5.5	Finite element (FE) modelling of woven fabrics	113
5.6	Future development of textile modelling	119
5.7	Acknowledgements	120
5.8	References and further reading	120

6   Digital technology and modeling for fabric structures   122
G. BACIU, E. ZHENG and J. HU, The Hong Kong Polytechnic
University, Hong Kong
6.1   Introduction   122
6.2   Background on woven fabric structure   123
6.3   Fabric geometry structure models   124
6.4   Fabric weave pattern   126
6.5   Description and classification of regular patterns   126
6.6   Description and classification of irregular patterns   128
6.7   Weave pattern and fabric geometry surface
appearance   131
6.8   Experimental pattern analysis   131
6.9   Methodology   132
6.10   Results and discussion   135
6.11   Acknowledgment   144
6.12   References   144

7   Modeling ballistic impact on textile materials   146
M. S. RISBY, National Defence University, Malaysia and
A. M. S. HAMOUDA, Qatar University, Qatar
7.1   Introduction   146
7.2   Computational aspects   150
7.3   Numerical modeling of single and multiple layer fabric   160
7.4   Conclusions   166
7.5   References and further reading   168

8   Modeling and simulation techniques for garments   173
S.-K. WONG, National Chiao Tung University, Taiwan
8.1   Introduction   173
8.2   Model development   174
8.3   Computer graphics techniques for garment structure
and appearance   178
8.4   Rendering of garment appearance and model
demonstration for garments   183
8.5   Considerations for real-time applications   188
8.6   Advanced modeling techniques   191
8.7   Future developments in simulating garment materials   192
8.8   Conclusions and sources of further information and
advice   193
8.9   References   195

Part III Computer-based technology for apparel   **201**

**9   Human interaction with computers and its use in the
          textile apparel industry**   **203**
          G. BACIU and S. LIANG, The Hong Kong Polytechnic
          University, Hong Kong
9.1   Introduction   203
9.2   Principles of human computer interaction (HCI)   204
9.3   Methods for improving human interaction with
          computers for textile purposes   207
9.4   Future trends   211
9.5   Conclusions   217
9.6   Acknowledgment   217
9.7   References   217

**10   3D body scanning: Generation Y body perception
          and virtual visualization**   **219**
          M.-E. FAUST, Philadelphia University, USA and
          S. CARRIER, University of Quebec at Montreal, Canada
10.1   Introduction   219
10.2   Literature review   221
10.3   Methodology   231
10.4   Results and findings   233
10.5   Conclusions and recommendations   239
10.6   Limitations   239
10.7   Future studies   240
10.8   References   240

**11   Computer technology from a textile designer's
          perspective**   **245**
          H. UJIIE, Philadelphia University, USA
11.1   Introduction   245
11.2   Role of computer technology in textile design   247
11.3   Main computer technologies in textile design   248
11.4   Benefits and limitations of computers for textile design   254
11.5   Future trends   256
11.6   Sources of further information and advice   257
11.7   References   257

12   Digital printing technology for textiles and apparel   **259**
          D. J. TYLER, Manchester Metropolitan University, UK
12.1   Introduction   259
12.2   Review of digital printing technology   260

12.3	Global developments in digital printing technology	263
12.4	Colour technology and colour management	266
12.5	Three stages of computing for digital printing	272
12.6	Future trends	277
12.7	Conclusions	279
12.8	Sources of further information and advice	279
12.9	Acknowledgements	280
12.10	References	280

13	**Approaches to teaching computer-aided design (CAD) to fashion and textiles students**	**283**
	P. C. L. HUI, The Hong Kong Polytechnic University, Hong Kong	
13.1	Introduction	283
13.2	Review of approaches to teaching computer-aided design (CAD)	283
13.3	Challenges presented by each approach	287
13.4	Case study	289
13.5	Areas for improvement in teaching computer-aided design (CAD)	293
13.6	Conclusions	294
13.7	References	294

14	**Three-dimensional (3D) technologies for apparel and textile design**	**296**
	C. L. ISTOOK, North Carolina State University, USA, E. A. NEWCOMB, North Carolina Agricultural & Technical State University, USA and H. LIM, Konkuk University, South Korea	
14.1	Introduction	296
14.2	Applications of three-dimensional (3D) human body modeling	297
14.3	Technologies of human body modeling in three-dimensions (3D)	299
14.4	Development of the body surface	304
14.5	Animation	308
14.6	Generic vs individualized body models	310
14.7	Virtual try-on technologies	313
14.8	Conclusions	319
14.9	References	320

15   Integrated digital processes for design and
        development of apparel   326
        T. A. M. LAMAR, North Carolina State University, USA
15.1   Introduction   326
15.2   Conventional design, development and production
        processes for apparel   327
15.3   Simultaneous design of textile and garment utilizing
        digital technology   332
15.4   Integrated processes in practice   343
15.5   Role of computer-aided design (CAD) and visualization
        technologies in integrated textile product design   344
15.6   The future of integrated digital apparel design and
        development processes   345
15.7   Conclusions   346
15.8   Sources of further information   346
15.9   References   347

        Index   351

Contributor contact details

(* = main contact)

Editor

J. Hu
Institute of Textiles and Clothing
The Hong Kong Polytechnic
   University
Hung Hom, Kowloon
Hong Kong
E-mail: tchujl@inet.polyu.edu.hk

Chapter 1

B. G. Xu
Institute of Textiles and Clothing
The Hong Kong Polytechnic
   University
Hung Hom, Kowloon
Hong Kong
E-mail: tcxubg@inet.polyu.edu.hk

Chapter 2

B. Xin
College of Fashion
Shanghai University of
   Engineering Science
333 Longteng Road
Songjiang District
Shanghai 210620
P.R. China
E-mail: xinbj@sues.edu.cn

Chapter 3

A. Kumar
Department of Computing
The Hong Kong Polytechnic
   University
Hung Hom, Kowloon
Hong Kong
E-mail: ajaykr@ieee.org

Chapter 4

Y. Zhong
Room 4047
College of Textiles
Donghua University
2999 North Renmin Road
Songjiang District,
Shanghai 201620
P.R. China
E-mail: zhyq@dhu.edu.cn

Chapter 5

X. Chen
School of Materials
University of Manchester
Manchester M13 9PL
UK
E-mail: xiaogang.chen@manchester.
   ac.uk

Chapter 6

G. Baciu
Department of Computing
The Hong Kong Polytechnic
    University
Hung Hom, Kowloon
Hong Kong
E-mail: csgeorge@inet.polyu.edu.hk

Chapter 7

M. S. Risby*
Protection and Sustainability
    Research Unit
Faculty of Engineering
National Defence University
    Malaysia
Kem Sg Besi
57000 Kuala Lumpur
Malaysia
E-mail: risby@upnm.edu.my

A. M. S. Hamouda
Mechanical and Industrial
    Engineering Department
College of Engineering
Qatar University
P.O. Box 2713
Doha
Qatar
E-mail: hamouda@qu.edu.qa

Chapter 8

S.-K. Wong
Department of Computer Science
National Chiao Tung University
1001 University Road
Hsin-Chu 30010
Taiwan, R.O.C.
E-mail: cswingo@cs.nctu.edu.tw

Chapter 9

G. Baciu* and S. Liang
Department of Computing
The Hong Kong Polytechnic
    University
Hung Hom, Kowloon
Hong Kong
E-mail: csgeorge@inet.polyu.edu.hk

Chapter 10

M.-E. Faust*
Fashion Merchandising, School of
    Business Administration
Philadelphia University
School House Lane and Henry
    Avenue
Philadelphia, PA 19144
USA
E-mail: faustm@philau.edu

S. Carrier
Department of Management and
    Technology
School of Management
University of Quebec at Montreal
315 St Catherine Street East
Montreal (Quebec)
Canada
H2X 3X2
E-mail: carrier.serge@uqam.ca

Chapter 11

H. Ujiie
Center for Excellence for Digital
    Inkjet Printing of Textiles
School of Design and Engineering
University of Philadelphia
4201 Henry Avenue
Philadelphia, PA 19144
USA
E-mail: ujiieh@philau.edu

Chapter 12

D. J. Tyler
Department of Clothing Design
 and Technology
Manchester Metropolitan
 University
Hollings Faculty
Old Hall Lane
Manchester M14 6HR
UK
E-mail: D.Tyler@mmu.ac.uk

Chapter 13

P. C. L. Hui
Institute of Textiles and Clothing
The Hong Kong Polytechnic
 University
Hung Hom, Kowloon
Hong Kong
E-mail: tchuip@polyu.edu.hk

Chapter 14

C. L. Istook*
College of Textiles, TATM
 Department
North Carolina State University
Campus Box 8301
Raleigh, NC 27695
USA
E-mail: cistook@tx.ncsu.edu

E. A. Newcomb
Department of Family and
 Consumer Sciences
North Carolina Agricultural &
 Technical State University
Benbow Hall
1601 E. Market St
Greensboro, NC 27411
USA
E-mail: eanewcom@ncat.cdu

H. Lim
Department of Textile Engineering
i-Fashion Technology Center
Konkuk University
1-Hwayang-dong
Gwangjin-gu
Seoul 143-701
South Korea
E-mail: maresea@gmail.com

Chapter 15

T. A. M. Lamar
College of Textiles, TATM
 Department
North Carolina State University
Campus Box 8301
Raleigh, NC 27695
USA
E-mail: traci_lamar@ncsu.edu

Woodhead Publishing Series in Textiles

1   **Watson's textile design and colour Seventh edition**
    Edited by Z. Grosicki

2   **Watson's advanced textile design**
    Edited by Z. Grosicki

3   **Weaving Second edition**
    P. R. Lord and M. H. Mohamed

4   **Handbook of textile fibres Vol 1: Natural fibres**
    J. Gordon Cook

5   **Handbook of textile fibres Vol 2: Man-made fibres**
    J. Gordon Cook

6   **Recycling textile and plastic waste**
    Edited by A. R. Horrocks

7   **New fibers Second edition**
    T. Hongu and G. O. Phillips

8   **Atlas of fibre fracture and damage to textiles Second edition**
    J. W. S. Hearle, B. Lomas and W. D. Cooke

9   **Ecotextile '98**
    Edited by A. R. Horrocks

10 **Physical testing of textiles**
    B. P. Saville

11 **Geometric symmetry in patterns and tilings**
    C. E. Horne

12 **Handbook of technical textiles**
    Edited by A. R. Horrocks and S. C. Anand

13 **Textiles in automotive engineering**
    W. Fung and J. M. Hardcastle

14   **Handbook of textile design**
      J. Wilson

15   **High-performance fibres**
      Edited by J. W. S. Hearle

16   **Knitting technology Third edition**
      D. J. Spencer

17   **Medical textiles**
      Edited by S. C. Anand

18   **Regenerated cellulose fibres**
      Edited by C. Woodings

19   **Silk, mohair, cashmere and other luxury fibres**
      Edited by R. R. Franck

20   **Smart fibres, fabrics and clothing**
      Edited by X. M. Tao

21   **Yarn texturing technology**
      J. W. S. Hearle, L. Hollick and D. K. Wilson

22   **Encyclopedia of textile finishing**
      H-K. Rouette

23   **Coated and laminated textiles**
      W. Fung

24   **Fancy yarns**
      R. H. Gong and R. M. Wright

25   **Wool: Science and technology**
      Edited by W. S. Simpson and G. Crawshaw

26   **Dictionary of textile finishing**
      H-K. Rouette

27   **Environmental impact of textiles**
      K. Slater

28   **Handbook of yarn production**
      P. R. Lord

29   **Textile processing with enzymes**
      Edited by A. Cavaco-Paulo and G. Gübitz

30   **The China and Hong Kong denim industry**
      Y. Li, L. Yao and K. W. Yeung

31   **The World Trade Organization and international denim trading**
      Y. Li, Y. Shen, L. Yao and E. Newton

32   **Chemical finishing of textiles**
      W. D. Schindler and P. J. Hauser

33   **Clothing appearance and fit**
     J. Fan, W. Yu and L. Hunter

34   **Handbook of fibre rope technology**
     H. A. McKenna, J. W. S. Hearle and N. O'Hear

35   **Structure and mechanics of woven fabrics**
     J. Hu

36   **Synthetic fibres: nylon, polyester, acrylic, polyolefin**
     Edited by J. E. McIntyre

37   **Woollen and worsted woven fabric design**
     E. G. Gilligan

38   **Analytical electrochemistry in textiles**
     P. Westbroek, G. Priniotakis and P. Kiekens

39   **Bast and other plant fibres**
     R. R. Franck

40   **Chemical testing of textiles**
     Edited by Q. Fan

41   **Design and manufacture of textile composites**
     Edited by A. C. Long

42   **Effect of mechanical and physical properties on fabric hand**
     Edited by Hassan M. Behery

43   **New millennium fibers**
     T. Hongu, M. Takigami and G. O. Phillips

44   **Textiles for protection**
     Edited by R. A. Scott

45   **Textiles in sport**
     Edited by R. Shishoo

46   **Wearable electronics and photonics**
     Edited by X. M. Tao

47   **Biodegradable and sustainable fibres**
     Edited by R. S. Blackburn

48   **Medical textiles and biomaterials for healthcare**
     Edited by S. C. Anand, M. Miraftab, S. Rajendran and J. F. Kennedy

49   **Total colour management in textiles**
     Edited by J. Xin

50   **Recycling in textiles**
     Edited by Y. Wang

51   **Clothing biosensory engineering**
     Y. Li and A. S. W. Wong

52   **Biomechanical engineering of textiles and clothing**
      Edited by Y. Li and D. X-Q. Dai

53   **Digital printing of textiles**
      Edited by H. Ujiie

54   **Intelligent textiles and clothing**
      Edited by H. R. Mattila

55   **Innovation and technology of women's intimate apparel**
      W. Yu, J. Fan, S. C. Harlock and S. P. Ng

56   **Thermal and moisture transport in fibrous materials**
      Edited by N. Pan and P. Gibson

57   **Geosynthetics in civil engineering**
      Edited by R. W. Sarsby

58   **Handbook of nonwovens**
      Edited by S. Russell

59   **Cotton: Science and technology**
      Edited by S. Gordon and Y-L. Hsieh

60   **Ecotextiles**
      Edited by M. Miraftab and A. R. Horrocks

61   **Composite forming technologies**
      Edited by A. C. Long

62   **Plasma technology for textiles**
      Edited by R. Shishoo

63   **Smart textiles for medicine and healthcare**
      Edited by L. Van Langenhove

64   **Sizing in clothing**
      Edited by S. Ashdown

65   **Shape memory polymers and textiles**
      J. Hu

66   **Environmental aspects of textile dyeing**
      Edited by R. Christie

67   **Nanofibers and nanotechnology in textiles**
      Edited by P. Brown and K. Stevens

68   **Physical properties of textile fibres Fourth edition**
      W. E. Morton and J. W. S. Hearle

69   **Advances in apparel production**
      Edited by C. Fairhurst

70   **Advances in fire retardant materials**
      Edited by A. R. Horrocks and D. Price

71   **Polyesters and polyamides**
      Edited by B. L. Deopura, R. Alagirusamy, M. Joshi and B. S. Gupta

72   **Advances in wool technology**
      Edited by N. A. G. Johnson and I. Russell

73   **Military textiles**
      Edited by E. Wilusz

74   **3D fibrous assemblies: Properties, applications and modelling of three-dimensional textile structures**
      J. Hu

75   **Medical and healthcare textiles**
      Edited by S. C. Anand, J. F. Kennedy, M. Miraftab and S. Rajendran

76   **Fabric testing**
      Edited by J. Hu

77   **Biologically inspired textiles**
      Edited by A. Abbott and M. Ellison

78   **Friction in textile materials**
      Edited by B. S. Gupta

79   **Textile advances in the automotive industry**
      Edited by R. Shishoo

80   **Structure and mechanics of textile fibre assemblies**
      Edited by P. Schwartz

81   **Engineering textiles: Integrating the design and manufacture of textile products**
      Edited by Y. E. El-Mogahzy

82   **Polyolefin fibres: Industrial and medical applications**
      Edited by S. C. O. Ugbolue

83   **Smart clothes and wearable technology**
      Edited by J. McCann and D. Bryson

84   **Identification of textile fibres**
      Edited by M. Houck

85   **Advanced textiles for wound care**
      Edited by S. Rajendran

86   **Fatigue failure of textile fibres**
      Edited by M. Miraftab

87   **Advances in carpet technology**
      Edited by K. Goswami

88   **Handbook of textile fibre structure Volume 1 and Volume 2**
      Edited by S. J. Eichhorn, J. W. S. Hearle, M. Jaffe and T. Kikutani

89   **Advances in knitting technology**
      Edited by K-F. Au

90   **Smart textile coatings and laminates**
      Edited by W. C. Smith

91   **Handbook of tensile properties of textile and technical fibres**
      Edited by A. R. Bunsell

92   **Interior textiles: Design and developments**
      Edited by T. Rowe

93   **Textiles for cold weather apparel**
      Edited by J. T. Williams

94   **Modelling and predicting textile behaviour**
      Edited by X. Chen

95   **Textiles, polymers and composites for buildings**
      Edited by G. Pohl

96   **Engineering apparel fabrics and garments**
      J. Fan and L. Hunter

97   **Surface modification of textiles**
      Edited by Q. Wei

98   **Sustainable textiles**
      Edited by R. S. Blackburn

99   **Advances in yarn spinning technology**
      Edited by C. A. Lawrence

100   **Handbook of medical textiles**
      Edited by V. T. Bartels

101   **Technical textile yarns**
      Edited by R. Alagirusamy and A. Das

102   **Applications of nonwovens in technical textiles**
      Edited by R. A. Chapman

103   **Colour measurement: Principles, advances and industrial applications**
      Edited by M. L. Gulrajani

104   **Textiles for civil engineering**
      Edited by R. Fangueiro

105   **New product development in textiles**
      Edited by B. Mills

106   **Improving comfort in clothing**
      Edited by G. Song

107   **Advances in textile biotechnology**
      Edited by V. A. Nierstrasz and A. Cavaco-Paulo

108   **Textiles for hygiene and infection control**
      Edited by B. McCarthy

109 **Nanofunctional textiles**
     Edited by Y. Li

110 **Joining textiles: principles and applications**
     Edited by I. Jones and G. Stylios

111 **Soft computing in textile engineering**
     Edited by A. Majumdar

112 **Textile design**
     Edited by A. Briggs-Goode and K. Townsend

113 **Biotextiles as medical implants**
     Edited by M. King and B. Gupta

114 **Textile thermal bioengineering**
     Edited by Y. Li

115 **Woven textile structure**
     B. K. Behera and P. K. Hari

116 **Handbook of textile and industrial dyeing. Volume 1: principles processes and types of dyes**
     Edited by M. Clark

117 **Handbook of textile and industrial dyeing. Volume 2: Applications of dyes**
     Edited by M. Clark

118 **Handbook of natural fibres. Volume 1: Types, properties and factors affecting breeding and cultivation**
     Edited by R. Kozlowski

119 **Handbook of natural fibres. Volume 2: Processing and applications**
     Edited by R. Kozlowski

120 **Functional textiles for improved performance, protection and health**
     Edited by N. Pan and G. Sun

121 **Computer technology for textiles and apparel**
     Edited by J. Hu

122 **Advances in military textiles and personal equipment**
     Edited by E. Sparks

123 **Specialist yarn, woven and fabric structure: Developments and applications**
     Edited by R. H. Gong

124 **Handbook of sustainable textile production**
     M. Tobler

Introduction

The textile and apparel industry comprises a complex network of interrelated sectors that produce fibers, spin yarns, fabricate cloth, and dye/finish/print and manufacture apparel. Computer technology is one of the most important tools contributing to the significant advancement of this industry. The transition towards digital solutions and computerization is an irreversible trend today and will accelerate in the future.

Scope of computer-based technology for textile applications

Just as in many other industries, computer-based technology in textile applications may be divided into many branches and sub-branches in terms of its applications. Generally speaking, there are three terms that are frequently used: (1) CAD (computer aided design), (2) CAM (computer aided manufacturing), and (3) CIM (computer integrated manufacturing). Furthermore, CAT (computer aided testing) is also used for applications of computer-based technology for quality evaluation and control as well as information management. Although it is difficult to summarize all the industrial applications based on computer technology alone, it is possible for us to describe some typical applications in these fields.

CAD has become particularly important for textiles and apparel design. It has been widely used in the design of yarns, fabrics and garments. CAD technology enables textile designers to develop and demonstrate virtual samples on the computer screen and to simulate the appearance of textile products without wasting materials and manufacturing processes. The development of CAD technology over the past decades has the advantages of lower product development costs and a greatly shortened design cycle with increased creative variation. These achievements have encouraged and simplified textile and garment manufacturing, material utilization, easier customization, and mass production.

CAM can be defined as the computer technology used for controlling textile manufacturing processes. Computer-based textile machines are used to support spinning, weaving, knitting, printing or finishing processes through programmable controllers, industrial computers, data gateways, cell controllers, data acquisition, batch controllers, and drive master controllers. CAM is one of the key parts of the computer integrated manufacturing (CIM) system.

CIM is the manufacturing approach of using computers to control the entire production process, typically relying on closed-loop control processes and based on real-time input from sensors. This allows individual processes to exchange information with each other and to initiate actions including planning, management and production. The final target of CIM is mainly to provide a digital platform of process control and information communication for textile design, manufacturing, testing, quality control and final product marketing/retailing. Through CIM, manufacturing can be faster and less error-prone, as well as easier to manage. Another term that may be related to CIM is CAE (computer aided engineering), which is a general term including all the engineering work based on computer equipment and technology used for the modernization of the textile industry.

CAT (computer aided testing) provides a digital and automatic solution for quality testing, evaluation and control of textile processing and products by using computer-related testing techniques, such as computer vision and artificial intelligence. In the textile industry, testing traditionally relies on heavily subjective estimation without objective testing instruments. Computer aided testing technology and methods have been used to replace these traditional subjective evaluation methods.

In addition, textile products can also be components of electrical devices and sensors used for data acquisition and information media, and could be developed as intelligent sensing, monitoring and control units worn on the human body. We call these E-textiles. For example, the development of wearable computers is one typical application, which integrates textile and computer technologies into one wearable and controllable device embedded in different functional garments. E-textiles constitute a new direction for computer-based technology which has undergone rapid development in recent years. The integration and embedding of electronic sensors and controlling units can offer clothes higher levels of functionality than common garments. It can provide information exchange between wearers and the monitoring system. Thus an E-textile solution can establish a wearable computational linked system and wearable electronic textiles can be worn in everyday situations. Computer and textile products will be merged seamlessly in the future.

Thus digital technology for textiles is broad in its scope. Loosely speaking, it refers to the utilization of computer hardware, software, networking, soft

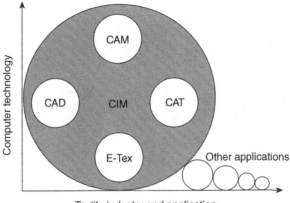

I.1 Scope of digital technology for textiles and apparel.

computing technology, robotics, and wearable sensing, monitoring and controlling technology directly or indirectly for textiles. Typical applications of computer technologies for textiles and apparel are illustrated in Fig. I.1, which includes the CAD for product design, CAM technology for product manufacture and CAT for the quality testing and evaluation of product, while E-Textiles are a new technology focused on the development of wearable computing devices directly used for the textile and apparel.

Objectives of the book

The history of computer-based technology for the textile industry can be traced back to 1801 when Joseph Marie Jacquard invented the first automatic weaving machine by introducing a series of punched paper cards which allowed the loom to weave intricate patterns. The invention of the Jacquard loom (Fig. I.2) was an important step in the development of computers. The use of punched cards to define woven patterns can be viewed as an early, albeit limited, form of programmability. Thus, the Jacquard loom could be considered as the grandfather of computer-based machinery.

Another interesting illustration of the close relationship between computers and textiles is that textile products such as woven materials have a binary interlacing format through weaving of warp and weft yarns. This binary interlacing format is based on the warp lifting which provides a channel for the weft inserting motion. This has its origins in the key motions of the weaving machine: shedding and weft insertion. From this point of view, the ancient weaving principle has led to the modern computing strategy and textile products could be considered as the prototype of binary materials as the media of information recording.

I.2 Jacquard loom on display at the Manchester Museum of Science and Industry. Photograph taken by George H. Williams in July 2004.

Now, however, computing technologies have become independent and have dominated human civilization over the past 50 years, while applications of computer technology to textiles and apparel have now extended to areas including design, machinery, manufacturing, integrated production control and management, information management, quality testing and evaluation, and tools for marketing and retailing. This is a typical example demonstrating the value, efficiency and effectiveness of computer science and technology while practical problems and market requirements presented by consumers, textile engineers and scientists trigger the development of new computer-based technologies. As the cost of computers has fallen dramatically, their functions have increased with powerful algorithms, and software systems, and academic works from both textile scientists and computer scientists have provided stronger support for the application of computer technology in the textile industry. Computer-based technology for textiles and apparel will therefore be even more important than ever before.

Thus, an understanding of the advances and principles of these computer-based technologies used in the modern textile industry is very necessary for both textile technologists and scientists to develop their abilities and utilize these techniques in their practical operations. A comprehensive review of computer-based technologies for textiles and apparel can provide professionals with guidelines. Thus this book has been intended to cater for the above needs. It has three parts and 15 chapters which cover various aspects of computer-based technologies for the textile industry including CAD, CAM and CAT. It consists of theories, principles, methodologies, applications, education, current status and future trends of computer-based technologies used in these fields. This book is beneficial to academics as well as practitioners in their research and daily work.

Arrangement of the contents

Part I deals with computer-based technology for textile materials. Quality assurance for textiles is vital for business competitive advantage. Traditional subjective evaluation methods and equipment have been found to be slow, fatiguing, subjective and inconsistent. To improve the automation and accuracy of quality testing and evaluation of textile products, objective methods and systems based on artificial intelligence and computer vision have been developed and utilized for the inspection of defects, appearance and other attributes. Thus in Part I of the book, we have three chapters which review the methods and theories for developing evaluation systems for defects and appearance attributes of yarns and fabrics. In particular, Chapter 1 presents the recent developments of digital technologies for yarn structure and appearance analysis. In Chapter 2, a new digital method based on the dual side scanning and active grid model (AGM) is introduced, with some experiments and discussions, which demonstrate the means of identifying the weave pattern of woven fabrics. The third chapter provides an overview of the open problems, the state-of-the-art in fabric defect detection and classification techniques, and points out the potential of the new technologies that could make low-cost robust defect detection a reality for textile industries.

Part II examines modeling and simulation of textiles and garments, which can help with increasing industrial competitiveness through reduction of design, manufacturing and validation cycle time. This can lead to the availability of virtual design assessment, or the ability to evaluate the problem entirely in a computer without carrying out time-consuming experiments. In many application areas, accurate simulations are routinely obtained using high performance computers while others are, at best, qualitative and capable of describing only trends in physical events. Modeling of textile structures coupled with computer technology has become an enabling discipline that has led to greater understanding of textile physics and structures and advances in computer software systems. In Part II, we have five chapters by leading experts in their areas:

- Key techniques for 3D garment design
- Modelling and simulation of fibrous yarn materials
- Digital technology and modeling for fabric structures
- Modeling ballistic impact on textile materials
- Modeling and simulation techniques for garments.

The first of these (Chapter 4) discusses three topics related to recent advances in computer technologies for textiles and apparel. These include sketch-based modeling, surface flattening and problems/solutions for an online garment-shopping system. The corresponding implementations are highlighted for evaluating the performance of the algorithms and/or techniques.

Theoretical modeling of yarn and fabric materials has contributed to the development of CAD software for textile materials by providing simulation methods for modeling of yarn segments. The research into this topic mainly concerns the geometrical definition of yarn and fabric structure and appearance based on their physical parameters. Chen's chapter introduces various methods for modeling and simulation of yarn and fabric structures.

In Chapter 6 by Baciu, Zheng and Hu, fabric geometry structure in two-dimensional space is modeled. The interlacing rules and their representation of the weave pattern are investigated. Experimental results are shown on 2D fabric weave pattern description in terms of main characteristic perception, indexing and classification in the fabric design domain and the information technology domain. Future research directions in this area are also noted.

For textile-based ballistic products under extreme environmental conditions, computational model experiments can be a cost-effective option compared to laboratory experiments; thus expensive live-fire tests can be culled and used only for validation purposes. The capability of these experiments to virtually simulate the physical impact event has been demonstrated in detail. Chapter 7, by Risby and Hamouda, presents methods for modeling and simulation to assess the ballistic impact of fabrics. The modeling and simulation of single-layer and multi-layers fabric provide a mathematical and virtual way to predict the ballistic impact of fabrics.

Chapter 8 reviews the three approaches for modeling garments: geometry-based, physically-based and hybrid. Computer graphics techniques and collision detection methods are then described for interactive garment simulation. The future development of modeling and simulation of garment materials are included at the end of the chapter.

In Part III, Computer-based technology for apparel, comprising Chapters 9 to 15, we have a more diversified content relating to soft and practical aspects of apparel, which include the views of strategists, practitioners and educators.

With the introduction of CAD and its many software capabilities, the entire process of designing a fabric has been revolutionized to replace traditional hand producing: where previously designers labored over graph paper and stencils, now they simply play with a mouse or stylus pen to produce innovative designs.

Baciu and Liang's chapter discusses the principles of human computer interaction and its usage in textiles as CAD systems. New features for the future development of garment design systems are highlighted, such as sketch interface, gesture-based interaction, shape retrieval and dynamic association, as well as garment database management. In Faust and Carrier's chapter, Hong Kong Generation Y females' body perception and virtual visualization is demonstrated. They believe, after being 3D body scanned, that a smart card with their body measurements would improve their garment shopping experience.

Chapter 11 by Ujiie discusses computer technologies used in the domain of textile design from the textile designer's perspective. In this case, computer technologies are tools for (1) information gathering, (2) designing, (3) design editing and (4) presentation in textile design workflows. The benefits and limitations are also explained from the initial design concept to final production and presentation.

We have selected digital printing as an example to demonstrate the application of CAM to textiles and apparel. Tyler's Chapter 12 reviews the role of textile digital printing from the perspective of product development and effective communication across the supply chain.

In addition to the technical chapters in the book, we have one chapter – Chapter 13 by Hui – which aims to review current pedagogic approaches in the teaching and learning of computer-based technologies, particularly CAD systems, in responding to the main challenges such as the rapid changes of industry needs and wants, computer technology, and the global working environment for fashion and textiles, as well as the non-standardization of jargon used in fashion and textiles.

Chapter 14 provides an overview of the technologies used for 3D apparel CAD systems. It points out that research continues in the development of more realistic and mechanically accurate models, and completed models may be integrated with 3D garment design, visualization and 3D fabric modeling to create a truly interactive 3D product development and marketing tool.

In the final chapter, by Lamar, the role of digital technology is examined in enabling a design and development process integrating the functions of both textile and apparel creation. First, digitally enabled processes integrating design of textiles and design of apparel simultaneously are discussed, including examples of garments developed through integrated processes. The focus of the chapter then shifts specifically to the role of computer aided design (CAD) and visualization technologies in enabling an integrated digital process, and concludes with discussion of future directions.

J. Hu
The Hong Kong Polytechnic University
Hong Kong

Part I
Computer-based technology
for textile materials

1
Digital technology for yarn structure and appearance analysis

B. G. XU, The Hong Kong Polytechnic University, Hong Kong

Abstract: Yarn structure and appearance are important characteristics in yarn quality assessment and assurance. This chapter presents the recent developments of digital technologies for yarn structure and appearance analysis. The chapter broadly reviews the latest advances that have been made in digital measurement and analysis of yarn evenness, yarn hairiness, yarn twist, yarn snarl, yarn blend and yarn surface appearance.

Key words: yarn structure, yarn appearance, yarn assessment, yarn image processing, digital signal processing.

1.1 Introduction

Yarn is used worldwide for making a wide range of textiles and apparel. Its structure and appearance have significant influence on the properties and performance of the yarn and its end-products. Therefore, the analysis of yarn structure and appearance is an important need and procedure in assessing yarn quality in the textile industry. Traditionally yarn structure and appearance are evaluated subjectively by manual methods, but some of the methods are subjective, less reliable and labour intensive. With the rapid development of computer technology, efficient and low-cost techniques have been established for the accurate image acquisition and massive image storage. At the same time, image processing, computer vision and pattern recognition have achieved their respective high levels of progress. Those developments in digital technology bring new data acquisition apparatus, new data analysis and recognition approaches, thus providing an alternative objective method for yarn feature analysis. A generic diagram of digitalized yarn feature analysis is shown in Fig. 1.1.

In this chapter, we will present the state-of-the-art digital technologies for yarn structure and appearance analysis. More specifically, we look into the latest developments in yarn evenness measurement, yarn hairiness analysis, yarn twist and snarl measurements, yarn blend analysis and yarn surface appearance grading (Sections 1.2–1.7). Afterwards, we will discuss the future trend of this area in Section 1.8. Concluding remarks are drawn in Section 1.9.

3

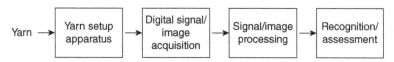

1.1 A generic diagram of digitalized yarn feature analysis.

1.2   Measurement of yarn evenness

Consistent yarn thickness is essential for the high quality of textile products. For many years, yarn irregularity has been measured by the capacitance evenness tester using two parallel capacitive sensors. The capacitance based method is accurate and stable in yarn mass measurement and has been well accepted in the textile industry for decades. Nevertheless this method can only give a rough description of yarn irregularity in diameter. Optical measurement alternatively provides a more accurate method in determining the yarn diameter and its variation by using optical sensors. As the diameter of a yarn is measured, the optical based method is not affected by moisture content or fibre blend variations in the yarn. The Uster evenness tester and the Zweigle G580 are two representative instruments commercially used in the textile industry for the capacitive and optical measurements of yarn evenness, respectively.

Recent developments in capacitive and optical measurements, along with the digital signal processing of analysis, make the yarn evenness results more practical and sensible. For instance, Rong and Slater (1995) developed a microcomputer system using digital signal processing for yarn unevenness analysis. In this system, the analogue signal of the diameters of a single yarn, measured by the Uster Tester, was converted into a digital form and then further analyzed by means of frequency spectrum analysis techniques. In addition to the traditional statistical parameters for characterizing irregularity, yarn unevenness could be assessed by using the probability density function, which was known to be closely correlated with fabric appearance quality. Based on the capacitive principle, Carvalho *et al.* (2006) also developed a new system for accurate yarn thickness and evenness measurement by using capacitive sensors and digital signal processing techniques. In comparison to commercial instruments, this system enabled direct measurement of yarn mass with a high resolution of 1 mm in yarn length. With the accurate measurements of yarn mass, some signal processing algorithms, such as FFT (Fast Fourier Transform) and FWHT (Fast Walsh–Hadamard Transform), were employed to detect a wide range of yarn faults in lengths of 1 mm and above.

In optical measurement, an optical signal processing system has been developed (Carvalho *et al.*, 2008) to measure yarn diameter. As shown in Fig. 1.2, a helium neon (HeNe) laser was used to emit a coherent light which

Image plan   Lens 4   Low-pass   Lens 3   Yarn sample   Lens 2   Lens 1   Diaphragm   HeNe
                      spatial filter   plan   laser

1.2 Experimental setup of the optical system with a low-pass spatial filter for yarn diameter measurement (adapted from Carvalho *et al.*, 2008).

was directed to pass through a diaphragm, a set of planoconvex lenses and a yarn sample and finally was received by a detector (photodiode). The novelty of this method is that a low-pass spatial filter was inserted in front of the optical sensor so as to eliminate the signal from yarn hairs and thus keep the entire core of the yarn remaining in the image for diameter evaluation. The working principle is based on the simple fact that the diameter of a spun yarn is usually one or more orders of magnitude larger than that of the small hairs (i.e. single fibres) protruding from its surface. Therefore, the low-pass spatial filter was able to eliminate all spatial frequencies above a predetermined value, covering the characteristic sizes of most textile fibres. As a result, the hairiness was almost completely removed in the yarn sample image before it was received by the optical detector. This system could also be easily adapted to measure the projection of the yarn diameter along a single direction and, consequently, to infer yarn irregularities.

1.3   Analysis of yarn hairiness

Yarn hairiness is defined as hairs protruding from the main body of a yarn. The amount of hairiness is important to both the textile operations and the appearance of fabrics and garments. Commercial instruments, such as the Uster Testers 3 and 4, the Shirley Hairiness Tester and the Zweigle G565 and G566, have been widely used in the textile industry for the assessment of yarn hairiness. So far, most commercial systems to quantify yarn hairiness have been based on the optical principle. Barella (1983) and Barella and Manich (2002) have systematically reviewed a wide range of measurement techniques and various commercial instruments designed for measuring yarn hairiness.

In addition to the traditional methods, digital analysis of high-quality images of a textile yarn can characterize hairiness. In this method, an image of high resolution or a microscope image with an appropriate magnification is usually adopted to ensure the clear presence of yarn hairs in the image.

In the captured image, the yarn can be divided into two basic components: yarn core and hairiness. Yarn core refers to the main body of the yarn, which is composed of a compact agglomeration of fibres. As the yarn core and yarn hairs all consist of constituent fibres, they may present similar image features (e.g. colour, texture). Therefore, the main task (also a challenge) of this method is the segmentation or separation of yarn hairs or the core of the yarn from the background image. For this purpose, various image processing techniques such as the edge detection and thresholding algorithms could be used.

For example, Cybulska (1999) proposed a digital-image processing method for estimating yarn hairiness. In this method, a microscope image is taken by a camera and then processed by the segmentation algorithms to separate the yarn core and yarn hairs as a whole from the background. Figure 1.3 shows the segmentation results in which the white areas represent all the yarn elements. In order to further separate the pixels of yarn hairs and those of the yarn core, two processing algorithms, namely a connected length detection method and a special correction condition, were implemented to determine the two actual edges (the upper and lower) of the yarn core, as shown in Fig. 1.3.

Finally, the yarn hairiness can be estimated by calculating the ratio of the width of the hairiness on both sides of the yarn to the width of the yarn

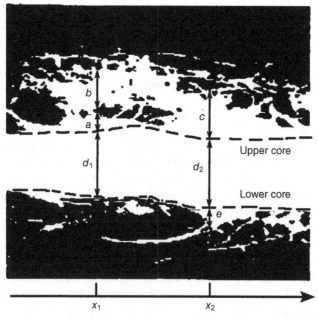

1.3 Estimation of yarn hairiness (adapted from Cybulska, 1999).

core along the direction orthogonal to the yarn axis. For example, at two axial points (x_1 and x_2) of the yarn length shown in Fig. 1.3, the values of yarn hairiness can be formulated as follows:

$$w(x_1) = \frac{a+b}{d_1} \times 100\%$$

$$w(x_2) = \frac{c+e}{d_2} \times 100\%$$ 1.1

where d_1 and d_2 are the yarn diameters at the two axial points, and a, b, c and e are the widths of yarn hairiness at the two axial points on both sides of the yarn.

More recently, a digital image processing method has been developed by Guha *et al.* (2010) for the automatic measurement of yarn hairiness. In this method, yarn images were captured by a camera from a moving yarn under a microscope. Different from the conventional way, the yarn images were captured under transmitted light (i.e. light shining from the back of the yarn). These kinds of images were easier to analyze because, under the transmitted light, the yarn core and hairs become dark areas or lines surrounded by the lighter background, as shown in Fig. 1.4. After the acquisition of a yarn image, a series of pre-processing algorithms were applied to convert it into a grey-scale image and then a binary image. In order to segment the yarn hairs in the image, a row-by-row scanning algorithm was adopted to identify the yarn core in the binary image. Supposing the yarn axis was horizontal, the image could be scanned row-wise and only those rows whose pixel values were greater than a certain threshold can be

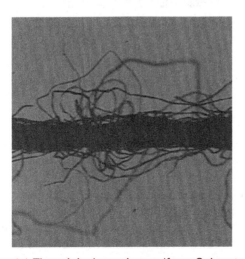

1.4 The original yarn image (from Guha *et al.*, 2010).

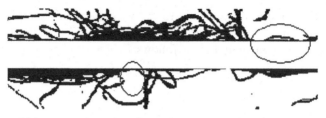

1.5 Yarn core detection (from Guha *et al.*, 2010).

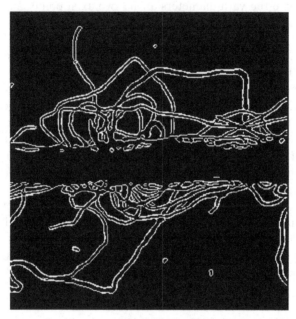

1.6 Yarn hairiness edge detection (from Guha *et al.*, 2010).

identified as constituting the yarn core. The determination of the threshold is crucial to this method. An extreme high or low value of the threshold may lead to a narrow yarn core (Fig. 1.5) or to faulty regions (like neps) being recognized as in the yarn core.

After the identification and removal of the yarn core, Canny's edge-detection algorithm was applied to identify the edge of every hair for the further assessment of yarn hairiness. Figure 1.6 shows the results of edge detection for each hair. Compared with the original image in Fig. 1.4, all individual hairs on the yarn surface have been well identified. In this method, the segmentation process of the yarn core was simple and physically reasonable. However, it was based on row-wise straight scanning, so the variations in the thickness of the yarn core may not be well identified. Thus the yarn sample should be divided into many small zones along the

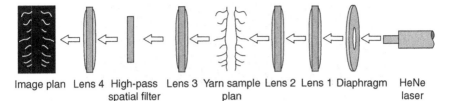

Image plan Lens 4 High-pass Lens 3 Yarn sample Lens 2 Lens 1 Diaphragm HeNe
                    spatial filter   plan   laser

1.7 Experimental setup of the optical system with a high-pass spatial filter for yarn hairiness measurement (adapted from Carvalho *et al.*, 2006).

yarn length. In each zone, the length of yarn under analysis should be short enough so that the yarn thickness variation in this short range can be ignored.

Because the yarn core and hairs are a pair of complements, the working principle discussed above in the evenness optical system can also be employed for the hairiness analysis (Carvalho *et al.*, 2006) with only minor modifications, as shown in Fig. 1.7. Compared with the above evenness system, a high-pass spatial filter instead of the low-pass spatial filter was adopted in the system for quantifying yarn hairiness. Contrary to the function of the low-pass spatial filter, the presence of the high-pass spatial filter removed all spatial frequencies below a predetermined value, permitting only the high spatial frequencies (covering the characteristic sizes of most fibres) in the image to propagate to the detector. This results in the contours of the edges of the yarn and associated hairs being highlighted in the image for further evaluation.

1.4   Measurement of yarn twist

Yarn twist is primarily used to hold the staple fibres together to form a strong and continuous yarn. The level of yarn twist has been traditionally measured on the yarn twist tester by using the untwisting or untwist–twist method. In a digital way, yarn twist can be estimated indirectly by measuring the angle between the tangent to the fibre helix and the axis of the yarn in images. According to the spatial location and colour, the fibres available for such evaluation can be classified into two categories: surface fibres and tracer fibres. Surface fibres are those laid along the twisting curves on the main body of a yarn, while tracer fibres are an extremely small percentage of colour fibres (preferably black) purposely introduced into a spun yarn during spinning. Different from the location of surface fibres, most tracer fibres are distributed inside the yarn structure with similar geometrical and migrational performances to other fibres. The tracer fibres were used to trace the actual fibre path inside a spun yarn and thus are very useful for

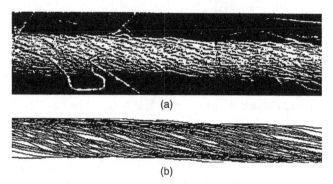

(a)

(b)

1.8 Estimation of yarn twist (from Cybulska, 1999): (a) image of the yarn surface; (b) set of lines characterizing yarn twist.

the structural analysis of spun yarns. In order to capture the tracer fibres, the yarn sample usually needs to be immersed in a trough of special solvent so as to optically dissolve other fibres based on their different refractive indexes. Compared with the yarn sample with tracer fibres, no such effort is needed in the image acquisition of yarn surface fibres.

When analyzing the image of a yarn, Cybulska (1999) observed some fragments of helices on the yarn surface, which were formed as a result of twisting the fibres in the process of yarn formation, as shown in Fig. 1.8(a). Based on this observation, Cybulska developed an image processing technique to estimate the yarn twist by identifying these helices or twist lines of surface fibres. After the image segmentation and filter processing of a yarn image, a black-and-white image was obtained in which the black points corresponded to the arrangement of individual fibres on the yarn surface. Thus the twisting lines formed by individual fibres (Fig. 1.8(b)) on the yarn surface could be recognized and used for estimating the surface helical angle and yarn twist.

Based on a similar principle, Jia (2005) also estimated the yarn twist by calculating the average angle of the yarn surface fibres. In this method, a yarn image captured by a CCD camera was first converted into a binary image for obtaining the preliminary fragments of helices. The resultant binary image was then transformed by using the 2D Fourier transform to obtain accurate information on these helices or twisting lines. Finally, the surface helical angle was determined by linear regression processing of the grey levels of the resultant Fourier spectrum image.

In the digital measurement of tracer fibres for assessing the structural parameters of spun yarns (Alagha *et al.*, 1994), a CCD or video camera coupled with a lens could be adapted to capture the images of a yarn when it is passing through a trough containing ethyl salicylate. Because the white fibres which formed the bulk of the yarn are optically dissolved in the

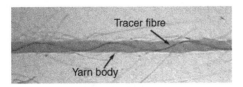

1.9 A tracer fibre image captured by CCD camera.

solvent, the black tracer fibres inside the yarn could be readily observed, as shown in Fig. 1.9. After the acquisition of yarn images, image segmentation methods such as grey level thresholding can be used to separate the yarn boundary from the background and then the tracer fibre from the body of the yarn. In addition to the twist level of the spun yarn, the geometrical information obtained by this method could also be used for the determination of other yarn structural parameters, such as fibre migration, yarn diameter, helix angle and helix diameter.

1.5   Recognition of yarn snarl

Yarn snarling occurs when a twist-lively yarn is given sufficient slack. As a result, the yarn will retire into itself and simultaneously twist in the opposite twist direction to form yarn snarls. This is also known as the torsional buckling effect (Primentas, 2003). In textile processing, yarn snarling due to yarn twist liveliness is considered to be a problem leading to yarn breakage, deterioration of yarn properties and equipment malfunction. Therefore recent developments on yarn quality assurance have included yarn snarl measurement as an effective evaluation method for yarn twist liveliness and yarn residual torque (Murrells *et al.*, 2007; Murrells, 2008).

From a geometrical point of view, the formation of yarn snarls is similar to that of yarn twists, because both are formed as a result of the twisting action. Therefore, the number of yarn snarls per unit length has also been manually measured by using the twist tester. To reduce the testing time and the amount of manual operations involved in counting the number of snarls, Xu *et al.* (2008, 2010) proposed a computerized method for automatic measurement of yarn snarls from a captured yarn image. This method adopts a fundamentally different measurement principle from the digital twist evaluation system discussed above. Instead of the fibre helical angle measurement, it uses a fluctuation profile measurement to determine the snarl features.

Figure 1.10 shows a simulated 2D projection of an ideal snarled yarn sample. It is obvious that the yarn snarls possess a periodic appearance and their top and bottom boundaries appear to be two fluctuating curves. As these fluctuations result directly from yarn snarls, it is feasible to recognize

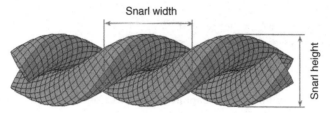

1.10 A schematic view of 2D projection of yarn snarls.

1.11 Original image of a snarled yarn sample.

1.12 Binary image of the yarn sample.

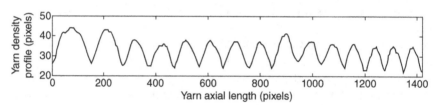

1.13 Yarn density profile extracted from the yarn sample.

a yarn snarl by calculating the fluctuating features along either of the fluctuating curves. In order to magnify this effect, a yarn density profile can be defined as the yarn cross-sectional width along its length and is used for the measurement of yarn snarl features.

Figure 1.11 shows an original image of a snarled yarn sample captured by a CCD camera. After the image is converted into a binary image (Fig. 1.12), the yarn density profile can be extracted, as shown in Fig. 1.13. Adaptive orientated orthogonal projective decomposition (AOP) is then employed for recognizing yarn snarl features. In the AOP algorithm discussed below, all fluctuations in yarn density profile can be adaptively simulated by the Gauss functions. The number of Gauss functions adopted for the simulation can be used to compute the number of yarn snarls in the sample. In addition to the number of yarn snarl turns, the method is also accurate for the detection of yarn snarl heights and widths, which are unobtainable by the manual operation method using the twist tester.

Adaptive signal decomposition (Bultan, 1999; Yin *et al.*, 2002; Qian and Chen, 1994; Qian, 2002) is a powerful method in signal modelling and processing. The principle can be described as follows (Yin *et al.*, 2002). For a given signal $s(t)$ and a set of predefined atom signals $G = \{g(t)\}$, select an atom signal $g_0(t)$ so that the distance between $s(t)$ and its orthogonal projection on $g_0(t)$ is minimized:

$$\min_{g} \|s_1(t)\|^2 = \min_{g} \|s_0(t) - \langle s_0(t), g_0(t) \rangle g_0(t)\|^2 \qquad 1.2$$

where $s_0(t) = s(t)$, $<>$ and $\|\ \|$ are the inner product and module operators, respectively, and $s_1(t)$ is the residual signal after subtracting the projection on $g_0(t)$ from the original signal $s(t)$.

Repeating this process, the following relationship can be obtained:

$$\begin{aligned} s_{k+1}(t) &= s_k(t) - \langle s_k(t), g_k(t) \rangle g_k(t) \\ &= s_k(t) - A_k g_k(t) \end{aligned} \qquad 1.3$$

Therefore, the original signal $s(t)$ can be decomposed by K atom signals as:

$$s(t) = \sum_{k=1}^{K} A_k g_k(t) + s_{k+1}(t) \qquad 1.4$$

The choice of atom signals G depends on the nature of the applications. In this case, a normalized Gaussian function can be employed as the atom signal:

$$g_l(z) = (\pi \sigma_l)^{-0.25} e^{-\frac{(z - \mu_l)^2}{2\sigma_l^2}}, l = 1, 2, \ldots \infty \qquad 1.5$$

The normalized Gaussian function is chosen because its shape can well match the morphological property of a yarn snarl. Figure 1.14 shows the decomposition results of the yarn density profile in terms of Gaussian functions. There are a total of 15 Gaussian functions used for the decomposition, therefore the number of yarn snarl turns in Fig. 1.11 is 15/2 = 7.5 turns.

With these Gaussian functions, it is also feasible to reconstruct a profile by adding these Gaussian functions for a comparison with the original yarn density profile, as shown in Fig. 1.15. The reconstructed profile generally

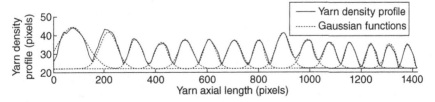

1.14 The Gaussian functions calculated by AOP algorithms.

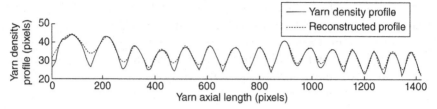

1.15 Profile reconstructed by the Gaussian functions.

coincides with the yarn density profile. With the magnitudes and variances of these Guassian functions, the snarl height and width can be calculated (Xu *et al.*, 2008, 2010).

1.6   Analysis of yarn blend

The combination of fibres from different types of material can give the resultant yarn certain characteristics which are unobtainable from a single component, such as strength, colour or effect. Therefore, blending is a common practice in yarn spinning. Traditionally, the blending composition of blended yarns has been analyzed by using human visual inspection, but such a method is time-consuming and subject to operation errors.

For the digital analysis of blend yarns, image processing techniques can be adopted for evaluating longitudinal or cross-sectional yarn images. For instance, Watanabe *et al.* (1992a, 1992b) carried out a series of investigations on the evaluation of blend ratio and irregularities for black-and-white yarns. In this model, a computer image processing system was developed to convert the longitudinal images of blended yarn into trivalued images (black fibres, white fibres and grey background), and then quantitatively to evaluate the blend irregularity on the surface of a short length of yarn (about 4 mm). Because the image conversion processing had to be involved in this process, the system was found to be time-consuming. Therefore, a more efficient image processing system (Watanabe *et al.*, 1995) was proposed to evaluate blend irregularities by directly measuring the line sense (a series of intensity values) near the axis on a surface image of a blended yarn. The new system enabled real-time evaluation of blend irregularity without converting the original images into trivalued images.

Line sense can be defined as a series of intensity values on a straight line around the centre along the yarn axis in the blended yarn (Watanabe *et al.*, 1995). An example of this measurement is shown in Fig. 1.16. An intensity value of 0 represents black and an intensity value of 255 represents white. The line sense shown in Fig. 1.16 is composed of 310 pixels, which corresponds to a 4 mm yarn length.

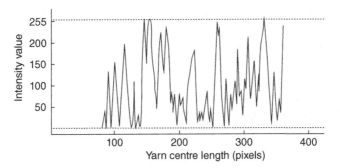

1.16 Line sense of pixels (from Watanabe *et al.*, 1995).

Towsend and Cox (1951) proposed a method of analyzing yarn thickness irregularity using a variance-length curve. The concept has also been applied by Watanabe *et al.* (1995) to analyze blend irregularity in slivers and yarns. In this method, the total variance of a blend irregularity curve is represented by

$$V_\infty = B_L + V_L = \text{constant} \qquad\qquad 1.6$$

where V_∞ is an overall variance, B_L is a relative variance among sample lengths L of the average intensity values of the pixels within the sample length L, and V_L is an average intensity value among sample lengths L of the relative variance of the pixels within the sample length L.

From the line sense, B_L V_L and V_∞ are computed and variance-length curves can be displayed. It has been revealed that V_∞ represents the total variance of all the intensity values and thus is proportional to the blend ratio of the component fibres. It is also noted that the magnitude of the intersection of the B_L and V_L curves represents a variance period of all the intensity values, and that it is proportional to the cluster sizes of the component fibres.

Besides the longitudinal yarn image, cross-sectional images of the yarn could also be used to characterize the blend ratio of blended yarns and fibre distribution in the cross-sections of blended yarns. Chiu *et al.* (1999) proposed a spatial-domain image processing technique to analyze the cross-sections of PET/rayon composite yarns. In this method, fibre features (area, perimeter, compactness and shape factor) were first extracted with image processing techniques and then used for the classification of filaments. Finally, the fibre distributions of the composite yarn can be determined with three index categories of radial, lateral and angular distributions. As there were a number of threshold values and processing steps involve in this method, it was not easy to predefine all suitable parameters in practice. Therefore, the method was further extended (Chiu and Liaw, 2005) by using some new image algorithms. In this method, a grey-level segmentation was first used to convert the captured image of a PET/rayon composite yarn into a binary image and then an edge image. The edge and binary images were then used to obtain single fibre locations and recognize the fibre patterns by implementing two voting techniques – connected component voting and circle parameter voting.

More recently, Xu *et al.* (2007) adopted a neural network to analyze the cross-sectional images of a wool/silk blended yarn. In this method, the original yarn cross-sectional images were processed by the algorithms of image enhancement and shape filtering, and then characteristic parameters (fineness, roundness, rectangle degree and shape features after thinning) were extracted for distinguishing wool and silk fibres. With these features as the input, a neural network computing approach with a single layer of perceptrons was used for recognition and classification of different fibres. The image analysis provided a method of characterizing the fibre blend ratio of blended yarns and the fibre distribution in the cross-section of blended yarns.

1.7 Grading of yarn appearance

The grading of yarn appearance is one of the important testing procedures in assessing spun yarn quality in the textile industry. In the standard test method presented in ASTM D2255–02 (2007), a yarn specimen is continuously wound onto a rectangular or trapezoidal black board in a specified length to allow the side-by-side comparison of a number of relatively short lengths of yarn. The board covered with the wound yarn is then examined and a visual appraisal of appearance is made based on the fuzziness, irregularity and visible foreign matter. Traditionally, the inspection is carried out by direct observation in which a skilled specialist visually compares the wound table with photographic standards labelled in Grades A, B, C and D, and then judges the quality of the yarn sample according to the standard definition.

Recently, attempts have been made to replace the conventional observation method with computer vision to resolve the limitation of human vision in yarn appearance grading. Compared with the objective evaluation of individual yarn properties (i.e. yarn evenness, hairiness, twist), the yarn appearance grading involves an additional operation of subjective judgement on the overall quality (i.e. yarn grades). Thus, in addition to the digital characterization of yarn surface appearance (i.e. fuzziness, irregularity and foreign matter), an artificial or classification algorithm has to be employed in the computerized method in order to resemble the human appraisal mechanisms. This is also a differentiating feature of the digital grading of yarn appearance. For this purpose, various methods such as splitting cluster analysis (Rong *et al.*, 1994) and artificial neural network (Lien and Lee, 2002; Semnani *et al.*, 2006) have been used for the objective classification of yarn grades.

In the digital characterizations of yarn surface appearance, Nevel *et al.* (1996a, 1996b) proposed an electronic system to read the diameter of a single yarn when it moved over a CCD camera. The whole length of yarn was then split into a number of shorter lengths so that they could be displayed electronically side-by-side to assist in manual grading. In the system, some yarn defects, such as thick places and neps, could also be digitally identified and counted, and the number of defects along the length of the yarn was used as a grading index for yarn surface appearance.

In the above methods, although it was possible to define a classification for yarn appearance, a grading method based on an image captured from a standard yarn board was found to be impossible. To address this issue, Semnani *et al.* (2006) have proposed a computerized grading method by using image analysis and an artificial intelligence technique. This method enabled the identification of yarn faults and the classification of yarn appearance directly from a yarn image captured from a standard yarn board. In this method, a threshold method was repeatedly used to separate thick places from the yarn body and then the background from the thick places. In order to reduce the negative influence of the inclination angle of yarns wound on the black board, which was an inevitable phenomenon due to the yarn winding operation, the image was separated into a large number of small square zones of equal size. The size of the small zones varied with yarn counts and could be experimentally optimized and determined. Based on the extracted fault features, a neural network was then trained and employed as the linear classifier for yarn surface appearance.

One of the recent advances in computer technology is the development of computational attention-driven modelling which enables the possibility of manual inspection simulation. When the operator is examining a yarn board, he or she will first consider all yarn segments on the board as a whole and then focus on certain abnormal areas by visually comparing yarn appearance among the segments. During this period, only the abnormal areas represented by yarn faults, such as thick places, thin places, neps and foreign matter, are visually highlighted and attract the attention of the operator. In the attention-driven model, a saliency map (Itti *et al.*, 1998; Itti and Koch, 2000) can be used to compute and indicate these abnormal or fault areas. The saliency map is an attention-driven image method and has been widely used in visual search, object detection and active vision (Avraham and Lindenbaum, 2006; Backer *et al.*, 2001). It can topographically identify the visual saliency or distinctive areas of a visual scene by

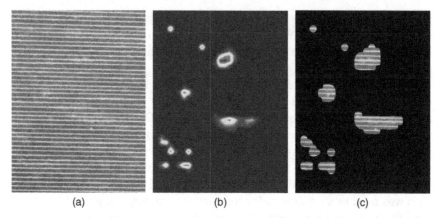

| (a) | (b) | (c) |

1.17 Attention-driven modelling for abnormality detection: (a) original image; (b) saliency map; (c) abnormal areas.

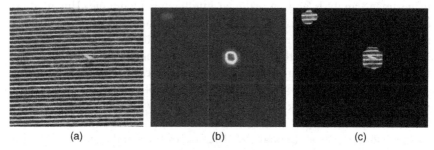

| (a) | (b) | (c) |

1.18 Attention-driven modelling for foreign matter detection: (a) original image; (b) saliency map; (c) foreign matter.

considering the centre–surround contrasts in terms of visual features, including intensity, colour and orientation.

Figure 1.17(a) shows an original image captured from a yarn board in which there appear to be a number of thick places and neps on the board. In order to identify the fault zones, the attention-driven method is applied by comparing the yarn segments. The formulated saliency map is shown in Fig. 1.17(b) in which the highlighted areas are identified as distinguished regions. Afterwards the identified fault areas on the board can be displayed by using the fault locations and sizes obtained from the saliency map and the results are shown in Fig. 1.17(c). The results are largely in agreement with human inspection. In addition, this method is also applicable to the detection of foreign matter and an example is shown in Fig. 1.18 in which foreign matter on the yarn board can be successfully identified.

1.8   Future trends

It is widely recognized and agreed that digitalized methods are benefiting the analysis of yarn structure and appearance. As discussed above, there is an increasing interest in the application of novel information technologies to facilitate and aid the analysis of yarn structure and appearance. This indicates a clear trend that yarn image acquisition will be improved in terms of image quality and modality, as well as by automation and standardization. High-resolution images will become mainstream so that more detailed characteristics of yarn structure and appearance will be available. This will also add new measures to the traditional yarn evaluation system. As an alternative to hardware improvement, super resolution image processing (Chaudhuri, 2001) can also be adopted to improve the resolution of a yarn image. In addition, some interest will probably be focused on the dynamic behaviour analysis of yarn when it is processed or tested, so high-speed cameras or line-scan cameras will be used to capture sequential pictures of yarn. Some characters that describe the dynamic properties of yarn will be studied and will become new features in its evaluation. There is no doubt that the acquisition of yarn images will become increasingly standardized and automated. Some important influences such as the luminance condition and the acquisition locus of yarn images should be well controlled and fixed in every process.

Similar to improvements in image quality and modality, research into yarn analysis will continue to take advantage of the latest advances in signal and image processing, pattern recognition, computer vision and other related areas. Some state-of-the-art classification approaches such as the support vector machine (SVM), artificial neural networks (ANN) and Bayesian networks may be helpful in automating yarn quality evaluation. With the increasing capacities of computer technology, massive yarn image acquisition and storage will be inevitable. Therefore, the efficient and accurate management of these images will also gradually become an important and specific topic. Some image retrieval methods such as content-based image retrieval (CBIR) will be especially useful.

Finally, the fundamental problem in this area will continue to be the acceptance of the new digital methods in the industry, which has been dominated by traditional manual methods for centuries. Although many valuable efforts have been made and several prototypes have been developed in recent years, most of the research work is still on a laboratory scale and techniques are less mature for providing testing standards for the textile industry. Under the trend of computerization, the application of digitalized yarn apparatus in textiles becomes prospective. Standardized and dedicated yarn devices with digitalized features will be devised and the corresponding standards will be established. Efforts have been and will continue to be made to develop and improve digital yarn technologies and devices for industrial application.

1.9   Conclusions

In this chapter, the latest advances in computer technology for the analysis of yarn structure and appearance have been presented. Generally, most digital evaluation methods involve three stages of processing: data acquisition, feature extraction and quantitative characterization. Data acquisition is the fundamental operation, in which the quality of the data is a major consideration. Electrical signals and digital images are two main types of data, such as the electrical signals from the yarn thickness, and images of yarn longitudinal or cross-sectional views that contain the yarn core, yarn hairs and/or tracer fibre. Feature extraction is the most essential and challenging task in digital yarn analysis. In this process, digital signal and image processing approaches are the main tools to extract the desirable yarn features (e.g. two edges of the yarn core, individual yarn hairs, yarn surface fibres and tracer fibres). For such purposes, some popular algorithms such as FFT (fast Fourier transform), FWHT (fast Walsh–Hadamard transform), spatial filtering, image segmentation, image filtering, edge detection and Fourier image transform have been employed. Quantitative characterization is the final stage in formulating the characteristic parameters based on the extracted features, such as thickness variation coefficient, hairiness index, helical angle of the fibre, blend ratio and irregularity. For grading yarn appearance, artificial intelligence and classification methods such as splitting cluster analysis and neural networks have also been adopted in the quantitative characterization stage for assigning the overall yarn grades.

1.10   Acknowledgement

The author wishes to acknowledge funding support from the Hong Kong Polytechnic University for the work reported here.

1.11   References

Alagha M J, Oxenham W and Iype C (1994), 'The use of an image analysis technique for assessing the structural parameters of friction spun yarns', *J Text I*, 85, 383–388.

ASTM International (2007), *Standard Test Method for Grading Spun Yarns for Appearance*, ASTM D2255–02, ASTM International, West Conshohocken, PA.

Avraham T and Lindenbaum M (2006), 'Attention-based dynamic visual search using inner-scene similarity: algorithms and bounds', *IEEE Trans Pattern Analysis and Machine Intelligence*, 28, 251–264.

Backer G, Mertsching B and Bollmann M (2001), 'Data- and model-driven gaze control for an active-vision system', *IEEE Trans Pattern Analysis and Machine Intelligence*, 23, 1415–1429.

Barella A (1983), 'Yarn hairiness', *Textile Progress*, 13, 1–57.

Barella A and Manich A M (2002), 'Yarn hairiness: a further update', *Textile Progress*, 31, 1–44.

Bultan A (1999), 'A four-parameter atomic decomposition of chirplets', *IEEE Trans Signal Processing*, 47, 731–745.

Carvalho V, Cardoso P, Belsley M, Vasconcelos R M and Soares F O (2006), 'Development of a yarn evenness measurement and hairiness analysis system', *IECON'06 Proceedings of the 32nd Annual Conference of the IEEE Industrial Electronics Society*, IEEE, Paris, 3621–3626.

Carvalho V, Cardoso P, Belsley M, Vasconcelos R and Soares F (2008), 'Yarn diameter measurements using coherent optical signal processing', *IEEE Sens J*, 8, 1785–1793.

Chaudhuri S (2001), *Super-resolution Imaging*, Boston, MA. Kluwer Academic Publishers.

Chiu S H and Liaw J J (2005), 'Fiber recognition of PET/rayon composite yarn cross-sections using voting techniques', *Text Res J*, 75, 442–448.

Chiu S H, Chen J Y and Lee J H (1999), 'Fiber recognition and distribution analysis of PET/rayon composite yarn cross-sections using image processing techniques', *Text Res J*, 69, 417–422.

Cybulska M (1999), 'Assessing yarn structure with image analysis methods', *Textile Res J*, 69, 369–373.

Guha A, Amarnath C, Pateria S and Mittal R (2010), 'Measurement of yarn hairiness by digital image processing', *J Text I*, 101, 214–222.

Itti L and Koch C (2000), 'A saliency-based search mechanism for overt and covert shifts of visual attention', *Vision Research*, 40, 1489–1506.

Itti L, Koch C and Niebur E (1998), 'A model of saliency-based visual attention for rapid scene analysis', *IEEE Trans Pattern Analysis and Machine Intelligence*, 20, 1254–1259.

Jia L F (2005), 'Determining the twist of yarns based on image analysis', *J Text Res*, 26, 39–40.

Lien H C and Lee S (2002), 'A method of feature selection for textile yarn grading using the effective distance between clusters', *Text Res J*, 72, 870–878.

Murrells C (2008), 'Twist liveliness of spun yarns and the effects on knitted fabric spirality', PhD thesis, Hong Kong Polytechnic University.

Murrells C, Wong K K, Tao X and Xu B (2007), *Yarn snarling testing apparatus and method*, US Patent 7219556 B2.

Nevel A, Avsar F and Rosales L (1996a), 'Graphic yarn grader', *Textile Asia*, 27, 81–83.

Nevel A, Lawson J B, Gordon Jr K W and Bonneau D (1996b), *System for electronically grading yarn*, US Patent 5541734.

Primentas A (2003), 'Direct determination of yarn snarliness', *Indian J Fibre Text Res*, 28, 23–28.

Qian S (2002), *Introduction to Time-Frequency and Wavelet Transforms*, Upper Saddle River, NJ, Prentice-Hall.

Qian S and Chen D (1994), 'Signal representation using adaptive normalized Gaussian functions', *Signal Process*, 36, 1–11.

Rong G H and Slater K (1995), 'Analysis of yarn unevenness by using a digital-signal-processing technique', *J Text I*, 86, 590–599.

Rong G H, Slater K and Fei R C (1994), 'The use of cluster analysis for grading textile yarns', *J Text I*, 85, 389–396.

Semnani D, Latifi M, Tehran M A, Pourdeyhimi B and Merati A A (2006), 'Grading of yarn appearance using image analysis and an artificial intelligence technique', *Text Res J*, 76, 187–196.

Towsend M W H and Cox D R (1951), 'The analysis of yarn irregularity', *J Text I*, 42, 107–113.

Watanabe A, Kurosaki S N, Konda F and Nishimura Y (1992a), 'Analysis of blend irregularity in yarns using image-processing. I. Fundamental investigation of model yarns', *Text Res J*, 62, 690–696.

Watanabe A, Kurosaki S N, Konda F and Nishimura Y (1992b), 'Analysis of blend irregularity in yarns using image-processing. II. Applying the system to actual blended yarns', *Text Res J*, 62, 729–735.

Watanabe A, Konda F and Kurosaki S N (1995), 'Analysis of blend irregularity in yarns using image processing. III. Evaluation of blend irregularity by line sense and its application to actual blended yarns', *Text Res J*, 65, 392–399.

Xu B, Dong B and Chen Y (2007), 'Neural network technique for fiber image recognition', *J Indust Text*, 36, 329–336.

Xu B G, Murrells C M and Tao X M (2008), 'Automatic measurement and recognition of yarn snarls by digital image and signal processing methods', *Text Res J*, 78, 439–456.

Xu B G, Tao X M and Murrells C M (2010), 'Evaluation of a digital image-signal approach on the automatic measurement of cotton yarn snarls', *Text Res J*, 80, 1157–1159.

Yin Q, Qian S and Feng A (2002), 'A fast refinement for adaptive Gaussian chirplet decomposition', *IEEE Trans Signal Processing*, 50, 1298–1306.

2

Digital-based technology for fabric structure analysis

B. XIN, Shanghai University of Engineering Science, China,
and J. HU, G. BACIU and X. YU,
The Hong Kong Polytechnic University, Hong Kong

Abstract: In this chapter, typical applications of image-based technologies used for the analysis of fabric structure and texture are summarized and discussed. Among them, a new digital method, based on the dual side scanning and active grid model (AGM), is introduced to demonstrate how to identify the weave pattern of woven fabrics. Some preliminary experiments and discussions are also included to validate this method.

Key words: image analysis, weave pattern, dual side scanning, active grid model, fabric structure.

2.1   Introduction

Generally, a fabric is a unique material with specific properties, such as flexibility, wearability, aesthetic quality and sewability. Both structure and appearance are important aspects for the quality evaluation and control of fabric products. Fabric structure defines the yarn construction and location inside the fabric material, including weave pattern and density, etc. Fabric appearance is a specific term more related to the visual aspects of the fabric surface; it pertains especially to wrinkling, pilling, fuzziness and surface roughness.

The textile industry has a long history, from manual operation to automatic machinery. From the very beginning of the textile industry, subjective human evaluation played a major role in quality assessment and control, fabric structure and appearance analysis. Although easy, simple and direct, this method suffered the human shortcomings of low speed, fatigue, subjectivity and inconsistency. With the development of the modern textile industry, the human eye could not satisfy the quick, objective and automated requirements now necessary for textile production. From the 1980s onwards, therefore, machine vision, based on image analysis and artificial intelligence, became one of the topics most investigated by researchers with a view to raising the standards of the textile industry and replacing the subjective human evaluation methods used previously.

23

Machine vision imitates the human eye–brain method using an automatic digital system. The behavior of the human eye is simulated by an electronic camera and the brain is replicated by a computer. Many successful attempts have been made to develop different image analysis algorithms or different objective evaluation systems for this purpose. This chapter will introduce the background and reference review of digital-based technology for fabric structure analysis and will present up-to date methods for systematic weave pattern recognition.

2.2   Background and literature review

The industrial identification of fabric weave construction is still dependent on the naked human eye and a teasing needle. It is very tedious and time-consuming work, and cannot satisfy the 'quick response' requirements of the modern textile industry. It has become necessary therefore to develop automatic digital artificial expert systems to replace the traditional eye and hand methods, thus avoiding fatigue and improving accuracy and reliability.

Since the 1980s, when research on the automatic identification of fabric structure started with the aid of computer technology, many related works have been reported. The research in this field can be divided into three periods: (1) the early and pioneer studies in Japan; (2) the period in which European and USA developments dominated; and (3) more recent advances in China and Asia.

2.2.1 Early Japanese studies

From the 1980s to the mid-1990s, Japanese scientist Akiyama and colleagues put forward a series of optical analyses based on digital image processing, pattern recognition [1–5] and yarn density measurement [6, 7]. They studied mainstream system design and digital testing concepts and tried to develop a method of transformation from optical analysis to digital image processing. They began by exploring the identification of the optical components and then measured the diffraction images on their window size and negative interval to identify the structural information on fabric weave. The relationship between the distribution patterns and their diffraction images was established by using plain, twill and satin fabrics [1]. Akiyama also applied the same digital image processing method to verify the above relationships in other derivatives of weave patterns [2].

In 1991, Nishimatsu [3] applied a charge coupled device (CCD) camera and frame grabber to digitalize the fabric surface at a magnification of 50×. Through a series of processing methods including image enhancement, threshold, image smoothing, filtering and so on, he was able to determine the floating-point position of yarns. Nishimatsu reported that the average measurement accuracy of this method for plain fabrics is 92.5%, while for twill fabrics it is only 79.1%.

In 1995, Ravandi and Toriumi [4] utilized the panel scanner and Fourier transform to process fabric images and discover their angular energy spectrum. They used the peak point extracted along the angular energy spectrum curve to determine the directional information of the fabric surface, and the Fourier inverse transformation and the autocorrelation function to measure the density of the fabric warp and weft. Ohta *et al.* [5] also obtained fabric images using scanning and adopted the rectangular bandpass filter and threshold to locate the warp and weft yarns. They then used the fan-shaped filter to determine the position of the interlacing point of the warp and weft yarns so as to determine the weave type. This method can determine the fabric structure simply and effectively using two bandpass filters and thresholding.

Prior to this, Imaoka *et al.* [6] in 1988 and Inui *et al.* [7] in 1992 and their research teams put forward two methods for analyzing the fabric of the optical image based on the Fourier transform and optical grating method. The accuracy of their methods is affected by the different types of fabrics, but the accuracy overall is about 80%. In 1991, Dusek [8] proposed a set of video image processing systems, named by the Visual Studio Industry Partner (VSIP), to measure the yarn density of knitted fabrics. Equipment included one or more digital cameras, a computer and a printer. The test results were accurate, but the yarn density of the fabric varied due to the irregular fabric texture.

2.2.2 Mainly European and US developments

From the 1990s to the early twenty-first century, American scientists Wood and Xu published their research works on the application of digital image processing methods to analyzing fabric construction. Their research into algorithm and system development brought studies up to a much higher level. In 1990, Wood [9] used a Fourier transform and an autocorrelation function to process the fabric image. He assumed that the arrangement of yarn fabric structure is periodic and available to be characterized by the Fourier transform method. He introduced the Fourier transform theory in detail and revealed the relationship between the Fourier transform and the autocorrelation function. In 1993, Yurgartis and his colleagues [10] at Clarkson University, New York, also used image processing methods to measure the yarn shape and arrangement of the fabric composites based on its microscopic image. The research concluded that the sine curve is not accurate for the description of a yarn tracing profile in fabric construction where the yarn profile can vary greatly.

In 1996, Sari-Sarraf and Goddard [11] developed an online visual system to monitor yarn fabric density. The system included linear CCD image acquisition, a digital signal processing (DSP) unit and a host computer. Through the real-time digital Fourier transform, the yarn density can be calculated from the distribution curve of the ring energy spectrum. Also in 1996, Xu [12] described procedures for applying fast Fourier transform (FFT) techniques in image processing to identify weave patterns, fabric count, the skew of the yarn, and other structural characteristics of woven fabrics. In his research, a color scanner was used to digitize fabric images (two-dimensional functions in a spatial domain), and a customized software package was used to apply the FFT to the images. Fabrics with various weave patterns and yarn counts were tested using the FFT techniques successfully. This study systematically expounded the application of the energy spectrum and established the corresponding relationship between the spectrum and the fabric spatial structure. It is now a representative research achievement in the field.

In 1997, De Castellar et al. [13] proposed an image processing method for evaluating the cover factor of yarn fabric, at the 78th Annual Textile Conference. In 1999, Kang et al. [14] designed a set of lighting systems with a digital camera which produced images of fabric samples using transmitted and reflective lighting sources respectively; the transmission images can be used for locating yarn using image gray-scale analysis. The reflective images of the fabric samples can be used to determine their weave type and yarn color configuration. A variety of image processing algorithms were used in this research, such as Gaussian filtering, threshold, histogram equalization, pattern recognition, etc.

In 2000 and 2001 in Taiwan, Huang and colleagues [15, 16] presented two image processing techniques which, respectively, could identify the weave type and yarn density and measure some geometrical parameters such as yarn twist angle. According to their experimental reports, these methods could identify three basic weave patterns. In 2001 and 2003, Escofet, Millán and Ralló [17, 18] also established a fabric structure analyis method based on the energy spectrum, using convolution and template matching to determine the minimum unit of the fabric texture and structure.

In 2003, Kuo et al. [19] in Taiwan used c-means clustering analysis for the automatic identification of fabric structure. They selected the panel scanner to digitalize the fabric surface, applied image gray-scale morphological processing to enhance the fabric image, and used the obtained c-means clustering analysis to recognize the floating-point of woven fabrics. The experimental results showed that such analysis can effectively recognize three basic weave types. In 2003, Lachkar and his colleagues [20] adopted the same tactics when using Xu, Escofet, Millán and Ralló's Fourier transform research method to perform fabric structure identification.

2.2.3 More recent advances in China and Hong Kong

From the beginning of the twenty-first century, scientific research personnel from China began to expand their work in this field. Experts from several famous Chinese textile universities and research institutes, including Donghua University, Yangtze Southern University, Wuhan University of Science and Technology, Shanghai University of Engineering Science, Tianjin Polytechnic University, China Textile Institute and Suzhou University, developed many related works to speed up digital system development for fabric construction analysis. A great deal of valuable research and exploratory work has been completed in the weave identification and density measurement field. Experts from Zhejiang University, Tianjin University, Huazhong University of Science and Technology, Shanghai University of Science and Technology, Jilin University, Zhejiang Engineering Institute and Southeast University also launched this research topic with their technical contributions of computer, photo, electrical and mechanical control.

Although much meritorious research has been accomplished in this area, most of the studies were limited to the adaptability of solid fabrics and basic weave types. Extending the work from the solid and simple woven fabrics to multi-colored and multi-patterned fabrics, even including jacquards, in the future will be difficult. A research team, led by Jinlian Hu and George Baciu in Hong Kong, recently made some good attempts to develop a digital system for fabric weave analysis. Their method, based on dual-side scanning, reportedly has higher recognition accuracy and reliability than other common image analysis methods [21, 22].

2.3   The digital system for weave pattern recognition

This newly developed method is based on dual-side scanning technology and the active grid model (AGM) and was created to identify the weave pattern of woven fabrics. The dual-side imaging system, based on the panel scanner, is illustrated in Fig. 2.1. It is able to create images of both the top and bottom of the woven fabric. It comprises three components: (1) a high-resolution panel scanner (1200 dpi or more), which is used to scan the reflective surface of the woven fabric; (2) a specific sample holder, which can hold and fix the sample with two uniformly structured plates; and (3) a personal computer with software for image capture and analysis, the software having been developed using the Microsoft Visual C++ 6.0 program language in a Windows XP operating system. The top and bottom plates in (2) are joined by a hinge, and on each plate there is a sample window on the four corners at which four reference points, with a predefined geometrical positioning relationship, are located. These reference points are used to

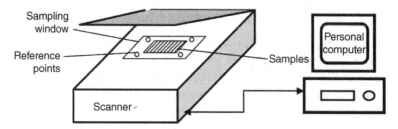

2.1 Dual-side imaging system.

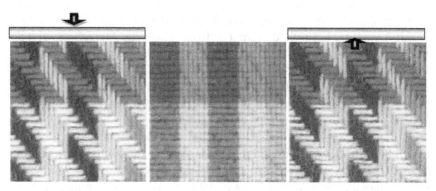

2.2 Dual-side images of a fabric after alignment (left: upper-side view; right: bottom-side view), and the merged image of dual-side images (center).

locate the mirrored yarn sections between the top and bottom images of the woven fabric.

The principle of dual-side imaging and matching relies on the alignment of the top image and the bottom image to locate the mirrored yarn sections between them, as illustrated in Fig. 2.2. In this case, the interlacing status of one yarn section, which can be viewed in both the top and bottom images, could be combined to classify the interlacing style of yarns. We used the following algorithm to implement the dual-side imaging technology: (1) locating the reference point, meaning locating all the reference points on the top and bottom plates; (2) affine transform to find the corresponding reference points on both plates; (3) merging the image pairs by transiting, scaling and rotating the images; and (4) image cropping to remove the boundaries of the merged image, only the fabric image patch inside the sampling windows being reserved. The dual-side sampling is helpful in improving the accuracy of yarn location, AGM self-adjustment and error correction. In this chapter, all the proposed woven pattern recognition algorithms are based on dual-side imaging technology.

An efficient digital imaging and sampling method is the basis for an AGM construction or weaving pattern recognition method. The dual-side imaging

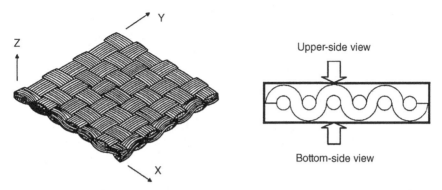

2.3 3D structure of plain woven fabric.

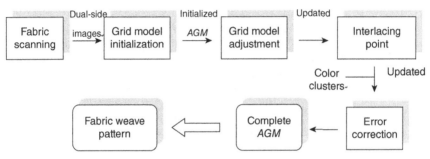

2.4 Systematic framework of dual-side scanning and AGM system.

system (Fig. 2.1) is one of the innovations presented in this chapter. The traditional fabric structure analysis methods consider only a single-side view of the fabrics. Information contained in a single-side image is not enough to provide an accurate analysis of weave pattern, which is naturally two-sided, especially when the woven fabric is made up of complex patterns. A single-side image contains only half the information of the interlacing status between the warp yarns and weft yarns, and the complete geometrical shape of a section of yarn cannot be observed. Only a dual-sided view of a fabric could illustrate the complete information and show the yarn interlacing from one side to the other side as illustrated in Fig. 2.3. The systematic framework of dual-side scanning and the AGM system is demonstrated in Fig. 2.4.

2.4   Theoretical background for weave pattern analysis

The AGM is a 2D grid mesh which represents the dual-side geometry and structural information of an arbitrary weave pattern. 'Active' means the grid structure is deformable and auto-adjustable to fit the actual geometrical

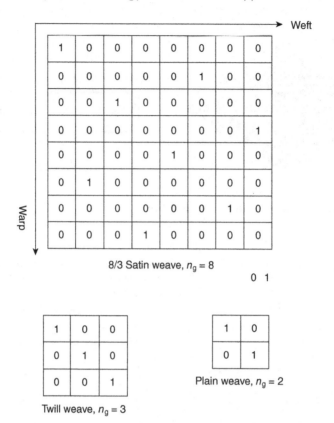

2.5 Point maps of three basic weave patterns (satin, twill and plain fabrics).

appearance of the woven fabric. The geometrical information on the yarns, such as curves and thickness variations, can be detected and represented. The concept of AGM is proposed based on three techniques: (1) the simple point map fabric model, (2) the dual-side scanning method, and (3) the active contour technique. In this section, we will introduce the concept of the AGM and the mechanism of its auto-adjustment.

The point map or weave diagram is a simple fabric model which forms the basis of AGM. The point map is a 2D diagram which is used to represent the fabric weave pattern. The columns and rows of the point map correspond to the warp and weft yarns of the fabric respectively. The dimensions of the grid are equal to the number of weave units along the warp or weft axes. For example, in Fig. 2.5, the weave unit of the satin fabric is 8 (wefts) × 8 (warps), so the dimensions of the point map grid are also 8 × 8. Each individual element of the grid represents a cross-point which is interlaced by one warp–weft pair. The state of the unit is set to 0 if the weft yarn is on the top at the corresponding cross-point and to 1 if the weft yarn is on the bottom at the corresponding cross-point. The point map/weave diagram could contain the following information: (1) the numbers of warp and weft yarns within one basic unit of the weave pattern; (2) the interlacing style of warp and weft yarns at the cross-points; and (3) the color arrangement of yarn dyed fabrics. The format of the point map/weave diagram is simple and readable and has been accepted by the textile industry for many years.

However, a fabric model which contains only the weaving pattern information is not enough to represent a real fabric. A fabric sample also has individual properties, such as the actual geometrical shapes of yarns and the local color information. These individual properties may vary from one sample to another. It is necessary to integrate the individual properties into the fabric model for a practical analysis of woven fabrics. For example, the cross-point classification algorithm could utilize the edge intensity information around each cross-point. A further predefined geometrical fabric model is useful when considering these properties. In order to establish a complete connection between a fabric model and a fabric sample, the point map should be extended to a more detailed and complete fabric model.

Based on the point map, the fabric AGM is proposed to represent both the weave pattern and individual properties of the woven fabric. The AGM inherits the basic 2D grid structure of the point map but has a more flexible and active structure. The characteristics of the AGM are defined as follows:

- The AGM has two extra components, a color palette and a color chart to store the color pattern information if the fabric is multicolored.
- The AGM adopts the 2D grid structure of the point map. Each AGM element is related to a short segment of the yarn. The geometrical shape of the yarn segment can be represented by a series of linked elements on the corresponding column or row of the AGM. For example, in Fig. 2.6, the element E_{ji} is related to the interlacing segment S_{ij} of warp W_i and weft W_j; and the weft yarn W_i can be represented by the linked elements $\{E_{0i}, E_{1i}, \ldots, E_{ji}, \ldots\}$ to those column i points;
- The shape of one yarn is determined by the geometrical information of all its AGM elements. The chain structure (CS) of the linked AGM elements makes the yarn flexible. By adjusting the positions and sizes of each AGM element they can be used to represent the shapes of different yarns. These CSs are the same as for the active contour which comprises a number of short segments and which can adjust its shape to fit an arbitrary object contour.
- One special feature of the AGM is that it is designed to represent the dual-side information of the woven fabrics. Woven fabrics are assembled

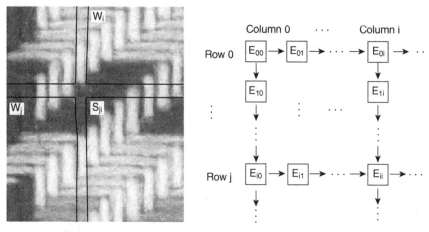

2.6 Demonstration of how to present a yarn by using a series of linked grid elements.

by interlacing yarns, so, from one side, only parts of the yarns can be observed. Besides the geometrical shape information, there are two other important properties on both sides of the woven fabric: the color information and the edge intensity information. By adopting the dual-side scanning, these two properties can be digitalized and integrated in the AGM model. For example, the E_{ji} in Fig. 2.6 can represent not only the upper warp segment but also the hidden weft segment.

The weave pattern of one woven fabric is also one component of the information carried by the AGM of the fabric. Other components include the fabric's geometry or color information. In an AGM, each element has a parameter to define the interlacing type, which also contains the relative positions of the warp and weft at this point. The combination of all these parameters within an AGM is the weave pattern. The extraction of these parameters is based on the geometrical and edge intensity information. So, in order to obtain the weave pattern, we need to initialize an AGM first.

2.4.1   Active contour and AGM self-adjustment

The active contours or 'snakes', first introduced by Michael Kass *et al.* and discussed in De Castellar *et al.* [13], are energy-driven curves that move within images to auto-adaptively deform and describe the shape of objects. Given an approximation of the boundary of an object, an active contour model can be used to locate the 'actual' boundary when an initial guess is provided by a user or by some other method. This model proposes a way to change a flexible structure to describe those complicated 2D objects, like the grid of a fabric structure, if there are observable boundaries to separate objects from the background or other parts of the objects.

An active contour is an ordered collection of n points in the image plane:

$$V = \{v_1, \cdots, v_n\}$$
$$v_i = (x_i, y_i), \ i = \{1, \cdots, n\} \tag{2.1}$$

The points in the contour iteratively approach the boundary of an object through the solution of an energy minimization problem. For each point in the neighborhood of v_i, an energy term is defined as:

$$E_i = \alpha E_{int}(v_i) + \beta E_{ext}(v_i) \tag{2.2}$$

where E_{int} is the internal energy function that depends on the shape of the contour, and E_{ext} is the external energy function that depends on the image properties, such as the gradient of the near point v_i; α and β are two constants providing the relative weighting of the energy terms. The internal energy function is intended to enforce a shape on the deformable contour and to maintain a constant distance between the points in the contour. The external energy gives the correction between the shape of the active contour and the actual boundary of the target object.

So, in order to make the AGM into one complete active energy-driven system, three things are needed: (1) an active contour, (2) an initial guess, and (3) an energy system including the external and internal energy. The AGM is a combination of a set of active contours. Each CS of an AGM comprises a series of linked and ordered elements and has an adjustable shape. A CS alone is an active contour, but its properties are under the constraints of its neighbors in the AGM. After the AGM initialization, each CS of an AGM will be moved to approximately overlay a yarn of the corresponding fabric sample. The initialization is a guess of the geometrical shape of the target. The AGM satisfies requirements (1) and (2) of an active system. The internal energy is the part that depends on the intrinsic properties of the active contour. The intrinsic properties here are usually the object-oriented constraints, such as the length or curvature. Considering the nature of the yarns and the fabrics, the following three properties are used to achieve the validation of the AGM.

• The width of a yarn should be within a particular scope. A yarn is not an even strip, but its width does not vary greatly. A maximum threshold $width_{max}$ and a minimum threshold $width_{min}$ are given to keep the width of an AGM CS within a rational scope. We can also use a deviation factor δ to define the maximum and minimum thresholds. Then, $width_{max} = m + \delta$, and $width_{min} = m - \delta$, where m is the statistical average of the yarn width.

- The yarns have small deformations along their length, but their curvature should be within a reasonable scope. The high density of the fabrics, either larger or smaller, forces the neighboring yarns to push against each other and thus keep approximately straight shapes. The yarn curvature is defined as the elasticity of the linked AGM elements. The distance between two neighboring linked AGM elements is restricted by an elasticity parameter E_{cur}.
- The weft or warp yarns in a fabric are parallel to one other. There is no crossing and no overlap. This property gives them a relatively restricted position among the CSs within an AGM. The boundary of a CS cannot crush into its neighbors' boundaries. For example, if the coordinate of the left boundary of a CS at position (i, m) is $l_{i,m}$, then $l_{i,m} < r_{i-1,m}$. Its right boundary coordinate $r_{i,m}$ should be smaller than the coordinate of the left boundary $l_{i+1,m}$ of the right neighbor.

The external energy of the AGM system is determined by the image of the fabric. The yarns are separated by the boundaries which are more numerous in the edge map. We want the CSs to latch on to bright structures in the edge map. Then the obvious energy function is the sum of the gray levels of the pixels that the CSs are on top of. The external energy of an AGM is the accumulation of the edge intensities for the pixels which lie under the left $EI_l(i)$ and right $EI_r(i)$ boundaries of all the CSs of an AGM:

$$E_{ext} = \sum_{c=1,\dots,H\times W} \sum_{i=0,\dots,n} EI_l(i) + EI_r(i) \qquad 2.3$$

where H and W are the size of the AGM along the warp and weft directions respectively. Enlarging this energy function (i.e. making it more negative) will move the CSs towards brighter parts of the image.

The potential energy between two arbitrary neighboring points produces the force necessary to push or lead the CS to stronger edge positions. If the edge map of the merged fabric image is considered as a function $I(x, y)$, where the domain of I is the set of point locations, and the range of I is the pixel intensity, the gradient of this edge map could be calculated, and then the potential directions at each point given by the image gradient. At each step of the self-adjustment, only one edge of an AGM element would be moved. The direction of an edge E_i would be:

$$\text{dir}(E_i) = \sum_{i\in E_i} \text{grad}(i) \qquad 2.4$$

At each step, the motion of the CSs would be calculated first, then the validation defined by the AGM properties and the global energy E_i checked. This process would be repeated until a local optimal solution was achieved.

2.5   Methodology for active grid model (AGM) construction and weave pattern extraction

The AGM construction includes four major steps. Firstly, a yarn detecting algorithm is applied to a merged fabric image based on dual-side images and the result is used to initialize the AGM. Secondly, the AGM self-adjusts to refine itself and achieve a more accurate result. In the third step, the type of yarn interlacing is classified based on the edge intensities and then corrected using the neighboring information. Finally, color information is sorted using color clustering and matching methods. A detailed fabric weave pattern analysis method based on AGM is described in the following sections.

2.5.1   Yarn detection and AGM initialization

The initialization of the AGM involves two major tasks. First, the size of the AGM needs to be determined. The numbers of the horizontal and vertical CSs of the AGM are equal to the numbers of the warp and weft yarns respectively. Second, the initialization of the AGM needs to be obtained. The corresponding position parameters of the AGM are updated by the approximate yarn positions. These two steps are based on the accurate yarn detecting algorithm.

The long stripe shape of the yarn is a distinct feature. In the edge map, the yarns are represented by two parallel lines. These features can be enhanced and used for yarn detection. In the edge map obtained by merging the front and back views of the fabric, the continuous yarn boundaries can be observed. The yarn detection problem can thus be solved by detecting the parallel lines in the edge maps of the fabric merged images. An innovative way of detecting objects with such a regular shape is to use template-based image matching. A yarn template can be designed to filter the edge image. The places containing the parallel line features will be enhanced. For example, a template with two vertical parallel lines can be used to detect the warp yarns by filtering the vertical edge map. The size of the template is decided by the fabric density which can be measured using a method based on Fourier analysis [15]. The match between the template and the floating sampling window on the edge map can be measured by the Hausdorff distance. The detection of the yarns is demonstrated in Fig. 2.7 and the detection result is shown in Fig. 2.8.

The positions of the yarns can be located using the yarn detection result image. Because the fabric is an approximately orthogonal grid, accumulating the yarn detection result along the vertical or horizontal direction can give a clearer view of the distribution of the yarns. In the accumulation histogram, each peak corresponds to a fabric yarn. By segmenting this

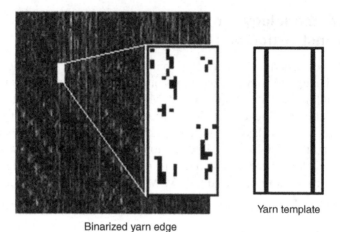

Binarized yarn edge   Yarn template

2.7 Demonstration of template-based (warp) yarn detection.

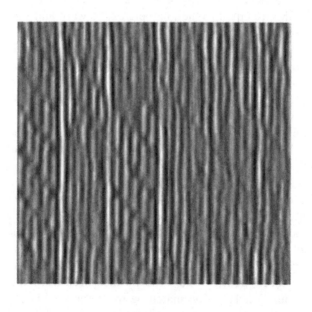

2.8 Warp yarn detection result and its intensity accumulation histogram along the vertical direction.

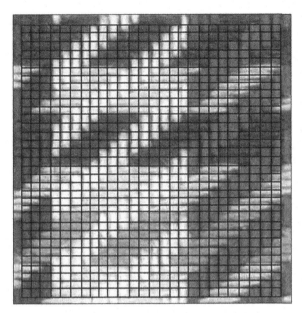

2.9 Initial grid model of woven fabric.

accumulation histogram, the yarns can be located. The initial AGM is obtained by establishing a one-to-one relationship between the CSs and the yarns (Fig. 2.9). When constructing the grid model, all the yarns are shown as straight stripes which follow either a vertical or a horizontal direction. The shapes and position values assigned to the grid columns are approximate values.

2.5.2   AGM self-adjustment

The AGM should accurately represent the geometrical information from the fabric sample such as small deformations and thickness variations of the yarns. In order to achieve this, the initial AGM should be self-adjustable using the method introduced in the previous section.

The self-adjustment of the AGM can be considered as the energy minimization process of a physical system. During self-adjustment, the edge should keep the continuity and a reasonable distance to adjacent edges. The results of the AGM self-adjustment contain the curves and thickness variations of the yarns. After the adjustment, the cross-points of the grid model are moved to the right positions on the yarn interlacing points of the fabric (Fig. 2.10). In our experiment, δ is set to be 3.0 and E_{cur} is set to 1.0. The setting of the two factors α and β is determined by the balance of the internal energy E_{int} controlling the shape adjustment and the external energy

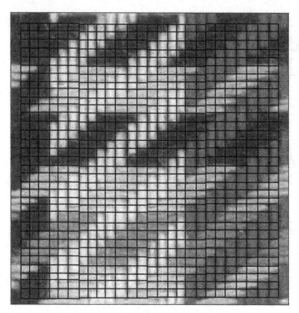

2.10 Fabric grid model after adjustment.

E_{ext} controlling the yarn locating adjustment. The α of fabrics with large shape vibration could be set larger than β; the β of fabrics with low boundary contrast could be set larger than α. In our experiment, the two factors α and β are both set to 0.5.

2.5.3 Classification of yarn interlacing

The type of yarn interlacing is determined by the relative position of the weft and warp yarns at the cross-point. The cross-point is defined as the warp cross-point if the warp yarn is on top, otherwise the cross-point is defined as the weft cross-point. In the AGM, a cross-point is represented by a quadrangular element. The intensities of the four edges of this element are determined mainly by the type of cross-point and, vice versa, the type of cross-point can be derived from the intensities of its four edges.

In an ideal case, the intensities of the edges would have two states: 0 and 1. The cross-points could then be partitioned into eight classes according to the different combinations of the intensities of the four edges (Fig. 2.11). Here, both the warp and weft cross-points can be further partitioned into four sub-classes. However, in practical cases, the intensities of the edges are not absolutely 1 or 0 but are relatively high or low. The classification of cross-points is a typical K-partition problem that has four input data and eight output states. We propose a BP neural network for this classification task.

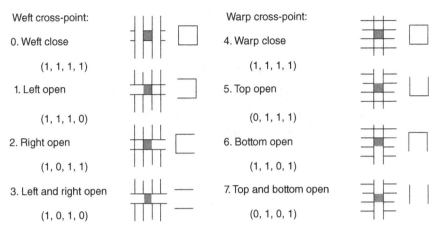

Weft cross-point:

0. Weft close

(1, 1, 1, 1)

1. Left open

(1, 1, 1, 0)

2. Right open

(1, 0, 1, 1)

3. Left and right open

(1, 0, 1, 0)

Warp cross-point:

4. Warp close

(1, 1, 1, 1)

5. Top open

(0, 1, 1, 1)

6. Bottom open

(1, 1, 0, 1)

7. Top and bottom open

(0, 1, 0, 1)

2.11 Cross-point classification according to the different combinations of the intensities of the four edges.

Table 2.1 Possible types of neighbor cross-points

Type	Left	Top	Right	Bottom
0	4,5,6,7	4,7	4,5,6,7	5,7
1	2,3	0,1,2,3,4,7	4,5,6,7	0,1,2,3,5,7
2	4,5,6,7	0,1,2,3,4,7	1,3	0,1,2,3,5,7
3	2,3	0,1,2,3,4,7	1,3	0,1,2,3,5,7
4	0,1,4,5,6,7	5,6	0,2,4,5,6,7	0,1,2
5	0,1,4,5,6,7	0,1,2,3	0,2,4,5,6,7	4,6
6	0,1,4,5,6,7	5,6	0,2,4,5,6,7	4,6
7	0,1,4,5,6,7	0,1,2,3	0,2,4,5,6,7	0,1,2

It is enough for the weave pattern extraction to partition the cross-points into two classes. The reason to partition the two classes into eight sub-classes is that the combinations of the edge intensities of one cross-point are related to combinations of the types of its neighbors, and the neighboring information can be used to correct the classification errors. For example, if the left neighbor of a cross-point is a weft left open cross-point (type 1), then there are only two choices as to the type of the cross-point itself: weft right open (type 2) or weft left and right open (type 3). There are two principles to follow for error correction based on neighboring information. First, if the front cross-point is a weft cross-point, the back cross-point should be a warp cross-point, and vice versa. Second, the types of two neighboring cross-points should comply with the relationships listed in Table 2.1. Figure 2.12 illustrates the result of cross-point classification and correction based on neighboring information with few errors.

2.12 Cross-point classification and result of correction based on neighboring information.

2.5.4 Color configuration

The color pattern should be considered if the fabric has more than one colored yarn. After the self-adjustment of the AGM, the local color information on the two sides of an AGM element and the CSs can be collected. This information can be used to (1) decide the color of the colors within the fabric, (2) decide the color pattern of the fabric, and (3) correct the cross-point classification errors.

The attempt to discover the number and properties of colors in a fabric can be defined as a color clustering problem. Unlike the requirements for color calibration and color matching, it is not necessary during color clustering to find out the reflection spectrum or to keep the consistency of the color. In this step, the colors are usually represented by the raw RGB values. There are two ways to obtain the color information for a fabric. The most reliable way is to manually sample the colors in the image. Alternatively a histogram-based automatic color clustering method [16] can be used. This method can quickly obtain the color information for the fabric, but may sometimes gives an incorrect result if the fabric includes colors which are adjacent or even overlap in the color space. If automatic clustering has errors, we will use manual operation to adjust it and discover the correct color information.

2.13 Final fabric grid model after the color assignment and the color-based correction.

The color pattern of a fabric is the arrangement of different color yarns along the warp and weft directions. The colors of the yarns are classified according to the color clusters. By matching the color values of a CS and the color clusters, the color properties of the corresponding yarns can be decided. After that, the color yarn arrangement of the fabric is obtained. The repeat of this arrangement is the color pattern of the fabric.

The color pattern can be used to correct the errors of cross-point classification. This correction is only applicable to cross-points which are produced by yarns of different colors. The types of these kinds of cross-points can be determined by the yarn colors. If the color of the front of the cross-point matches the color of the weft yarn, the cross-point is a weft cross-point, and if vice versa it is a warp cross-point. If the cross-point classification result is contrary to the classification result obtained using colors, it can be corrected. Figure 2.13 shows the final fabric grid model after the color assignment and the color-based correction.

In this chapter, four kinds of fabric with different colors/textures were used for the validation and demonstration of this proposed method. The results of the fabric simulation using the extracted weaving pattern are shown in Fig. 2.14.

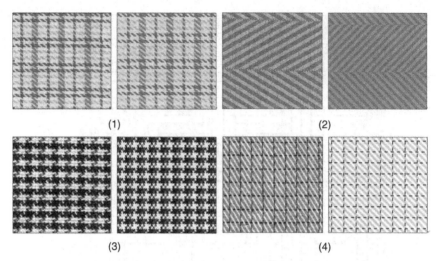

2.14 The real fabric images (left) and simulated images based on the AGM (right).

2.6   Conclusions

Digital-based technology for fabric structure analysis is reviewed in this chapter. A new method based on the active grid model (AGM) is used to identify the weave pattern of woven fabrics. This proposed method utilizes dual-side scanning technology to merge dual-side images of a fabric at yarn level. It describes a four-step method for constructing an AGM. Firstly, a yarn detection algorithm is applied on the dual-side scan images to initialize the AGM. Secondly, the AGM self-adjustment scheme is used to adjust the AGM accurately. Then, the types of interlacing yarn are classified based on the edge map and the result is refined using the information from neighboring yarns. Finally, the color pattern is determined using color clustering and matching; error correction is also made based on its color configuration. Preliminary experiments show that AGM is effective for the classification of fabric weaving patterns.

2.7   References

[1] Ryuichi, A., *et al.*, Detection of weave types in woven fabrics by observing optical diffraction pattern. *Sen'i Gakkaishi (Journal of the Society of Fiber Science and Technology, Japan)*, 1986, **42**(10): 574–579 (in Japanese).
[2] Kinoshita, M., *et al.*, Determination of weave type in woven fabric by digital image processing. *Journal of the Textile Machinery Society of Japan (English Edition)*, 1989, **35**(2): 1.

[3] Nishimatsu, T., Automatic recognition of the interlacing pattern of woven fabrics. Part 1. Determination of the pattern by using image information analysis (in English). *Journal of the Textile Machinery Society of Japan (Japanese Edition)*, 1991, **44**(6): T126.

[4] Ravandi, S.A.H. and K. Toriumi, Fourier transform analysis of plain weave fabric appearance. *Textile Research Journal*, 1995, **65**(10): 676.

[5] Ohta, K.I., Y. Nonaka, and F. Miyawaki, Automatic analyzing of a weaving design with the spatial frequency components. *Proceedings of the 3rd IEEE-ICSC'95 (International Computer Science Conference)*, 1995, 516–517.

[6] Imaoka, H., *et al.*, Trial on automatic measurement of fabric density (in English). *Sen'i Gakkaishi (Journal of the Society of Fiber Science and Technology, Japan)*, 1988, **44**(1): T32.

[7] Inui, S., *et al.*, Development of automatic measurement system for fabric density (in English). *Bulletin of the Research Institute for Polymers and Textiles (Yokohama)*, 1992, **169**: 7.

[8] Dusek, Z., Contactless thread density measurement of woven and knitted fabrics. *Melliand Textilberichte – International Textile Reports*, 1991, **72**(11): 917.

[9] Wood, E.J., Applying Fourier and associated transforms to pattern characterization in textiles. *Textile Research Journal*, 1990, **60**(4): 212–220.

[10] Yurgartis, S.W., K. Morey, and J. Jortner, Measurement of yarn shape and nesting in plain-weave composites. *Composites Science and Technology*, 1993, **46**(1): 39.

[11] Sari-Sarraf, H. and J.S.J. Goddard, On-line optical measurement and monitoring of yarn density in woven fabrics, in *Photonics China '96, Symposium on Automated Optical Inspection for Industry: Theory, Technology, and Application*, Beijing, 1996, 444–452.

[12] Xu, B., Identifying fabric structures with fast Fourier transform techniques. *Textile Research Journal*, 1996, **66**(8): 496.

[13] De Castellar, M.D., *et al.*, New finding of fabric cover factor determination by image analysis. *Textiles and the Information Society*, papers presented at the 78th World Conference of the Textile Institute in Association with the 5th Textile Symposium of SEVE and SEPVE, 1997: 245.

[14] Kang, T.J., C.H. Kim, and K.W. Oh, Automatic recognition of fabric weave patterns by digital image analysis. *Textile Research Journal*, 1999, **69**(2): 77.

[15] Huang, C.C., S.C. Liu, and W.H. Yu, Woven fabric analysis by image processing. Part 1: Identification of weave patterns. *Textile Research Journal*, 2000, **70**(6): 481.

[16] Huang, C.C. and S.C. Liu, Woven fabric analysis by image processing. Part 2: Computing the twist angle. *Textile Research Journal*, 2001, **71**(4): 362.

[17] Escofet, J., M.S. Millán, and M. Ralló, Modeling of woven fabric structures based on Fourier image analysis. *Applied Optics*, 2001, **40**(34): 6170–6176.

[18] Ralló, M., J. Escofet, and M.S. Millán, Weave-repeat identification by structural analysis of fabric images. *Applied Optics*, 2003, **42**(17): 3361–3372.

[19] Kuo, C.-F.J., Chung-Yang Shih and Jiunn-Yih Lee, Automatic recognition of fabric weave patterns by a fuzzy c-means clustering method. *Textile Research Journal*, 2004, **74**(2): 107–111.

[20] Lachkar, A., *et al.*, Textile woven-fabric recognition by using Fourier image-analysis techniques: Part I: A fully automatic approach for crossed-points detection. *Journal of the Textile Institute*, 2003, **94**(3): 194–201.

[21] Binjie Xin, Jinlian Hu, George Baciu, and Xiaobo Yu, Investigation on the classification of weave pattern based on an active grid model. *Textile Research Journal*, 2009, **79**(12): 1123–1134.

[22] Xiaobo Yu, Micro structure information analysis of woven fabrics, PhD thesis, Hong Kong Polytechnic University, 2008.

3

Computer vision-based fabric defect analysis and measurement

A. KUMAR,
The Hong Kong Polytechnic University, Hong Kong

Abstract: Quality assurance for textile fabrics generated from production lines is vital for competitive advantage. The computer vision-based automated inspection of textile fabrics has invited significant attention from researchers to develop effective techniques for fabric defect detection and classification. In this chapter we provide an overview of the open problems, the state-of-the-art on fabric defect detection and classification techniques, and also the promises of the new technologies that could make low-cost robust defect detection a reality for textile industries. This chapter also provides a summary of the research efforts for the automated classification of dynamically populated fabric defects. Finally we discuss the evolving new technologies which promise to significantly enhance the state-of-the-art in fabric defect analysis for industrial usage.

Key words: fabric defect detection, textile inspection, fabric defect classification, quality assurance.

3.1   Introduction

Automated quality assurance for textile fabric materials and products is one of the most challenging computer vision problems in real-world applications. The random variations in the knitting process, quality of fabric yarn, production and operating conditions often result in dynamically populated defects that vary in size, shape, appearance and color. The economic benefits resulting from the visual inspection for textile product quality are huge and justify the investment in automated computer vision-based solutions for product quality assurance. It is estimated that even the most highly trained inspectors can detect only about 70% of fabric defects, and that fabric defects reduce the value of produced fabrics by about 45–65% [41]. Recently some commercially available fabric inspection machines have entered the market. However, their cost is significantly high and the ranges of defects that can be detected are quite limited. The increasing availability of low-cost high-speed computers, high-resolution digital cameras and low-cost storage has generated much promise for robust automated textile inspection solutions to become popular in the near future.

45

3.2   Fabric inspection for quality assurance

The inspection of textile fabrics is traditionally performed by human inspectors who are specifically employed and trained for this task. While trained inspectors locate the fabric defects offline with their prior knowledge, their judgment is highly subjective and often influenced by expectations. The subjectivity, high cost, labor and fatigue associated with human fabric inspection can be alleviated by automated computer vision-based inspection. Such inspection can potentially achieve high speed, consistency, efficiency, low cost and more objective quality assurance of textile fabrics. There are several kinds of fabric defects which can also be corrected for their repetitive occurrence if detected early during the online inspection. Such defect correction from the *automated* adjustment of process parameters is highly desirable and helps to reduce waste.

The production lines in textile industries generate a variety of fabrics: apparel fabrics, tie cord, industrial fabrics, piece-dyed fabrics, finished fabrics, denim, upholstery fabrics, etc. The range of commonly occurring defects in each category of these fabrics is large and the same class of fabric defect can appear quite differently on the same or different fabrics. Therefore the algorithmic requirements for the fabric defect classification are highly constrained by challenges from large intra- and inter-class variations in the fabric defects. The appearance of some class of fabric defects is a very poor guide to their true nature, and reference [42] details the response of such defects for human inspectors using physiological methods.

3.2.1 Fabric defects

The textile industry produces a variety of fabrics, e.g. woven, knitted, bondage, braided, upholstery, etc., for civilian and industrial applications. Each of these fabrics is essentially composed of yarns, which can be simple or complex, and characterized by their strength, twist and density. The woven fabric defects have been largely categorized into yarn defects and weave defects. Figure 3.1 illustrates nine different types of common defects on plain weave fabric samples and three on twill weave samples which are visible for low-resolution imaging. The photographs of the samples shown in Fig. 3.1(a)–(c) were acquired using front lighting, while those in Fig. 3.1(d)–(l) were acquired using back lighting. It may be noted that some defects (e.g. *start marks*) can be better observed in transmitted light while others (e.g. *oil spots*) can be seen more clearly in reflected lighting.

These defects can be generally categorized into horizontal, vertical and block types of weaving defects. Due to the nature of the weaving process, the majority of fabric defects are likely to appear in the vertical and/or horizontal directions. Some defects can be quite ambiguous, such as (e) and

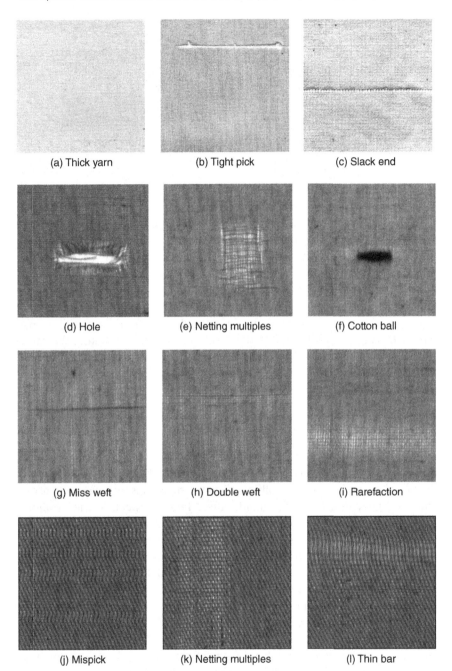

3.1 Image samples from common fabric defects in plain weave fabric (a)–(i) and twill weave fabric (j)–(l).

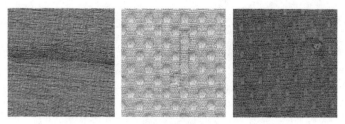

3.2 Image samples from fabric defects in patterned fabrics.

(f). They seem to have no exact distinction simply from color or shape. Therefore the classification of several categories of such defects can be highly subjective even for human inspectors. There are several established standards which provide details on the classes of fabric defects and help to impart more objectivity in the categorization of fabric defects, e.g., *Manual of Standard Fabric Defects in the Textile Industry* [38]. Figure 3.2 illustrates samples of patterned (textured) image samples acquired using high-resolution imaging (200 pixels per inch). The current state-of-the-art in textile inspection is more mature for the detection of plain/twill weave fabrics (Fig. 3.1) and the defect detection algorithms can detect the most popular defects with high accuracy. However, the detection of patterned fabric defects (Fig. 3.2) is quite challenging and is inviting a lot of research interest in developing effective defect detection approaches.

3.2.2 Automation for fabric inspection

The automated inspection of the textile web is often performed offline primarily for two reasons: the slow speed of the textile web from the production lines and the unfavorable environment because of excessive noise and fiber heap. On the other hand, online fabric inspection can be more suitable for high-resolution imaging and inspection as the slow speed of the moving fabric can allow complex computations on high-resolution images. Another advantage of online inspection is related to the possibility of online defect correction, i.e., some class of fabric defects (e.g. running warp defects, recurring filling defects) can be eliminated by correction of weaving parameters. The automation requirements for high-speed textile web inspection are summarized in Fig. 3.3. As shown in this figure, an array of digital cameras is synchronized with the fabric movement and covers the entire area of the moving fabric. Each of the acquired image frames is analyzed for the defect and the identified defects (if any) are classified into one of the many known fabric classes. The scaled map of the textile web is marked with the location and type of defects and used for grading/segmenting the quality of the fabric at different locations on the web.

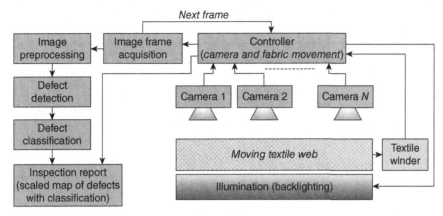

3.3 Automation for quality assurance using computer vision-based inspection.

There are several commercially available solutions [46] for automated fabric defect detection and classification. The IQ-TEX system from Elbit Vision Systems [43], Fabriscan from Zellweger Uster [44] and Cyclops from BarcoVision [45] are some popular examples of such solutions. IQ-TEX and Fabriscan inspect the textile web offline or at the exit end of the inspection machine. Cyclops contains a traveling scanning head which can be deployed on the weaving machine itself for online inspection as it is generated from the weaving machine. The weaving process is therefore automatically stopped if online inspection detects any running or serious defects. These systems are designed to perform high speed, offline, fabric defect detection and classification. The high cost and limitations on the range of detected fabric defects have limited the large-scale deployments of currently available commercial solutions in the textile industries. There are also several efforts to develop (low-cost) fabric inspection solutions, using fuzzy wavelets and fractal scanning of fabric [47], multiscale wavelet decomposition [51] and Markov random field models [14, 49]. These efforts have attracted a lot of attention and continue to drive further research efforts to develop robust, low-cost and effective solutions for textile inspection.

3.3   Fabric defect detection methods

Most of the research efforts in the literature of fabric inspection have been focused on the detection of fabric defects. Reference [1] provides an extensive summary of fabric defect detection approaches, their limitations and advantages, presented in the literature (up to 2005). Table 3.1 summarizes some of the best-performing approaches for fabric defect inspection. The fabric defect detection approaches that can analyze the information from

Table 3.1 Best-performing approaches for plain and twill fabric inspection

Feature extraction	Reference	Comments
Gabor filters	[48]	Spatial and spatial-frequency domain local features
Wavelets	[47]	Wavelet analysis of 1D data from fractal scanning
Regularity	[50]	Texture similarity using localized regularity
Optimal filters	[52]	Linear finite impulse response filters

local regions have shown the most promising results in the literature. The fabric defect detection approaches that perform global analysis of the gray levels in the acquired image frames are more suitable for the detection of coarse defects. However, such appearance-based approaches can also be used to extract localized feature details, e.g. using kernel PCA, kernel LDA, etc., but have yet to be investigated in the literature. The localized extraction of information in the spatial and spatial-frequency domain has been shown to offer the best overall experimental results. Such localized feature extraction approaches, e.g. Gabor filters [48], wavelets [47], optimal filters [52], etc., are highly suitable for online fabric inspection. In addition, localized feature extraction using texture regularity [50] has been shown to offer the most promising results on the publicly available TILDA database and is highly effective for detecting defects from complex texture backgrounds.

Another localized feature extraction approach recently presented in the literature uses localized binary patterns (LBP) for the feature extraction. Such an approach effectively extracts phase information from the localized window regions to characterize the textile image. Tajeripour *et al.* [28] demonstrated that the LBP phase features showed promise in detecting fabric defects in patterned fabrics. These authors applied LBP masks to the local windows of each fabric image, and this LBP feature from a sub-window containing defects was compared with that of the reference image (normal fabric) to determine whether a defect can exist.

Mak *et al.* [29] presented another promising approach for fabric defect detection using local texture analysis based on morphological operators. The authors used a Gabor wavelet network (GWN) for the extraction of localized texture features from the defect-free fabric images. The GWN is constructed from the imaginary Gabor wavelets and used to construct a structuring element for morphological filters. Finally, the authors employed a few such tuned morphological filters and illustrated promising defect detection results on 32 commonly appearing textile defects on twill, plain and denim fabrics. One of the notable advantages of this approach lies in its smaller computational complexity resulting from the tuned morphological operators, which makes it an attractive choice for online textile inspection.

3.3.1 Fabric defect detection in patterned fabrics

The detection of defects in patterned fabrics is quite challenging and is inviting increasing interest among researchers. The patterns of textures in textile fabrics can be more effectively analyzed from the spatial arrangement of gray levels. We can consider patterned fabrics as being constructed from the repetition of a unit pattern [7]. The analysis of such textures involves four key steps [31]: texture segmentation, texture classification, texture synthesis and shape studies. Some properties such as uniformity, coarseness, roughness, directionality, regularity, etc., have been widely studied in the texture segmentation literature. Among these properties, regularity analysis of patterned textures has received growing attention, which is an important feature because it is an invariant but perceptually motivated feature. Ngan and Pang [32] also used the wide existing regularity in textured images and proposed an approach they called the regular bands method, which is based on the idea of periodicity. The key idea of the regularity approach is to study or represent signal generation for each vertical and horizontal line of the defect-free region. Any defect in a defective region would correspond to an irregularity in the signal.

Kuo et al. [30] developed an automated visual inspection approach, based on wavelet transform (WT), for the detection of defects on a patterned fabric. Wavelet transform had been successfully applied earlier [47, 51] for the detection and classification of defects on plain fabrics. The authors compared three methods on the WT decomposed sub-images and showed that the method of wavelet pre-processed golden image subtraction (WGIS) has the best rate, which is up to 96.7%. Actually, this method is the combination of WT and the technique of golden image subtraction (GIS). WT is used as a tool to reduce the noise in the texture image, which is a key problem in defect detection. Then GIS is applied to the WT decomposed image. In this process, the authors acquired the golden image, which is bigger than the pattern unit in the textures from a reference image. Then, for every test image, this golden image moves over and the energy measure is computed for each pixel. The thresholding of the entropy values for each pixel determines the location of defects in the textured test images.

Defect detection on patterned textures using hash functions has been investigated by Baykal et al. [39]. Hash functions were originally used in cryptography to ensure integrity of files. A family of special hash functions was developed in this work. The authors applied these functions to textured images in the horizontal and vertical directions respectively. The location of defects in the texture image can be ascertained by thresholding the output from the two directions. These functions were shown to be quite effective and sensitive for defect detection in the texture. The key motivation for using this approach is that the hash functions are immune to variations in illumination and contrast.

3.4   Fabric defect classification

Most of the literature on automated textile inspection is predominantly focused on the detection of fabric defects [1–6, 8, 15] and little attention has been paid to the automated classification of localized or detected defects. Automated fabric defect detection is required to automatically segment defect-free fabric from fabric with defects and precisely compute the location of defects on the moving textile web. The classification of such localized defects into one of the several known categories is necessary to further ensure quality assurance and for the economic usage of the textile product. Some of the detected fabric defects are severe and accordingly have to be eliminated (trimmed) from their localized position on the textile web. However, some defects can be very subtle, i.e., least severe, and result in lowering the quality of fabric for its economic usage. In summary, fabric defect detection is required to ensure freedom from defects and a high level of quality assurance, while the classification of such detected defects is required for the economic usage of products and for the defect analysis to reduce or eliminate the occurrence of such defects.

Fabric defect classification is largely treated as a popular pattern classification problem and therefore researchers prefer to use mature machine learning approaches for defect classification [9–12]. The supervised pattern classification methods require categorization of multiple classes of detected or presented fabric defects into one of the possible patterns in the training database. Due to the large number of fabric defect classes, large intra-class and small inter-class variations, the selection of training samples and classes largely determines the upper limit to the achievable performance. There has been no effort, to the best of our knowledge, by researchers to train the classifier for more than 100 classes of fabric defects and to carry out the performance analysis. Plausible reasons for this could be the absence of such a large database and/or the large inter-class similarity of fabric defect classes, which is likely to present less discriminating (or more confusing) samples for effective training. Several fabric defects look quite similar (or the same) while belonging to different defect classes, and several defect classes may appear differently at different locations or frequencies on the textile web. A rigorous comparative experimental evaluation of the performance for the fabric defect classification on a very large number of defect classes, from several machine learning approaches, has not yet been attempted and would be of high interest for system development.

The entire process of fabric defect classification largely consists of three stages: feature extraction, feature analysis and training the classifier. For the classification task, it has to be ensured that enough representative fabric defect samples are available. These samples are used for training the classifier. The classifier can then learn to discriminate among those defects of known class using the discriminant features. Therefore the selection of feature representation scheme is equally important in achieving accurate defect classification. The classifier learning process depends greatly on the training samples, the feature representation, the training algorithm and the parameters of the training process. The most promising fabric defect classification results have been achieved from the neural network (NN) and support vector machine (SVM) based classifiers. Such learned classifiers often have little complexity and therefore are highly suitable for online classification of fabric defects. We now briefly discuss these classifiers.

3.4.1 Neural network-based fabric defect classification

A simplified structure of a neural network with one hidden layer is shown in Fig. 3.4. The two-layer structure illustrated here could be the simplest one, which includes an input layer and an output layers. The more complex

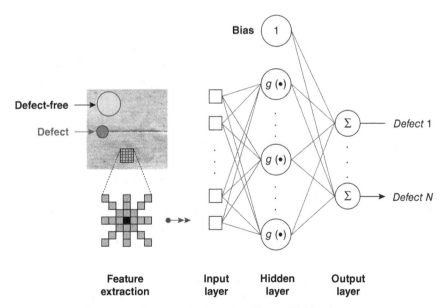

3.4 Structure of a neural network with one hidden layer.

structure, often required for the effective classification of large classes of fabric defects, additionally has one or more hidden layers between the input and output layers. A neural network with a two-layer structure is too simple and lacks generalization capability for the classification problem. Therefore we prefer to use three-layer neural networks, which contain one hidden layer, and can provide more effective classification capability from representative feature vectors. Each layer has several nodes which play different roles. The nodes in the input layer connect to the real world, where the discriminant features extracted from the unknown or input image are made available. The number of these nodes highly depends on the number of the dimension of feature vectors. The number of the output nodes is related to the nature of the classification problem. Each node can be seen as a class output from neural networks, which means that if the value of this node is the maximum, then the unknown input image or defect can be classified in its corresponding class. The exact number of the nodes in the hidden layer is a critical parameter which highly affects the generalization capability of the neural network. Empirically, it can be estimated from the number of input and output nodes. If N_i, N_h and N_o represent the number of input, hidden and output nodes respectively, then they are related by $N_h = \sqrt{N_i N_o}$. This is a popular rule of thumb and provides a reasonable estimate to determine the number of nodes. The values of the nodes in the hidden layer and the output layer are related as follows:

$$\varphi_j^l = \sum_{i=1}^{N_{l-1}} w_{ij}^{l-1,l} y_i^{l-1} \qquad\qquad 3.1$$

$$y_j^l = g(\varphi_j^l) \qquad\qquad 3.2$$

where the sum of weighted inputs for the jth ($j = 1, \ldots, N_{l-1}$) neuron in the lth layer is represented by φ_j^l. The weights from the ith neuron at the $(l-1)$th layer to the jth neuron in the lth layer are denoted by $w_{ij}^{l-1,l}$ and y_i^l is the output for jth neuron in the lth layer. The popular practice is to provide the values of −1 and 1, corresponding to 'defect' and 'defect-free' responses, to the network during training as the correct output responses expected for the two-class classification (pixel gray-level values) from the training.

Reference [16] uses a two-layer network with a hyperbolic tangent sigmoid activation function for learning the local arrangement of gray levels to detect fabric defects. Back-propagation (BP) is used to minimize the training error between the output and the expectation. In this iteration process [16], the BP algorithm adjusts the weights between layers automatically. Such a trained neural network can be well adapted for the two-class fabric classification problem.

The approach detailed in [16] uses the feature vector formed by the gray-level values of neighboring pixels for training the neural networks. The trained network successfully segments the defects from the fabric under

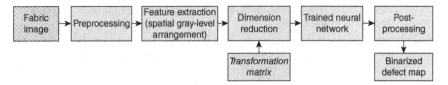

3.5 Flow diagram of fabric defect segmentation using neural network.

Table 3.2 Classification accuracy of neural network classifier for five leather defects

	Line	Stain	Hole	Knot	Wear	Total
Accuracy (%)	100	100	95	100	90	96.25

inspection and generates a binarized defect map as illustrated in Fig. 3.5. In this method, principal component analysis (PCA) has been applied to feature vectors to reduce the dimension of the feature vector. The experimental results suggest that this method is robust for the segmentation of 11 different fabric defects, including twill weave fabric and plain fabric samples. The work detailed in [16] is quite effective and details the effectiveness of a neural network in learning the gray-level arrangement in a fabric structure to find defects. However, the key purpose of our discussion, to detail fabric defect classification, has not been attempted in [16]. The classification of detected defects into one of the known defect categories has, however, been attempted in the literature for quality assurance applications with promising success. Kwak *et al.* [17] employed a multi-layered neural network to classify five different types of leather defects: lines, holes, stains, wears and knots. They extracted geometrical information from the localization of defects detected through the local thresholding and morphological operations. In the experiment, the authors used 140 image samples with defects, 60 of which were employed for the training (12 for each class) with the other 80 used for the test (10 for each of line and hole, 20 for each of the other three). The trained neural network with a single hidden layer was used for the defect classification and the results are summarized in Table 3.2. The authors showed that the neural network can achieve better classification accuracy than when a decision tree is employed for the classification.

Liu *et al.* [18] used a PSO-BP (Particle Swarm Optimization – Back Propagation) neural network with orthogonal wavelet transform as the feature for the classification of fabric defects. In this work [19], the authors attempted to alleviate two key problems with the use of a neural network: large training complexity and local minimum. The strategy is to use PSO

Table 3.3 Classification accuracy of neural network classifier for five stitching fabric defects

	Pleats	Puckers	Tension	Skipped-stitches	Holes
Accuracy	100	100	100	100	93

for the global optimization with a high search speed to achieve more effective optimization in a given time. The authors evaluated this approach on the classification of five different types of fabric defects: warp direction defect, weft direction defect, particle defect, hole and oil stain defect. They claimed to have achieved better performance than the BP neural network from the output of the classifier, and to shorten the training time and complexity simultaneously.

The classification and detection of stitching fabric defects using wavelet decomposition and neural networks has been investigated by Wong *et al.* in [13]. The classification results achieved in this paper demonstrate that the proposed method can identify five classes of stitching defects effectively. The five stitching defects are pleats, puckers, tension, skipped-stitches and holes. Table 3.3 summarizes the classification accuracy achieved in [13].

Kuo *et al.* [30] have also illustrated the classification of fabric defects using 1×4096 line scan camera imaging. This is a simplified approach that uses projective summation of fabric image gray levels for the neural network-based classification. The authors claimed to have achieved classification accuracy of 90% for weft-lacking and warp-lacking fabric defects, while oil stains were more effectively identified with an accuracy of 95%.

3.4.2 Support vector machine (SVM)-based fabric defect classification

In statistical learning theory, SVM is the representative of a set of methods referred to as kernel methods. In this approach, the SVM estimates the required support vectors from the training samples. In the two-class case (Fig. 3.6), these support vectors are the points from which two parallel lines pass to discriminate the feature points of the two classes and also have the maximum distance between them. However, not all the features from different classes are well separated in practice. In order to eliminate the overlap among them, features are first projected into a space with higher dimension through a nonlinear transform. In this new space, features tend to be more effectively separated from those of different classes. In this process, the inner product of new features will be computed, which can instead be solved in the initial space using a nonlinear function. There are

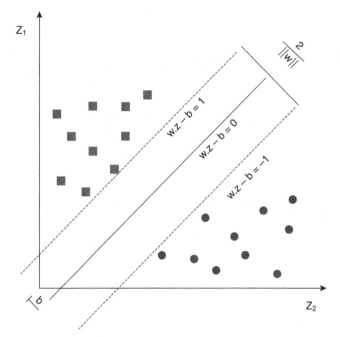

3.6 Estimation of decision boundary for a two-class SVM.

several choices for the nonlinear functions: Gaussian, polynomial, tan-hyperbolic function, etc. The decision surface using the Gaussian function is estimated as follows:

$$f(x) = \text{sgn}\left\{ \sum_{i=1}^{N} w_i \exp\left(-\frac{\|x - x_i\|^2}{c_i} \right) + b \right\}$$ 3.3

where c_i and b are constants, and x_i is the centers of support vectors.

Hou and Parker [20] investigated a promising approach for texture defect detection using SVM and Gabor features. In this paper, the authors adopted the one-against-all strategy for a binary SVM classifier to discriminate between defective and non-defective pixels. They selected 50 images from a Brodatz texture album [21] and the classification accuracy for the two cases is summarized in Table 3.4.

Kumar and Shen [22] used SVM and gray-level values of neighboring pixels as the feature to inspect fabric defects. A polynomial kernel of order 3 was employed in the SVM. This work details comparison between the SVM and the neural network performance for fabric inspection. The results in [22] suggest that the two schemes are both computationally similar for the inspection but SVM does not suffer from the problem of a local minimum and is computationally simple to train. Kim *et al.* [23] investigated

Table 3.4 Classification accuracy from two different cases (%)

Texture images	Proportion of training data (%)	Full Gabor filters	Optimal Gabor filters
D77 vs others	2	92.37	87.36
	4	92.57	87.31
	6	92.66	87.39
D4 vs others	2	81.01	77.84
	4	81.75	77.92
	6	84.99	78.09

Table 3.5 Summary of related defect classification methods from the literature

Reference	Year	Classifier	Database	Fabric quality	Accuracy (%)
[20]	2002	ANN	2 (defect-free and with-defect)	Plain and twill	NA
[21]	2000	ANN	5	Leather	96.25
[22]	2008	ANN	5	Plain	NA
[24]	2009	ANN	5	Stitching defect	98.6
[25]	2005	SVM	2 (positive and negative)	Brodatz texture	92.66 (best)
[27]	2002	SVM	2 (defect-free and with-defect)	Plain and twill	NA
[28]	2002	SVM	2 (positive and negative)	Brodatz texture	83.9 (best)
[30]	2006	GMM	12	Brodatz texture	99.61
[31]	2009	GMM	9	Plain	85
[32]	2003	K-NN	21491 (15 defect classes)	Wood surface (parquet slabs)	91.4% (best)

texture classification using SVM rigorously on several image databases including a Brodatz album [21] and an MIT vision texture database [24]. The gray-level values of neighboring pixels were used as the feature for classification. Different kernel functions and different multi-classification schemes for SVM were compared in this work. On the multi-texture images (four Brodatz textures), the authors achieved the best error rate of 16.1% when a 17×17 median filter was employed for the post-processing.

Although neural networks and SVM have emerged as the most promising classifiers for fabric defect classification (refer to Table 3.5), some other classifiers have also been investigated for texture classification. Kim and Kang [25] used a Gaussian mixture model (GMM) and a wavelet packet frame to extract texture features. The texture features from each class are evaluated by GMM and observed to be quite distinct from each other. In the test phase, the features extracted from each of the unknown images are employed to estimate the probability and find the model with maximum likelihood. The class of maximum probability model is assigned to the unknown test image. For 12 different textures from a Brodatz album [21], the authors achieved a classification accuracy of 99.61%.

Zhang et al. [26] have used GMM and Gabor filters to effectively demonstrate fabric defect classification. The Gabor-filtered responses from each of the fabric images are statistically quantified to extract feature vectors. The mean and covariance of these responses are used to estimate the model parameters for each GMM. The key advantage of GMM is that it can approximate any unknown distribution in a weighted sum of several multi-variance Gaussian distributions with different parameters. Therefore the GMM can be effectively used to characterize the distribution of features from the different fabric defects. The GMM is derived using an expectation maximum (EM) criterion. Generally the number for GMM has to be estimated first. However, the precision of the estimated model highly influences the performance. Zhang et al. [26] also employed the minimum description length (MDL) estimator to estimate the appropriate model order for each of the GMMs. They illustrated experimental results for the nine fabric defect classes and achieved an average classification accuracy of more than 85%, which is a little better than SVM in their work.

Mäenpää et al. [56] presented one of the most promising efforts for the inspection of wooden surface images that employed a large number of test samples for evaluation on 15 different classes of defects. They explored a variety of features and developed a methodology for combining color and texture features. They achieved their best results, i.e., an error rate of 8.6%, using signed differences quantized with 128 code vectors along with LBP features. The experimental results indicate that such an approach is likely to achieve high performance if attempted for the classification of fabric defects.

3.5   Fabric properties and color measurement using image analysis

Computer vision-based analysis of fabric samples is not limited to quality assurance but is increasingly employed for the objective/automated assessment of color, pattern and physical measurement [27, 40]. The manual measurement of fabric density, i.e., number of warp and weft yarns per unit distance, is time consuming and prone to errors. Therefore computer vision-based analysis of fabric images to automatically compute the actual fabric density has been developed. The Fourier spectra of the acquired (high resolution) fabric image can be used to ascertain warp and weft yarn density. Escofet *et al.* [54] have used such spectral domain energy analysis to identify fabric structure in the spatial domain. However, such an approach is not suitable for patterned fabrics as the periodicity of patterns, rather than individual yarns, appears in the combined power spectrum. The gray-level co-occurrence matrix-based fabric image analysis is more complex but has been attempted in [36] with little success. Jeong and Jang [34] and Kuo *et al.* [37] proposed fabric density measurement using measurements of the gray-level fabric profile. The experimental results in these efforts are promising and illustrate success on plain and patterned fabrics. The fabric image analysis can also be used to automate the measurements of yarn thickness and weave characteristics of underlying fabric.

The weave patterns for double-layer weft woven (necktie) fabric can also be ascertained from the autocorrelation analysis of the fabric image [53]. However, the common necktie patterns are complicated and this algorithm cannot work for complex fabrics of more than double-layer or even special single-layer weaves. Woven fabrics can also be geometrically modeled to describe their structure. The identification of fabric weave patterns using such *active grid models* that can contain their geometrical, structural and color details with a high success rate is proposed in Xin *et al.* [55]. Another emerging application of fabric image analysis is for fabric color analysis, i.e., color density, color composition, etc. Pan *et al.* [33] present fabric color analysis using a fuzzy clustering algorithm. The comparative fabric color analysis using a chromatography, spectroscopy and mass spectrometry has a range of applications in criminal investigation. The forensic analysis of textile fabrics is primarily employed to trace the materials that can identify the suspects, victims and crime scenes to scientifically establish the contact between individuals and objects. For example, visible microspectrophotometry can effectively distinguish between colored fibers, which may appear to be visually similar, from their spectral characteristics. There has been a significant amount of research into the forensic analysis of dyed textile fibers using mathematical analysis of spectroscopic data, and reference [35] summarizes these efforts.

3.6   Conclusions and future trends

Computer vision-based textile inspection is inviting a lot of interest, primarily for the objective assessment of textile quality and non-destructive measurements of the physical properties. The majority of research efforts in the literature [41] have evaluated fabric quality assessment on either plain or twill fabrics. The defects in such fabrics have a comparatively clear background texture to distinguish from the defect-free regions. Therefore the defect inspection approaches tested on such uniform colored and textured fabrics have very limited applications for the textile industry. Most of the textile fabric products for consumer use are colored and patterned. Such fabrics have complex patterns which are quite challenging for defect inspection and present a common limitation for the popular fabric defect detection approaches. Therefore the next generation of textile inspection approaches should effectively be able to detect defects from the complex patterns on the moving textile web. Such capabilities use scale, translation and rotation invariance techniques to match given regular pattern(s) whose repetition can be random rather than uniform. Research efforts should therefore be focused on the use of advanced feature extraction and matching techniques which can offer scale, translation and rotation invariant capabilities. Secondly, a higher level of imaging setups should be explored, since the textile fabric is a 3D surface. The 3D imaging of a textile surface can allow us to detect defects which are not visible using conventional imaging. Such an approach can therefore ensure a higher level of quality assurance and therefore should be pursued for the next generation of textile fabric inspection systems. The classification of fabric defects using a novel neural network based on the principle of *Autonomous Mental Development* (AMD) [57, 58] can be a promising tool for addressing the problem of dynamic defect populations that has not yet attracted attention from researchers. The AMD detector can be judiciously designed to autonomously develop its classification skills according to its learning environment during the online fabric inspection.

It is widely expected that sensing, storage and computational capabilities of automated computer vision-based fabric inspection systems will continue to improve. The development of smart sensing technologies will allow researchers to effectively exploit extended fabric features, i.e. structure, texture and color, and develop high-performance feature extractors. While this will significantly improve the throughput and usability, there are still some fundamental issues related to the representation of complex structured (patterned) fabrics, robust defect detection, and accurate classification of dynamically populated fabric defects. Therefore, future textile inspection systems must overcome many hurdles and challenges to meet a wide range of application requirements for the textile industry at low cost.

3.7   Acknowledgments

The author thankfully acknowledges support from Yu Zhang and reviewers for their valuable suggestions.

3.8   References

1. P. Ott, 'Procedure for detecting and classifying impurities in longitudinally moving textile inspection material', US Patent No. 7,292,340, November 2007, http://www.patentstorm.us/patents/7292340.html
2. H.-F. Ng, 'Automatic thresholding for defect detection', *Pattern Recognition Letters*, vol. 27, no. 14, pp. 1644–1649, 2006.
3. H. A. Abou-Taleb and A. T. M. Sallam, 'On-line fabric defect detection and full control in a circular knitting machine', *AUTEX Research Journal*, vol. 8, pp. 21–29, 2008.
4. R. Perez, J. Silvestre and J. Muñoz, 'Defect detection in repetitive fabric patterns', *Proc. Conf. Visualization, Imaging and Image Processing*, Marbella, Spain, September 2004.
5. N. Uçar and S. Ertuğrul, 'Prediction of fuzz fibers on fabric surface by using neural network and regression analysis', *Fibres & Textiles in Eastern Europe*, vol. 15, pp. 58–61, June 2007.
6. X. Yang, G. Pang and N. Yung, 'Robust fabric defect detection and classification using multiple adaptive wavelets', *IEEE Proc. Image Signal Processing*, vol. 152, no. 6, pp. 715–723, December 2005.
7. H. Y. T. Ngan and G. Pang, 'Novel method for patterned fabric inspection using Bollinger bands', *Optical Engineering*, vol. 45, no. 8, 087202, 2006.
8. H.-G. Bu, J. Wang and X.-B. Huang, 'Fabric defect detection based on multiple fractal features and support vector data description', *Engineering Applications of Artificial Intelligence*, vol. 22, no. 2, pp. 224–235, March 2009.
9. S. Liu, J. Liu and L. Zhang, 'Classification of fabric defect based on PSO-BP neural network', *Proc. Second Int. Conf. Genetic and Evolutionary Computing*, WGEC'08, pp. 137–140, September 2008.
10. X. Yang, G. Pang and N. Yung, 'Fabric defect classification using wavelet frames and minimum classification error training', *Proc. 37th IAS Annual Meeting*, vol. 1, Pittsburgh, PA, 2002.
11. E. Shady, Y. Gowayed, M. Abouiiana, S. Youssef and C. Pastore, 'Detection and classification of defects in knitted fabric structures', *Textile Research Journal*, vol. 76, no. 4, pp. 295–300, April 2006.
12. C. Kwak, J. A. Ventura and K. Tofang-Sazi, 'Automated defect inspection and classification of leather fabric', *Intelligent Data Analysis*, vol. 5, no. 4. pp. 355–370, 2001.
13. W. K. Wong, C. W. N. Yuen, D. D. Fan, L. K. Chan and E. H. K. Fung, 'Stitching defect detection and classification using wavelet transform and BP neural network', *Expert Systems with Applications*, vol. 36, no. 2, pp. 3845–3856, March 2009.
14. R. Stojanovic, P. Mitropulos, C. Koulamas, Y. A. Karayiannis, S. Koubias and G. Papadopoulos, 'Real-time vision based system for textile fabric inspection', *Real-Time Imaging*, vol. 7, no. 6, pp. 507–518, 2001.
15. S.-W. Li, L.-W. Guo and C.-H. Li, 'Fabric defects detecting and rank scoring based on Fisher criterion discrimination', *Proc. Int. Conf. Machine Learning and Cybernetics*, vol. 5, pp. 2560–2563, July 2009.
16. A. Kumar, 'Neural network based detection of local textile defects', *Pattern Recognition*, vol. 36, pp. 1645–1659, October 2003.

17. C. Kwak, J. A. Ventura and K. Tofang-Sazi, 'A neural network approach for defect identification and classification on leather fabric', *Journal of Intelligent Manufacturing*, vol. 11, pp. 485–499, 2000.
18. S. Y. Liu, L. D. Zhang, Q. Wang and J. J. Liu, 'BP neural network in classification of fabric defect based on particle swarm optimization', *Proc. Int. Conf. Wavelet Analysis and Pattern Recognition*, Hong Kong, 30–31 August 2008.
19. R. C. Eberhart and Y. Shi, 'Particle swarm optimization: Developments, applications and resources', Piscataway, NJ, Seoul, Korea, *Proc. IEEE Congress on Evolutionary Computation*, IEEE Service Center, 2001, pp. 81–86.
20. Z. Hou and J. M. Parker, 'Texture defect detection using support vector machines with adaptive Gabor wavelet features', *Proc. 7th IEEE Workshop on Applications of Computer Vision* (WACV/MOTION'05).
21. P. Brodatz, *Textures; a Photographic Album for Artists and Designers*. New York: Dover Publications, 1966.
22. A. Kumar and H. C. Shen, 'Texture inspection for defects using neural networks and support vector machines', *Proc. Int. Conf. Image Processing*, ICIP 2002, vol. 3, pp. 353–356, August 2002.
23. K. I. Kim, K. Jung, S. H. Park and H. J. Kim, 'Support vector machines for texture classification', *IEEE Trans. PAMI*, vol. 24, pp. 1542–1550, November 2002.
24. MIT Vision and Modeling Group, 1998.
25. S. C. Kim and T. J. Kang, 'Texture classification and segmentation using wavelet packet frame and Gaussian mixture model', *Pattern Recognition*, vol. 40, no. 4, April 2007.
26. Y. Zhang, Z. Lu and J. Li, 'Fabric defect detection and classification using Gabor filters and Gaussian mixture model', *Proc. 9th Asian Conference on Computer Vision*, Xian, China, 2009.
27. A. Cherkassky and A. Weinberg, 'Electro-optical method and apparatus for evaluating protrusions of fibers from a fabric surface', US Patent No. 10/601858, 2004.
28. F. Tajeripour, E. Kabir and A. Sheikhi, 'Defect detection in patterned fabrics using modified local binary patterns', *EURASIP Journal on Advances in Signal Processing*, Article ID 783898, 12 pages, 2008.
29. K. L. Mak, P. Peng and K. F. C. Yiu, 'Fabric defect detection using morphological filters', *Image and Vision Computing*, vol. 27, pp. 1585–1592, 2009.
30. C. F. J. Kuo, C.-J. Lee and C.-C. Tsai, 'Using a neural network to identify fabric defects in dynamic cloth inspection', *Textile Research Journal*, vol. 73, no. 3, pp. 238–244, 2003.
31. H. Y. T. Ngan, G. K. H. Pang, S. P. Yung and M. K. Ng, 'Wavelet based methods on patterned fabric defect detection', *Pattern Recognition*, vol. 38, pp. 559–576, 2005.
32. H. Y. T. Ngan and G. K. H. Pang, 'Regularity analysis for patterned texture inspection', *IEEE Trans. Automation Science and Engineering*, vol. 6, pp. 131–144, 2009.
33. R. Pan, W. Gao and J. Liu, 'Color clustering analysis of yarn-dyed fabric in HSL color space', *Proc. World Congress on Software Engineering*, pp. 273–278, Xiamen, China, May 2009.
34. Y. J. Jeong and J. Jang, 'Applying image analysis to automatic inspection of fabric density for woven fabrics', *Fibers and Polymers*, vol. 6, no. 2, pp. 156–161, 2005.

35. J. V. Goodpaster and E. A. Liszewski, 'Forensic analysis of dyed textile fibers', *Analytical and Bioanalytical Chemistry*, vol. 394, pp. 2009–2018, June 2009.
36. J. J. Lin, 'Applying a co-occurrence matrix to automatic inspection of weaving density for woven fabrics', *Textile Research Journal*, vol. 72, pp. 486–490, 2002.
37. C. J. Kuo, C. Y. Shih and J. Y. Lee, 'Automatic recognition of fabric weave patterns by a fuzzy c-means clustering method', *Textile Research Journal*, vol. 74, pp. 107–111, 2004.
38. A. G. Blackmon, *Manual of Standard Fabric Defects in the Textile Industry*, Graniteville Co., Graniteville, SC, 1975.
39. I. C. Baykal, R. Muscedere and G. A. Jullien, 'On the use of hash functions for defect detection in textures for in-camera web inspection systems', *IEEE International Symposium in Circuits and Systems*, ISCAS 2002, vol. 5, pp. 665–668, 2002.
40. M. L. Gulrajani (ed.), *Colour Measurement: Principles, Advances and Industrial Applications*, Cambridge, UK: Woodhead Publishing, Textiles Series No. 103, August 2010.
41. A. Kumar, 'Computer vision based fabric defect detection: a survey', *IEEE Trans. Industrial Electronics*, vol. 55, no. 1, pp. 348–363, January 2008.
42. R. Mcgregor, D. H. Mershon and C. M. Pastore, 'Perception, detection, and diagnosis of appearance defects in fabrics', *Textile Research Journal*, vol. 64, no. 10, pp. 584–591, 1994.
43. Elbit Vision Systems, http://www.evs-sm.com/home/evs/Textile_Fabric.aspx, accessed 2 April 2010.
44. Uster Technologies, Inc., Charlotte, NC.
45. BarcoVision, http://www.visionbms.com/textiles/en/pressreleases/show.asp?index=2062, accessed 2 April 2010.
46. R. Guruprasad and B. K. Behara, 'Automatic fabric inspection system,' *Indian Textile Journal*, June 2009, http://www.indiantextilejournal.com/articles/FAdetails.asp?id=2131
47. J. G. Vachtsevanos, M. Mufti and J. L. Dorrity, 'Method and apparatus for analyzing an image to detect and identify defects', US Patent No. 5,815,198, September 1998.
48. J. Escofet, R. Navarro, M. S. Millan and J. Pladelloreans, 'Detection of local defects in textiles webs using Gabor filters', *Optical Engineering*, vol. 37, pp. 2297–2307, August 1998.
49. A. Baykut, A. Atalay, A. Erçil and M. Güler, 'Real-time defect inspection of textured surfaces', *Real-Time Imaging*, vol. 6, pp. 17–27, 2000.
50. D. Chetverikov and A. Henbury, 'Finding defects in texture using regularity and local orientation', *Pattern Recognition*, vol. 35, pp. 2165–2180, 2002.
51. H. Sari-Sarraf and J. S. Goddard, 'Vision system for on-loom fabric inspection', *IEEE Trans. on Industry Applications*, vol. 35, pp. 1252–1259, November–December 1999.
52. A. Kumar and G. Pang, 'Defect detection in textured materials using optimized filters', *IEEE Trans. on Systems Man and Cybernetics: Part B, Cybernetics*, vol. 32, pp. 553–570, October 2002.
53. H. Wu, M. Zhang, Z. Pan and H. Yin, 'Automatic identifying weave patterns for double-layer weft woven fabric', *Proc. Int. Conf. Computer Graphics, Imaging and Vision: New Trends* (CGIV'05), 2005.

54. J. Escofet, M. S. Millán and M. Ralló, 'Modeling of woven fabric structures based on Fourier image analysis', *Journal of Applied Optics*, vol. 40, pp. 6171–6176, 2001.

55. B. Xin, J. Hu, G. Baciu and X. Yu, 'Investigation on the classification of weave pattern based on an active grid model', *Textile Research Journal*, vol. 79, pp. 1123–1134, December 2009.

56. T. Mäenpää, J. Viertola and M. Pietikäinen, 'Optimizing color and texture features for real-time visual inspection', *Pattern Analysis and Applications*, vol. 6, no. 3, pp. 169–175, 2003.

57. N. J. Butko and J. R. Movellan, 'Infomax control of eye movement', *IEEE Trans. Autonomous Mental Development*, vol. 2, no. 2, pp. 91–107, June 2010.

58. C. M. Vigorito and A. G. Barto, 'Intrinsically motivated hierarchical learning in structured environments', *IEEE Trans. Autonomous Mental Development*, vol. 2, no. 2, pp. 132–143, June 2010.

Part II
Modelling and simulation of textiles and garments

4

Key techniques for 3D garment design

Y. ZHONG, Donghua University, China

Abstract: This chapter introduces three topics related to recent advances in computer technologies for textiles and apparel, including sketch-based modeling, surface flattening and some of the problems and solutions surrounding shopping for garments online. The mainstream techniques developed in these regions are discussed in view of their applications towards garment design and online garment shopping. The corresponding implementations are highlighted for evaluating the performance of these algorithms and/or techniques. A section on future trends and challenges, along with a list of supplementary reading material, is also provided at the end of this chapter.

Key words: sketch-based modeling, surface flattening, shape morphing, virtual try-on.

4.1   Introduction

Scientific investigation involved in the apparel industry always contains efforts to transfer human behavior to computational tasks. In this chapter, we will focus on several mainstream requirements. The first requirement is how to find a method of designing a 3D garment from sketches. The second requirement is how to flatten a 3D garment into 2D patterns. The third requirement is how to try on a 3D garment in cyberspace. These requirements are believed to be the key techniques in 3D garment modeling and implementation.

The first requirement actually demands an effort to prototype the 3D garment directly from either 2D or 3D drawings. The key to both is the generation of a 3D surface that follows the shape of a mannequin. Two different approaches to this topic can be found in Section 4.2.

The heart of the second requirement is mesh parameterization, which is to build a one-to-one correspondence (mapping) between the 3D surface and a 2D domain (in garment design, a 2D pattern). This technique can be used wisely in both reverse engineering and texture mapping, which is covered in Section 4.3.

The third requirement is a very challenging task since the virtual garment cannot be physically tried on. Unlike the traditional model of sewing 2D patterns to predict a garment's 3D configuration, it is rather a procedure which fits a given 3D garment model onto an avatar of the user, most likely

69

an online shopper. Problems concerning avatar generation, virtual try-on and the corresponding solutions are depicted in Section 4.4.

Challenges and future trends involved in these topics are outlined in Section 4.5, while sources of further information and references are provided in Section 4.6.

4.2   Sketch-based garment design

4.2.1   Method 1: 2D sketches associated with 3D garment prototypes

There are two methods involved in sketch-based garment design. The first one is called 'sketch-based interface and modeling' (SBIM), as compared to the traditional WIMP (Window, Icon, Menu, Pointer) paradigm. For garment designers, sketching is a natural way to express their ideas. With few strokes, the complete 3D configuration can be evoked in viewers. This is because human eyes have the capacity to interpret a 2D drawing into its 3D counterpart seamlessly. A recent research direction in computer modeling is to mimic this 'interpretation' procedure as a manipulating tool in garment design. The goal is to assist users in generating 3D prototypes adapted to existing mannequins, meaning that sewing 2D patterns for the 3D configuration is no longer necessary.

The sketch-based 3D garment design was first introduced by Turquin *et al.* [1] in 2004. The user draws an outline of the front or back of the garment, and the system makes reasonable geometric inferences about the overall shape of the garment (ignoring constraints arising from physics and from the material of the garment). Thus both the garment's shape and the way the character is wearing it are determined. Despite the details of sketch acquisition, the distance from the 2D garment silhouette to the character model is the key to inferring the variations of the distance between the remainder of the garment and the character in 3D.

The perception of the 3D shape from 2D sketches is usually ill-posed and thus only an optimized answer can be reached. A plausible guess of the intentions of drawing can be decoded via the relationships between the silhouettes and the enclosed body area. The undefined z-depth of the edge can be roughly realized by finding the nearest point on the body providing that the garment surface is of an isotropic configuration. With suitable interpolations, the interior part of the garment can be predicted.

Firstly, the user initializes the 2D sketches under a given software interface where the segments of the curve are classified as border lines (which crossed the character's body) or silhouette lines. The next step is to infer the 3D position of silhouette and border lines via the minimum distance to the body, and this value is recorded as the basis for interpolating the z value in the 3D space. The third step is to generate the interior shape of the garment via following procedure:

- Meshing the interior area with a rectangular grid in the (x, y) plane
- Using the interpolation of distances to the body as the main clue for roughly inferring the 3D position of the interior of the garment
- Adjusting this tentative assignment of z values by surface tension of the garment between two limbs of the character
- Re-triangulating the grids and the boundary to form a complete garment model.

The major advantage of this method is its capacity for fast prototyping, since most of the computation is linear interpolation, which can be accelerated via a predefined distance field. Although this naive approach can only generate certain types of simple garment, it is believed that directly generating 3D garments from 2D sketches remains an interesting and valuable challenge.

4.2.2 Method 2: Fit and tight-fit garment design based on parameterized 3D sketching

The second method of sketch-based garment design is that the user draws the outline of the 3D garment directly on the mannequin, and the system generates the 3D surface from a bumped collision patch constrained by the boundary of the 3D sketches [2].

The user is required to drop down a series of points onto the mannequin. These points are consequently connected to represent the initial boundary. Each segment is then projected on the surface of the mannequin, as shown in Fig. 4.1(a).

After the projection, the resulting boundary forms a closed 3D contour which can be approximated by Bezier curves. The end point of each segment is regarded as the anchor of the Bezier curve, while the user-inserted points are regarded as the control points (one or two, representing the cubic and quadratic Bezier curves respectively). These Bezier contours provide the user with an editable option to fine-tune the original design, as shown in Figs 4.1(b) and (c).

For an interactive 3D garment design system, the sketching line only confines the boundary of the design, to finalize the corresponding surface. Delaunay triangulation is employed to generate a local plane and then it is mapped back onto the mannequin, as shown in Fig. 4.2.

To form a seamed virtual garment, the user should draw each patch of the garment and sew them together. With suitable left-to-right mapping and component reuse, the resulting 3D garment can be finalized without tedious effort, as shown in Fig. 4.3.

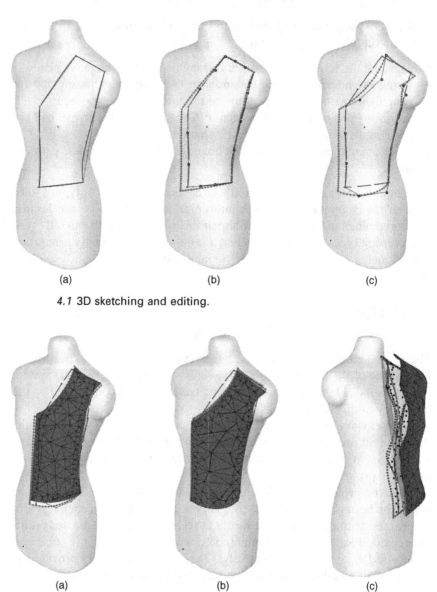

4.1 3D sketching and editing.

4.2 Generating the 3D garment patch from the 3D sketch.

The garment surface can be further adjusted through an extended free-form deformation (EFFD) procedure [3]. The boundary of the 3D garment patch can be either included in the deforming lattices or isolated from the lattices according to the design interest.

Based on the aforementioned pipeline, the curvature of the 3D garment surface is determined by either the shape of the mannequin or the EFFD

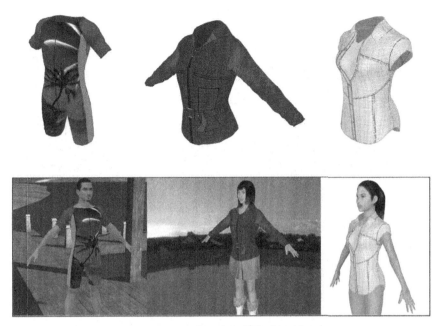

4.3 Resulting garment surface from 3D sketching.

adjustment. In case the smoothness of the 3D surface is highly preferred, a global smoothness can be performed upon the resulting surface, such as Loop's subdivision algorithm [4] to smooth the surface curvature, which is commonly adopted in 3D authorizing software.

The major purpose of this method is to provide a useful tool for professionals in 3D garment tailoring. These people have a better understanding of how to make a garment from 2D patterns. Obviously, tight-fit (swimsuit) and fitted (jacket and shirt) garments can be designed more directly and efficiently with this method, since the designers are drawing and interpreting their ideas in 3D space.

4.3   Surface flattening for virtual garments

Technically, surface flattening is referred to as 'mesh parameterization', which is widely studied by those in the fields of CAD and computational geometry. The common procedure for flattening a triangulated 3D surface with non-zero Gaussian curvature (which means it is undevelopable) involves decomposing the surface to discrete patches, building the correspondence between 3D meshes and their isomorphic counterparts in a (u, v) space through piecewise mapping, and minimizing the introduced distortions (anisotropy) via a linear and non-linear solver. Fig. 4.4 shows examples of a triangulated surface in both 3D and parametric (u, v)space.

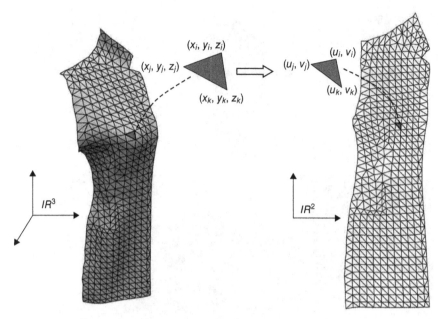

4.4 Illustration of mesh parameterization.

For the triangulated representation, the parameterization can be defined as a piecewise linear function:

$$f(u, v) = \lambda_1 \mathbf{p}_i + \lambda_2 \mathbf{p}_j + \lambda_3 \mathbf{p}_k$$

where (i, j, k) denotes the index triplet such that the triangle (u_i, v_i), (u_j, v_j), (u_k, v_k) in parameter space contains the point (u, v). The triplet $(\lambda_1, \lambda_2, \lambda_3)$ denotes the barycentric coordinates at point (u, v) in that triangle. Basically, constructing a parameterization of a triangulated surface means finding a set of couples(u_i, v_i) associated with each vertex i. An unconstrained flattening method may cause self-intersections and/or anisotropy, i.e., an elementary circle becomes an elementary ellipse after flattening.

There are two types of applications for flattening the 3D surface of a virtual garment. The first is to generate a 2D pattern for future cutting and plotting, and the second is to generate the texture atlas for rendering. For implementations falling into the first category, boundary shape in (u, v) space is of paramount importance. Thus the key point of surface flattening is to promise at least a conformal parameterization, i.e., an elementary circle becomes an elementary circle after flattening, as shown in Fig. 4.5.

The anisotropy or distortion introduced by mapping a 3D surface from IR^3 to IR^2 involves computation of the gradients of u and v. In the case of a triangulated surface, the gradients are constant in each triangle, which can be expressed by providing each triangle with an orthonormal basis X, Y

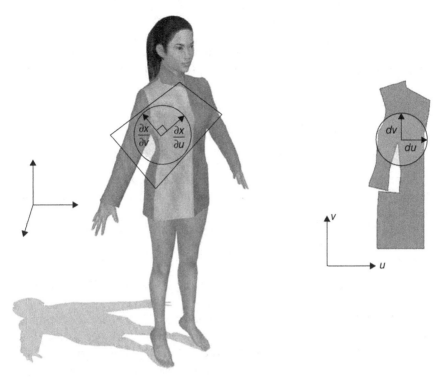

4.5 Conformal parameterization with darts insertion.

(defined in the local domain of the triangle). The gradient of mapping a point (X, Y) of the triangle to a point (u, v) can be expressed as:

$$\nabla u = \begin{pmatrix} \partial u/\partial X \\ \partial u/\partial Y \end{pmatrix} = \frac{1}{2|T|} \begin{pmatrix} Y_j - Y_k & Y_k - Y_i & Y_i - Y_j \\ X_k - X_j & X_i - X_k & X_j - X_i \end{pmatrix} \begin{pmatrix} u_i \\ u_j \\ u_k \end{pmatrix} = M_T \begin{pmatrix} u_i \\ u_j \\ u_k \end{pmatrix}$$

where M_T is constant over triangle T, and $|T|$ denotes the area of T.

The most popular conformal parameterization for the triangulated surface is the LSCM method (least squares conformal maps) introduced by Levy *et al.* [5] in 2002. For a single triangle, the conformality can be expressed as (refer to Fig. 4.5):

$$\nabla v = \text{rot}_{90}(\nabla u) = \begin{pmatrix} 0 & -1 \\ 1 & 0 \end{pmatrix} \nabla u$$

where rot_{90} denotes counter-clockwise rotation of 90 degrees. This means

$$M_T \begin{pmatrix} v_i \\ v_j \\ v_k \end{pmatrix} - \begin{pmatrix} 0 & -1 \\ 1 & 0 \end{pmatrix} M_T \begin{pmatrix} u_i \\ u_j \\ u_k \end{pmatrix} = \begin{pmatrix} 0 \\ 0 \end{pmatrix}$$

As depicted in [5], this equation cannot be strictly enforced for the unde-velopable triangulated surface. Hence the violation of the conformality condition is minimized in the 'least squares' sense, i.e., to minimize the discrete conformal energy corresponding to the 'non-conformality', which is given by:

$$E = \sum_{T=(i,j,k)} |T| \left\| M_T \begin{pmatrix} v_i \\ v_j \\ v_k \end{pmatrix} - \begin{pmatrix} 0 & -1 \\ 1 & 0 \end{pmatrix} M_T \begin{pmatrix} u_i \\ u_j \\ u_k \end{pmatrix} \right\|^2$$

Another similar method, the ABF method (angle based flattening), developed by Sheffer and Sturler [6], reformulates the flattening as a con-strained optimization problem in terms of angles, i.e., the parameterization is to find out the angles α_i^t at the corner of triangle t incident to vertex i. The energy under minimization is given by:

$$E(\alpha) = \sum_{t \in T} \sum_{k=1}^{3} \frac{1}{w_k^t} (\alpha_k^t - \beta_k^t)^2$$

where the α_k^t's are the unknown 2D angles, and the β_k^t's are the 'optimal' angles, measured on the 3D mesh. The weights w_k^t are set to $(\beta_k^t)^{-2}$ to measure a relative angular distortion rather than an absolute one.

The shape of the garment usually contains undevelopable manifolds with high Gaussian curvature, which have to be decomposed into discs or charts to efficiently minimize the energy accumulated during flattening. The decomposition (partitioning) must be carefully evaluated to avoid the occurrence of a large number of flattened discs. More particularly, this means that the distortion should be handled either individually or collec-tively. In this context, McCartney et al. [7] presented a method of flattening nearly planar surfaces while evaluating the local distortion energy of par-ticle movements. The optimal local positioning of projected nodes was based on a sequential addition of the nodes. This method did not guarantee preservation of the metric structure of the two-dimensional mesh or even its validity. Later, the same authors proposed an orthotropic strain model to mimic the flattening problem of orthotropic materials [8]. The strain energy was minimized through gradient optimization.

Based on the same methodology, Wang et al. [9] transferred the whole surface into a mass-spring system and applied a penalty function to recover the overlapped area. The deformation produced in the flattening procedure was dissipated through energy release and surface cutting.

A common feature of these methods is that they first need to establish the 2D isomorphic planar configuration through a one-to-one geometrical mapping and then to minimize the introduced distortion energy through various approaches. Geometrical mapping usually requires a specially

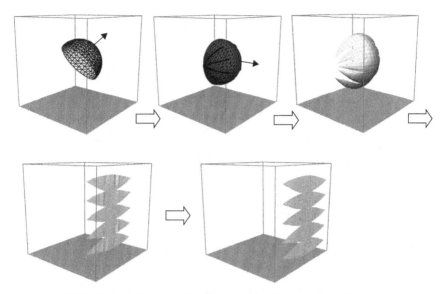

4.6 Surface flattening via releasing bending configuration.

designed data structure and rules to determine the best 2D coordinates. To bypass this procedure, Zhong and Xu [10] proposed a method to release the bending configuration between the winged triangles. As shown in Fig. 4.6, the unwrapping force field shapes the half-sphere into a 'blossoming' configuration until it flattens.

The core of this method is the physical opening of the bending configuration between two winged triangles via forces. This process is defined as a combination of unfolding and spreading, which is illuminated by the works of Baraff and Witkin [11] and Bridson *et al.* [12] in calculating the bending force for cloth simulation.

As shown in Fig. 4.7, an unfolding element is a winged pair of triangles that consist of four particles x_1, x_2, x_3 and x_4. If the original dihedral angle of the winged triangle pairs is $\pi - \theta$, the rest dihedral angle is set as π to reflect the parameterized position in (u, v) space. The unwrapping force is given by

$$F = F_i^e + F_i^d = k_e \frac{|\mathbf{E}|^2}{|\mathbf{N}_1| + |\mathbf{N}_2|} (\sin(\theta/2)) \mu_i - k_d |\mathbf{E}| (\mu \otimes \mu) v$$

where $i = 1, 2, 3, 4$ indicate the positions of the vertices, $\mu = (\mu_1, \mu_2, \mu_3, \mu_4)$ represents a motion mode that changes the dihedral angle but does not cause any in-plane deformation or rigid body motion, and

$$\mathbf{N}_1 = (\mathbf{x}_1 - \mathbf{x}_3) \times (\mathbf{x}_1 - \mathbf{x}_4)$$
$$\mathbf{N}_2 = (\mathbf{x}_2 - \mathbf{x}_4) \times (\mathbf{x}_2 - \mathbf{x}_3)$$

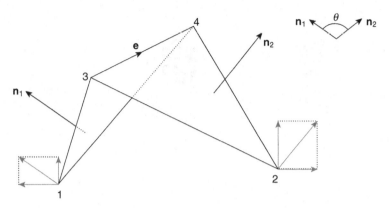

4.7 An unfolding element with dihedral angle.

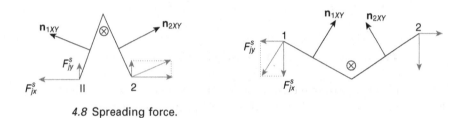

4.8 Spreading force.

where $\mathbf{N}_1$ and $\mathbf{N}_2$ are the area weighted normals and $\mathbf{E} = \mathbf{x}_4 - \mathbf{x}_3$ is the common edge.

The spreading is fulfilled by applying the spreading forces to particles 1 and 2. Basically, the spreading force F^s is the projection of a force in the triangle normal direction onto the XY-plane (Fig. 4.8), i.e.

$$F^s_{jx} = \pm k_s n_{jx}, \; F^s_{jy} = \pm k_s n_{jy}, \; F^s_{jz} = 0$$

where $j = 1, 2$ and k_s is the coefficient. As shown in Fig. 4.8, $\mathbf{n}_{1XY}$ and $\mathbf{n}_{2XY}$ are the projections of $\mathbf{n}_1$ and $\mathbf{n}_2$ in the XY-plane, respectively, and $\otimes$ indicates the direction of $\mathbf{e}_z$. The signs ($\pm$) of F^s_{jx} and F^s_{jy} are the same as the sign of $(\mathbf{n}_1 \times \mathbf{n}_2) \cdot \mathbf{e}$.

Finalization is reached by filtering the velocity (which is equal to the strain control) of the vertices to reduce the curvature difference and then pressing the planar-like surface onto (u, v) space (the 'front wall') via colliding. Fig. 4.9 demonstrates the result of flattening a 3D pattern.

The boundary shape of the flattened pattern is highly dependend on the cutting path – in the sense of garment design, the darts which will be inserted. Technically, the cutting path usually indicates linkage between the domain boundary and the vertices where the Gaussian curvature is high. In practice, the optimized path to decompose a given 3D surface may

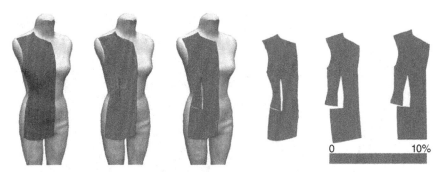

4.9 Physically-based method in surface flattening.

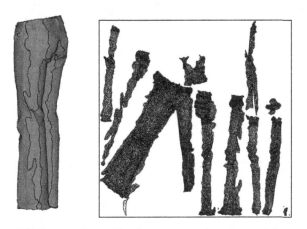

4.10 Poor texture atlas for a scanned pants model.

not be the best candidate for dart insertion, which is confined by the rules of fashion design, so the result of surface flattening may not be as expected. Successful flattening always demands human interference to determine the best cutting path, which is actually a balance between optimized energy minimization and dart insertion. Recent advances in flattening 3D surfaces for application in the clothing industry are also focused on this topic.

The same issues happen to the implementations where generating a texture atlas is of great concern. When artists are preparing the texture atlas for a given 3D object, they would like to choose those easy-to-manipulate techniques. For instance, the texture atlas shown in Fig. 4.10 is very cumbersome to work with, since too many pieces have to be processed individually.

Sometimes, the artist is in favor of the texture atlas as shown in Fig. 4.11. Although the overall shape is very different from the 2D patterns for pants,

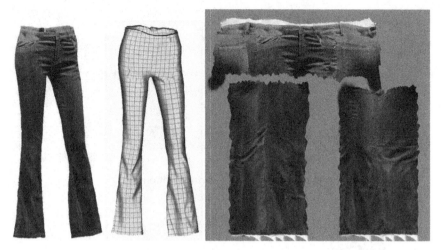

4.11 Rendering the pants model based on an appropriate texture atlas.

its rectangle-like silhouette makes this type of segmentation a popular choice since it is easier to fill and manipulate.

4.4   Online garment-shopping system: problems and solutions

To mimic the behavior of human clothing, an online shopping system has to solve two problems. One is to provide an avatar for the customer as a replacement in cyberspace. Another is to provide a method of evaluating fit associated with the visual effects. Theoretically, the body shape of various individuals can be found in a shape library containing thousands of human models. However, such a library would in practice need to include the majority of the human race, which is a great challenge even if a range of scanned data has been gathered. The answer to this question is shape inter-polation, or shape morphing, which was introduced by Allen *et al.* [13]. To model body shape variation across different people, the authors morphed a generic template shape from a corpus of whole-body 3D laser range scans of 250 different people in the same pose. A well-meshed template model was employed as the base for matching with the scans. Three types of error function were taken into account to control the quality of matching: data error, smoothness error, and marker error. The data error emphasized the criterion that the template surface should be as close as possible to the target surface. The smoothness error constrained the smoothness of the actual deformation applied to the template surface among local neighbor-hoods. When the template and example mesh were initially very close to each other, these terms would be sufficient. Otherwise, the optimization might stick in local minima, for instance, if the left arm were aligned with the right arm when the poses between the template model and the target model are very different. To overcome the undesirable minima, they identified a set of points on the example surface corresponding to known points on the template surface. This marker error minimized the distance between each marker's location on the template surface and its location on the example surface. The combination of these error constraints yields an objective function which can be solved by quasi-Newton optimization (L-BFGS-B).

Since each resulting mesh has the same topology, the morph between any two subjects can be performed by taking linear combinations of the vertices, and the texture maps can also be transferred between any pair of meshes. The variability of the human shape was captured by performing principal component analysis (PCA) over the displacements of the template points. The model was used for hole-filling of scans and fitting a set of sparse markers for people captured in the standard pose. Obviously, the synthesized individuals have their own character, distinct from those of the original scanned individuals.

Once the human space has been built, finding a similar avatar from it becomes a problem of data matching (Fig. 4.12) or shape analysis, which

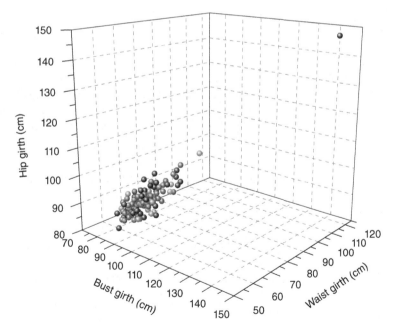

4.12 Finding out the body shape that can represent the online shopper from the human space.

can be solved by finding the minimum value compared to the input. After this, the focus can be moved into the second problem: performing virtual try-on with fit/ease evaluation. Before further examining the solutions to this problem, let's take a look at the current situation of online clothing sales. Although they have boomed in recent years, the rate of returns has also increased. The main reason for returning clothing items purchased online is due to problems with the fit. To improve customer satisfaction, the following criteria must be emphasized [14]:

- The virtual garment should be exactly the same as its counterpart (the real product) in the real world to comply with the principle of 'What You See Is What You Get'.
- The system should be able to provide the fit/ease information for the customer, since the virtual garment cannot be physically tried on.
- The speed of 3D garment 'manufacturing' should be fast enough to satisfy the requirements of industrialization.

Although virtual garment simulation has been investigated for years, the commercial software available in the market and the cutting edge modeling algorithms still cannot provide a three-dimensional garment with the same size as the real one, and the speed of generating a three-dimensional garment is highly dependent on the skills of the operator. It is usually necessary to plot the two-dimensional patterns and position them around the human body. A manual setup of the sewing relationship is inevitably necessary before the virtual sewing can be engaged, and an iteration of numerical integration is required to reach the final draping result. Using this method, it is a tedious job to produce the three-dimensional garments for an online shopping store in accordance with the aforementioned criteria. Despite the low productivity, the size of the virtual garment usually diverges from the original product, since the physical model cannot constrain the strain and strain rate under the threshold as the real fabric does. Hence the virtually sewn garment may not be a suitable candidate for online apparel retailing. However, methods for solving numerical integration, the collision detection/response scheme, and the strain control technique, can be employed in computing the drape of an already formed garment model. If approaching shape generation via range data scanning, the quality of the shape will be guaranteed according to the nature of the scanning. The speed of three-dimensional garment 'manufacturing' may be massively accelerated if the points cloud is well recovered. A great challenge is how to refit the garment obtained from a source human model to a target human model (i.e., the avatar selected from the shape space), especially when the poses of these two models are different.

One avenue of solving this problem is to copy the pose of the source model to the target model to facilitate the garment redressing. In the animation area, the character can be manipulated through an underlying hierarchical skeleton. The character mesh geometry is then attached to the underlying skeleton so that, as the skeleton deforms, the mesh synchronizes the deformation appropriately. When using the same methodology for pose duplication, a key problem is how to automatically generate the skeleton for human models. Most skeleton extraction algorithms require a volumetric discrete representation of the input model. This type of approach is inconvenient when transferring the polygonal meshes that are widely adopted by body scanners to volumetric representation, before extracting the skeleton and then transferring it back to the polygonal meshes. The potential difficulty in minimizing the discretization error is also an issue. The state of the art in this region is to extract the skeleton directly from the mesh contraction while maintaining the key features of the original mesh. Implicit Laplacian smoothing was employed to generate a zero-volume skeletal shape. The final skeleton is obtained through a series of edge collapses while removing collapsed faces from the degenerated mesh until all faces had been removed. The major drawback of this purely geometrical approach is that the generated skeleton did not provide any anthropometric information, which is critical in virtual garment refitting.

During pose duplication, another technique called 'smooth skinning' is also required for visual satisfaction. Smooth skinning indicates convincing skin deformations, including subtleties and details like muscle bulges. Some popular methods, including Pose Space Deformation [15], Shape by Example [16], and EigenSkin [17], introduced radial-based interpolation of corrections to blend skins linearly. These methods started with a simple skin combined with sparse data interpolation to rectify errors between the skin and a set of examples.

The skeleton can also be extracted from automatic body measurement [18], where the anthropometric principles have been taken into account, as shown in Fig. 4.13. The joints are defined as the geometrical center of the contour at the anthropometric position. As illustrated in Fig. 4.14, the coordinates of a joint **O** can be defined as $O_x = (X_{min} + X_{max})/2$, $O_y = y$, $O_z = (Z_{min} + Z_{max})/2$, where y is the height of the anthropometric contour.

If we want to fit the scanned garment to the target model, as shown in Fig. 4.15, the automatic body measurement must be performed for both models to generate two skeletons that can represent the pose of these two human bodies (Fig. 4.15(a)–(c)). The difference between these two skeletons is reduced through the pose duplication procedure (Fig. 4.15(d)). Once the pose of these two human models matches, the extracted 3D garment from the source model will be fitted onto the target model (Fig. 4.15(e)). The fit/ease information can then be evaluated by cutting the garment and the body through a series of planes. The drape of the refitted garment can

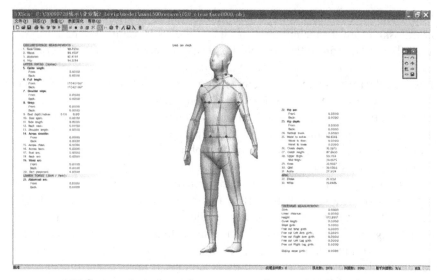

4.13 Screen shot of a software system for automatic body measurement.

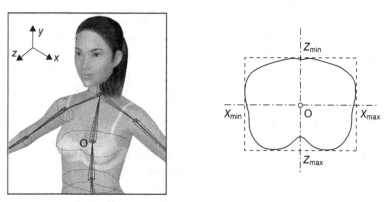

4.14 Calculating the joint as the center of the contour at the anthropometric position.

also be calculated by converting the geometrical model to a mass-spring system solved by an implicit integrator.

Duplicating the pose between two different human models can now be accomplished by recursively applied affine transformations and a smooth skinning procedure. There are two types of affine transformations. Type I is applied to the joints, such as right thigh root, left thigh root, abdomen, right arm and left arm. These joints usually have only one parent joint and more than one child joint. The type II affine transformation is applied to the joints

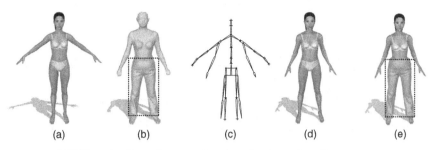

4.15 The pipeline of garment redressing based on the pose duplication.

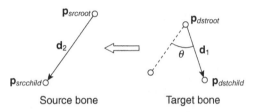

Source bone   Target bone

4.16 Type I affine transformation.

sub-chained with the type I joints that only have one parent and one child (or those that have no children at all, such as the end joints) among the hierarchical map. To copy the pose from one skeleton to another skeleton, these two types of affine transformations should be performed as follows.

- *Step 1*. Calculate the world coordinates for each joint.
- *Step 2*. For the type I joints, perform a type I affine transformation (Fig. 4.16), which is a pure rotation from $\mathbf{p}_{dstroot}\mathbf{p}_{dstchild}$ to $\mathbf{p}_{srcroot}\mathbf{p}_{srcchild}$.
- *Step 2a_1*. Compute the incline angle between $\mathbf{d}_1$ and $\mathbf{d}_2$ to get the rotation matrix $\mathbf{R}$ from $\mathbf{d}_1$ to $\mathbf{d}_2$.
- *Step 2a_2*. Apply $\mathbf{R}$ to $\mathbf{p}_{dstroot}$ to obtain the translation matrix $\mathbf{T}$.
- *Step 2a_3*. Apply the affine transformation matrix to the bones:

$$\mathbf{M} = \mathbf{TR}$$

$$\mathbf{p}'_{dstchild} = \mathbf{M}\mathbf{p}_{dstchild}$$

For the type II joints, perform a type II affine transformation (Fig. 4.17) as follows:

- *Step 2b_1*. Compute the incline angle between $\mathbf{d}_1$ and $\mathbf{d}_2$ to get the rotation matrix $\mathbf{R}$ from $\mathbf{d}_1$ to $\mathbf{d}_2$.
- *Step 2b_2*. Apply $\mathbf{R}$ to $\mathbf{p}_{dstroot}$ and compute the translation matrix $\mathbf{T}$:
- $\mathbf{p}'_{dstroot} = R \cdot \mathbf{p}_{dstroot}$, $\mathbf{t} = \mathbf{p}'_{dstchild} - \mathbf{p}'_{dstroot}$

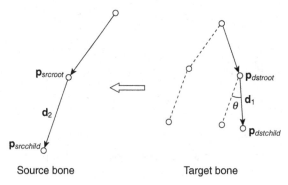

Source bone   Target bone

4.17 Type II transformation.

- *Step 2b_3.* Compose the affine transformation matrix and apply it to the bones:

  M = TR

- *Step 3.* Set the influence and the weights.

To reach a satisfied effect, each vertex is assigned a set of influencing joints and a blending weight for each influence. Computing the deformation in a new pose involves rigidly transforming each vertex of the target model by all of its influencing joints. Then, the blending weights are used to combine these rigidly transformed positions. For a vertex **p**, the deformed vertex position at a new pose $\mathbf{p}_{new}$ can be computed as

$$\mathbf{p}_{new} = \sum_{i=1}^{n} w_i \mathbf{M}_{i\_new} \mathbf{M}_{i\_original}^{-1} \mathbf{p}$$

where w_i are the weights ($0 < w_i < 1$) for the affine transformation, **p** is the original location, $\mathbf{M}_{i\_new}$ is the transformation matrix associated with the ith joint in the new pose and $\mathbf{M}_{i\_original}^{-1}$ is the inverse of the original pose matrix associated with the ith influence. Actually, $\mathbf{M}_{i\_original}^{-1}\mathbf{p}$ represents the location of **p** in the local coordinate of the ith influence. As explained in [19], this skinning algorithm is very fast and is widely accepted by commercial applications. For the posed target model, its skin is most heavily influenced by those bones to which the skin vertices are geometrically attached. The weights of the skin vertices are assigned by regarding the following criteria:

$w_i = 0$ if the vertex is not included within the binding area of a bone

$w_i = 1$ if the vertex is included within the binding area of a bone

The binding area of a bone is defined by the geometrical features of the body parts, shown as the dotted cylinder area in Fig. 4.18.

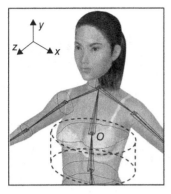

4.18 Bounding area attached to a bone.

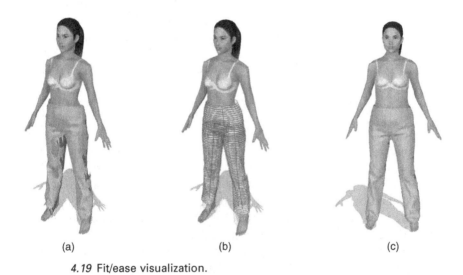

(a)   (b)   (c)

4.19 Fit/ease visualization.

- *Step 4*. Loop through the skeleton recursively to apply the transformation matrix to the vertices whose weight against the current bone is 1.

After pose duplication, it is more convenient to try the garment originally worn by the source model. The redressing procedure can be facilitated by using the crotch point, acromions and armpits as the benchmarks, since these positions can be regarded as the common constraints of dressing.

Taking refitting of pants as an example, it is not uncommon for penetration to occur. This does not always indicate that the source garment cannot be worn by the candidate human body, as shown in Fig. 4.19(a). To reveal the actual fit/ease, the girth comparison can be performed around the whole garment area by contour slicing the garment and the human body. As shown

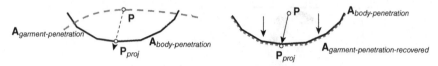

4.20 Penetration recovery.

in Fig. 4.19(b), there is a significant gap around the waist zone. This indicates that to achieve fit for these pants, another body shape with a larger waist girth may be more suitable, as shown in Fig. 4.19(c).

Although the penetration area shown in Fig. 4.19(a) does not jeopardize the girth comparison between the garment and the human body, for a better visual effect as illustrated in Fig. 4.19(c), the intersections among the garment and the human body can be solved by a penetration recovery scheme.

As shown in Fig. 4.20, let $\mathbf{A}_{garment-penetration}$ and $\mathbf{A}_{body-penetration}$ be the penetrated area on the pants and the target body, respectively. For each $\{\mathbf{P}|\mathbf{P} \in \mathbf{A}_{garment-penetration}\}$, there will be a projection point $\mathbf{P}_{proj}$ onto $\mathbf{A}_{body-penetration}$, noted as $\{\mathbf{P}_{proj}|\mathbf{P}_{proj} \in \mathbf{A}_{body-penetration}\}$. Then, the recovered position of $\mathbf{A}_{garment-penetration}$ can be performed by:

$$\mathbf{P} = \mathbf{P}_{proj}$$

which means $\mathbf{A}_{garment-penetration} \rightarrow \mathbf{A}_{body-penetration}$. In practice, this projection is directed by the normal of each $\{\mathbf{P}|\mathbf{P} \in \mathbf{A}_{garment-penetration}\}$. It is true that this kind of arbitrary shape recovery might change the girth at the intersected slices. Hence, the deformation rate (i.e. the strain of the garment) must be controlled under a given threshold, depending on the properties of the fabric. Otherwise, the fit/ease evaluation will take 'unfit' as the answer even if the visual effect is satisfied. This actually implies that to achieve fit on the given garment, either the fabric should be of excellent resiliency, or the body shape should be slimmer.

In the real world, if refitting the garment onto a different individual with a different pose, the drape will be changed. To generate the drape configuration, the refitted garment can be re-meshed by regular subdivision of the irregular meshes for the stability of the numerical integration (Fig. 4.21(c)). The vertices of the triangles are regarded as the masses, and the edges of the triangles are taken as the non-linear springs. For drape simulation, the following differential equation must be solved:

$$\frac{d}{dt}\begin{pmatrix} \mathbf{x} \\ \dot{\mathbf{x}} \end{pmatrix} = \frac{d}{dt}\begin{pmatrix} \mathbf{x} \\ \mathbf{v} \end{pmatrix} = \begin{pmatrix} \mathbf{v} \\ \mathbf{M}^{-1}\mathbf{f}(\mathbf{x}, \mathbf{v}) \end{pmatrix}$$

where $\mathbf{M}$ is the mass matrix of the garment model, and $\mathbf{f}$ is the matrix of resultant forces applied on the masses. In most cases, the external force is

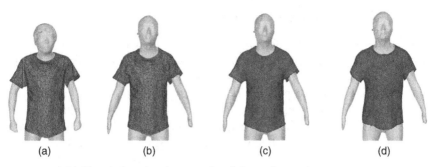

(a)   (b)   (c)   (d)

4.21 Simulating the drape style of the redressed garment.

gravity, and the internal force involves stretching, bending and shearing resistance, which can be simulated by non-linear springs:

$$\mathbf{f}_{\text{internal}} = \mathbf{f}_i + \mathbf{d}_i = -k\,\frac{\partial \mathbf{C}(\mathbf{x})}{\partial \mathbf{x}_i}\,\mathbf{C}(\mathbf{x}) - k_{\text{d}}\,\frac{\partial \mathbf{C}(\mathbf{x})}{\partial \mathbf{x}_i}\,\dot{\mathbf{C}}(\mathbf{x})$$

where $\mathbf{f}_i$ and $\mathbf{d}_i$ are the elastic force and damping force applied onto the ith mass, and $\mathbf{C}$ is the conditional function. k and k_{d} are the hook and the damping coefficient respectively, the value of which usually depends on the physical properties of the fabric. By defining $\mathbf{C} = |\mathbf{x}_{ij}| - L$, where $\mathbf{x}_{ij} = \mathbf{x}_j - \mathbf{x}_i$, L is the rest length between the ith and jth masses connected by a spring (the edge of a triangle), and setting Δt as the time step, the task is to solve the position $\mathbf{x}(t_0 + \Delta t)$ and the velocity $\dot{\mathbf{x}}(t_0 + \Delta t)$ at $t_0 + \Delta t$ for each mass, i.e.,

$$\begin{pmatrix} \Delta \mathbf{x} \\ \Delta \mathbf{v} \end{pmatrix} = \Delta t \begin{pmatrix} \mathbf{v}_0 + \Delta \mathbf{v} \\ \mathbf{M}^{-1}\mathbf{f}(\mathbf{x}_0 + \Delta \mathbf{x}, \mathbf{v}_0 + \Delta \mathbf{v}) \end{pmatrix}$$

The dynamics equation can be reformulated by first-order backward Euler integration as:

$$\left(\mathbf{I} - \Delta t \mathbf{M}^{-1}\frac{\partial \mathbf{f}}{\partial \mathbf{v}} - \Delta t^2 \mathbf{M}^{-1}\frac{\partial \mathbf{f}}{\partial \mathbf{x}} \right)\Delta \mathbf{v} = \Delta t \mathbf{M}^{-1}\left(\mathbf{f}_0 + \Delta t \frac{\partial \mathbf{f}}{\partial \mathbf{x}}\mathbf{v}_0 \right)$$

where $\mathbf{I}$ is the identity matrix. This is a well-studied formulation and can be solved by the conjugate gradient method [11]. Once we have $\Delta \mathbf{v}$, we can get $\Delta \mathbf{x} = \Delta t(\mathbf{v}_0 + \Delta \mathbf{v})$ to provide the new position and velocity for each mass.

For this approach, the strain limitation is set at around 1% to maintain size stability and prevent wrong fit/ease prediction. This is reached by an initial velocity filtering scheme, which adjusts the velocity of the masses to prevent over-deformation caused by extra impulses [20].

4.5   Challenges and future trends

The topics covered in this chapter reflect the challenges in computational textiles and the clothing industry. Although sketch-based application has become a special branch of interest, it is still hard to reproduce human-like shape perception. The capacity of modeling a range of objects with low complexity has limited the applications of this type of approach. One of the trends is to employ other shape cues such as hatching, scribble lines, stippling, shading, and suggestive contours to convey 3D forms.

As for surface flattening, many outstanding algorithms have been proposed during the first decade of this century. However, 3D garment models usually have creases which are not suitable for methods based on differential geometry where the differential quantities must be defined. Therefore, most of them cannot provide the accuracy required by the clothing industry in reverse engineering. The popularity of this technique falls in the domain of texture atlas generation. It is believed that designing numerically robust methods for industrial applications remains a huge challenge.

The last topic, trying on a virtual garment online, is actually a fantasy of virtual reality, but one that is demanded by both the customer and the industry. Most websites related to online clothing retail are currently relying on high quality 2D pictures for sales promotion. The challenges and costs of mass-preparation of high-fidelity 3D garments have set a barrier that is hard to breach. Designing fast and robust treatment in virtual garment 'manufacturing' might be a future trend in this area. As shown in Figs 4.22 and 4.23, the reusable 3D garment model might be another dawn for both virtual garment try-on and product demonstration in cyberspace.

4.22 Dressing the same items by various avatars of online shoppers.

4.23 3D garment demonstration for online shopping store.

4.6   Sources of further information

For topics on sketch-based modeling, we suggest that readers refer to a review paper written by Olsen *et al.* [21]: 'Sketch-based modelling: a survey', and the conference books of 'Proceedings of the Eurographics Workshop on Sketch-based Interfaces and Modelling (SBIM)'.

For topics on surface flattening, we recommend the Eurographics 2008 course notes: 'Geometric Modeling Based on Polygonal Meshes – Chapter 8: Mesh Parameterization'.* A more detailed version on this topic (with the proofs of theorems and formulae) is also available in the SIGGRAPH 2007 course notes: 'Mesh Parameterization, Theory and Practice'. †See also the surveys conducted in references [22] and [23]. Some source code is also available from http://alice.loria.fr/software.

For topics on human body measurement, shape morphing, virtual garment try-on and fit/ease evaluation, the detailed techniques are mainly reported by published research papers. References [13]–[20] provide some related works.

4.7   References

1. Turquin, E., Cani, M. and Hughes, J.F., 2004. Sketching garments for virtual characters. *Proc. Eurographics Workshop on Sketch-Based Interfaces and Modeling*, T. Igarashi and J.A. Jorge (eds), *Eurographics*, pp. 175–182.
2. Zhong, Y.Q. and 2009. Fit and tight-fit garment design based on parameterized 3D sketching. *The Fiber Society 2009 Spring Conference – International Conference on Fibrous Materials*, Shanghai, China, pp. 1320–1324.
3. Coquillart, S., 1990. Extended free-form deformation: a sculpturing tool for 3D geometric modeling. *Proc. SIGGRAPH 90, Computer Graphics*, 24(4), pp. 187–196.

* http://www.agg.ethz.ch/publications/course_note
† http://www2.in.tu-clausthal.de/~hormann/parameterization/index.html

4. Loop, C., 1987. Smooth spline surfaces based on triangles. Master's thesis, University of Utah, Department of Mathematics.
5. Levy, B., Petitjean, S., Ray, N. and Maillot, J., 2002. Least squares conformal maps for automatic texture atlas generation. *SIGGRAPH 2002, ACM Trans. Graph.*, 21(3), pp. 362–371.
6. Sheffer, A. and Sturler, E., 2001. Parameterization of faceted surfaces for meshing using angle based flattening. *Engineering with Computers*, 17(3), pp. 326–337.
7. McCartney, J., Hinds, B.K. and Seow, B.L., 1999. The flattening of triangulated surfaces incorporating darts and gussets. *Computer-Aided Design*, 31(4), pp. 249–260.
8. McCartney, J., Hinds, B.K. and Chong, K.W., 2005. Pattern flattening for ortho-tropic materials. *Computer-Aided Design*, 37(6), pp. 631–644.
9. Wang, C.C.L., Smith, S.S.F. and Yuen, M.M.F., 2002. Surface flattening based on energy model. *Computer-Aided Design*, 34(11), pp. 823–833.
10. Zhong, Y.Q. and Xu, B.G., 2006. A physically based method for triangulated surface flattening. *Computer-Aided Design*, 38(10), pp. 1062–1073.
11. Baraff, D. and Witkin, A., 1998. Large steps in cloth simulation. *Proc. ACM SIGGRAPH*, pp. 43–54.
12. Bridson, R., Marino, S. and Fedkiw, R., 2003. Simulation of clothing with folds and wrinkles. *Proc. ACM/Eurographics Symposium on Computer Animation*, pp. 28–36.
13. Allen, B., Curless, B. and Popovic, Z., 2003. The space of human body shapes: reconstruction and parameterization from range scans. *ACM Trans. Graph.*, 22(3), pp. 587–594.
14. Zhong, Y.Q., 2010. Redressing 3D garments based on pose duplication. Textile Res. J., 80(10), pp. 904–916.
15. Lewis, J.P., Cordner, M. and Fong, N., 2000. Pose space deformations: a unified approach to shape interpolation and skeleton-driven deformation. *Proc. ACM SIGGRAPH 2000, Annual Conference Series*, pp. 165–172.
16. Sloan, P.P.J., Rose III, C.F. and Cohen, M.F., 2001. Shape by example. *Proc. the 2001 Symp. on Interactive 3D Graphics*, pp. 135–143.
17. Kry, P.G., James, D.L. and Pai, D.K., 2002. Eigenskin: Real time large deforma-tion character skinning in hardware. *ACM SIGGRAPH Symp. on Computer Animation*, pp. 153–160.
18. Zhong, Y.Q. and Xu, B.G., 2006. Automatic segmenting and measurement on scanned human body. *Int. J. Clothing Sci. and Technol.*, 18(1), pp. 19–30.
19. Mohr, A. and Gleicher, M., 2003. Building efficient, accurate character skins from examples. *ACM Trans. Graph. (Proc. SIGGRAPH)*, 22(3), pp. 562–568.
20. Zhong, Y.Q. and Xu, B.G., 2009. Three-dimensional garment dressing simulation. *Textile Res. J.*, 79(9), pp. 792–803.
21. Olsen, L., Samavati, F.F., Sousa, M.C. and Jorge, J.A., 2009. Sketch-based model-ling: a survey. *Computers & Graphics*, 33(1), pp. 85–103.
22. Sheer, A., Praun, E. and Rose, K., 2006. Mesh parameterization methods and their applications. *Foundations and Trends in Computer Graphics and Vision*, 2(2), pp. 105–171.
23. Floater, M.S. and Hormann, K., 2005. Surface parameterization: a tutorial and survey. In *Advances in Multiresolution for Geometric Modelling*, N.A. Dodgson, M.S. Floater and M.A. Sabin (eds), Springer-Verlag, Heidelberg, pp. 157–186.

5

Modelling and simulation of fibrous yarn materials

X. CHEN, University of Manchester, UK

Abstract: In general, textiles are perceived as soft materials and structures and textile fibres can be organised into fabrics of different complexities to form new materials and structures. Textiles are widely used in many applications, ranging from clothing to advanced technical applications such as textile composites for the aerospace and other industries. Accurate understanding of the relationships between the construction and technical behaviour of textiles is the key in engineering textile materials for the intended applications. This chapter introduces the modelling techniques for different forms of textile materials and structures and reports on the latest progress in modelling textiles for different applications.

Key words: textiles, structures, behaviour, modelling, applications.

5.1   Introduction

Textile materials and products are used extensively nowadays for both domestic and technical applications (Chen, 2009). In both areas, precision in structural design and properties is of primary importance as it is associated with the behaviour of the textiles. Textile composites, for example, are used in advanced applications such as aircraft, and the ability to engineer the textile preform and the composite components is vital in achieving the overall aircraft efficiency and safety. Modelling textiles is an important step towards better understanding of how textiles work and behave, and therefore forms an advanced tool for investigation and exploration of the textile assemblies.

It is necessary to realise that textiles as a type of materials are special when compared to materials such as metal. Textile fabrics are far from homogeneous and isotropic because they are assemblies of fibres, which themselves are long and thin entities. In addition, fibres are made of a wide range of different chemical compositions, and when different fibres are used for making textiles, the physical and chemical properties can be vastly different. Because of all these special features, modelling of textile structures and behaviour has always been a focus of attention.

The textile hierarchy ranges from fibre as the basic element. Fibres are the construction units of yarns and some non-woven fabrics. The yarns are used as components for making fabrics based on their weaving, knitting and

93

braiding technologies. It is accepted that the behaviour of yarn depends on the fibre property and the yarn construction, and that fabric behaviour is determined by the properties of the yarn from which it is composed and the construction of the fabric. A fabric contains tremendous amounts of fibres of the same or different types, and there are many different ways in which a fibre is configured individually or collectively in a fabric. Phenomena such as these make the modelling of textiles very challenging.

The most important application of modelling of textiles is for the analysis of the performance and behaviour of textiles under different loading conditions, whether mechanical, thermal or fluid. Much work has been done on modelling different forms of dry textiles and textile-reinforced composites for their mechanical properties, for instance. Such work has enabled people's understanding of how the textiles would react when a mode of mechanical load is applied to the textiles and products.

5.2   Modelling and simulation of yarns

5.2.1 Types of yarn

Monofilament yarns may be regarded as the simplest form of yarn that is used in forming fabrics. As a filament, they are usually much coarser than the filaments involved in multifilament yarns. Monofilament yarns are mostly used for making fabrics for special purposes, such as polymer monofilament fabrics for filtration and metal wire fabrics for architectural decoration.

The major forms of yarn are those that involve a large number of fibres in their cross-sections. If the fibres are continuous throughout the length of the yarn, then the yarn is known as a continuous multifilament yarn. On the other hand, if the fibres are short staples, the yarn is known as a staple yarn. The filaments in a continuous multifilament yarn are lightly interlaced or lightly twisted as they come from a fibre producer. If more inter-filament cohesion is required, these yarns are twisted. For bulk, stretch and appearance, which are necessary for most domestic end-uses, continuous filament yarns are usually textured. The most important technique for texturing is false-twist texturing, either with a single heater to give stretch yarns or with two heaters to give set yarns (Hearle et al., 2001). Unlike the continuous filament yarns used for domestic applications, those used for technical applications, such as carbon, glass and aramid yarns, are mostly employed without adding twists due to considerations of properties. In case inter-fibre cohesion is needed, size is usually applied to hold the fibres together.

For short (staple) fibres, such as natural cotton and wool and cut manufactured fibres, means must be found of holding the separate fibres axially in order to form yarns. This can be achieved physically by twisting, wrapping, entangling or chemical bonding. Ring-twisting is the dominant process. There are some older methods, notably hand-spinning and mule spinning, and in the 1970s and 1980s many other methods were developed. Rotor spinning and air-jet spinning are extensively used; some others have survived in specialist applications. More information on staple fibre spinning was described and discussed by Grosberg and Iype (1999), Lawrence (2003) and Lord (2003). The structure and mechanics of twisted continuous filament and staple yarns are covered by Hearle *et al.* (1969).

Single yarns may be twisted together to give ply yarns. Ply yarns may then be twisted together to give cords and ropes. Usually twist direction alternates between levels.

5.2.2 Yarn geometry

Twist and twist factor

An idealised model of twisted yarn geometry is shown in Fig. 5.1(a). It consists of concentric helices with constant twist period h and has yarn lengths in one turn of twist, l and L, at an intermediate radius r and the external radius R. By opening out the cylinders into flat rectangles, as shown in Fig. 5.1(b) and (c), the relations between helix angles θ and α and the other quantities can be derived. Among these, helix angle α can be expressed as

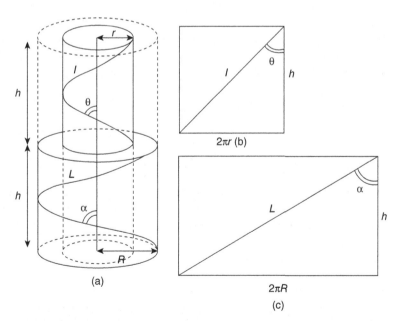

5.1 Idealised helical yarn geometry.

$$\tan \alpha = \frac{2\pi R}{h} = 2\pi R T_w \qquad\qquad 5.1$$

where $T_w = 1/h$ is the number of twists added in the yarn length h.

Suppose that the yarn linear density is T (tex) and the yarn specific volume is $1/\rho$ (cm^3/g), where ρ is the specific density of the yarn. According to the definition, $T = \pi R^2 \rho \times 10^5$, and hence

$$R = \sqrt{\frac{10^{-5}T}{\pi\rho}} \qquad\qquad 5.2$$

Substituting equation 5.2 into 5.1 gives

$$\tan \alpha = 2\pi \sqrt{\frac{10^{-5}T}{\pi\rho}} \cdot T_w = 0.0112 \sqrt{\frac{1}{\rho}} \cdot \left(\sqrt{T} \cdot T_w\right)$$

Let

$$\alpha_t = \sqrt{T} \cdot T_w \qquad\qquad 5.3$$

be defined as the twist factor, which reflects the twist effect to the yarn by taking into account both the twist number and the yarn count.

Fibre packing factor

Another feature of yarn geometry consists of the fibre packing, which influences yarn diameter and specific volume. The fibre packing factor (f) is defined as the ratio of the specific volume of yarn to the specific volume of fibres, $1/\rho$, namely,

$$f = \frac{\rho}{\rho_f} \qquad\qquad 5.4$$

In equation 5.4, ρ_f is the fibre specific density (g/cm^3). The tightest form is hexagonal close packing of circular fibres; the packing factor is then 0.92, but many factors lead to lower values. In twisted continuous filament yarns, values of 0.7–0.8 are common. In spun yarns of weakly crimped fibres, such as cotton and most manufactured staple fibres, values are nearer 0.5; in wool and in textured continuous filament yarns, values are much lower. Packing becomes tighter as twist is increased.

Fibre migration

In real yarns, the fibres do not maintain a constant radial position but migrate between inner and outer zones. The changing position is illustrated in Fig. 5.2(a); the radially expanded view of Fig. 5.2(b) shows the change

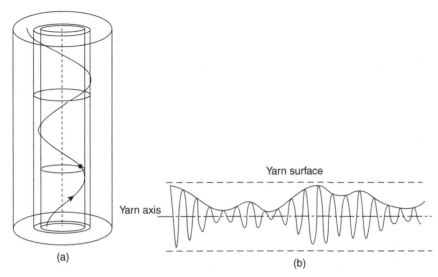

5.2 Yarn migration: (a) changing position of fibres; (b) radially expanded view.

over longer lengths. In modelling continuous filament yarns, migration can be neglected, because the structure is close to ideal over a short yarn element. In staple fibre yarns, migration has a more important role.

Ring-spun cotton and similar yarns have a helical structure, which is not too far from ideal, but there is an irregular partial migration with a period of around three turns of twist. The structure is more open and less regular with crimped wool fibres. In set textured continuous filament yarns, the fibre paths are alternating left- and right-handed helices; in unset stretch yarns, the filaments can contract into pig-tail snarls. In rotor-spun yarns, there is less migration, but wrapper fibres have an important role.

5.2.3 Yarn mechanics

Continuous filament yarns

The tensile stress–strain curve of twisted continuous filament yarns is the most successfully modelled example of the mechanics of a textile material, based on the idealised geometry of Fig. 5.1. According to the definition, the strain of yarn is expressed as

$$\varepsilon_y = \frac{\delta h}{h}$$

5.5

where δh is the extension of yarn length h corresponding to one turn of twist. From Fig. 5.1(b),

$$l^2 = h^2 + 4\pi^2 r^2 \qquad\qquad 5.6$$

Assuming that the yarn radius R does not change under the tensile loading leads to

$$\frac{dl}{dh} = \frac{h}{l} \qquad\qquad 5.7$$

In this process of tensile loading, the strain of a filament can be described as

$$\varepsilon_f = \frac{\delta l}{l} = \frac{1}{l}\cdot\frac{dl}{dh}\cdot\delta h = \frac{h^2}{l^2}\cdot\frac{\delta h}{h}$$

i.e.,

$$\varepsilon_f = \varepsilon_y \cdot \cos^2\theta \qquad\qquad 5.8$$

In this simplest approximate form, modelling goes from the strain distribution in fibres, which reduces approximately as $\cos^2\theta$, to give a contribution to stress reduced by a further factor of $\cos^2\theta$ through the axial component of tension and the oblique area on which fibre tension acts. The ratio of yarn to fibre modulus is thus the mean value of $\cos^2\theta$, which is $\cos^2\alpha$ for the ideal helical geometry. For a more exact model, an energy method is both simpler than the use of force methods and easily adapted to cover large strains and lateral contraction. An important simplification in modelling is that most fibres extend at close to constant volume, i.e. Poisson's ratio being 0.5. This means that the deformation energy is the same when the fibre is extended at zero lateral pressure in a tensile test as it is when the fibre is subject to combined tension and pressure within a yarn. Consequently, the measured fibre stress–strain curve is a valid input. Except for a small deviation at low stresses where some buckling of central filaments reduces their contribution to tension, there is good agreement between the theory and experimental results.

Twisted yarns break when the central fibres, which are the most highly strained, reach their break extension. Consequently, yarn and fibre break extensions are similar and the strength reduction is similar to the modulus reduction.

In the case where the yarn radius reduces under tensile loading, the strain of the filaments is expressed as follows, where σ_y is the yarn stress:

$$\varepsilon_f = \varepsilon_y \left(\cos^2\theta - \sigma_y \sin^2\theta\right) \qquad\qquad 5.9$$

Staple fibre yarns

For staple fibre yarns, the resistance to extension is reduced by slip at fibre ends, as illustrated in Fig. 5.3. The slippage factor *SF* equals (area

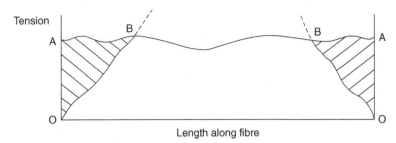

5.3 Slip at fibre ends.

OBBO/area OAAO). If the gripping pressure is too weak, the fibre is nowhere gripped. A low twist sliver or roving can thus be drafted with fibres sliding past one another. Above a critical gripping force when the gripped zone BB is present, the yarn is self-locking. The slippage factor, which depends on the fibre aspect ratio, the gripping force and the friction between fibres, can be represented by the following basic relation:

$$SF = 1 - \frac{r}{2\mu JL} \qquad\qquad 5.10$$

where r = fibre radius, L = fibre length, μ = coefficient of friction, and J is an operator determining how fibre tensions are converted into transverse stresses.

The length between A and B in Fig. 5.3 is the fibre length that is held by the yarn. Let it be l_h. This length can be determined by equalising the friction the fibre encounters in the yarn and the breaking tension of the fibre. That is,

$$2\pi r \cdot l_h \cdot q \cdot \mu = \pi r^2 \sigma_b$$

Hence,

$$l_h = \frac{r\sigma_b}{2q\mu} \qquad\qquad 5.11$$

In equation 5.11, r is the radius of the fibre, q is the lateral pressure received on the fibre surface, μ is the frictional coefficient between fibres, and σ_b is the breaking stress of the fibre.

Tension in helically twisted fibres clearly leads to inward pressures. However, twist alone is not enough. With the ideal structure of Fig. 5.1, a fibre on the outside of the yarn would be nowhere gripped and in turn could not grip fibres in the next layer. Slippage would be complete. The situation is saved by fibre migration, which occurs naturally in ring-spinning. The structure is self-locking, because fibres are gripped when they are near the centre of the yarn and provide gripping forces when they are near the outside. This was confirmed in an idealised model by Hearle (1965), which also yielded the following approximate relation for the ratio of yarn modulus to fibre modulus, γ:

$$\gamma = \cos^2 \alpha \left(1 - \frac{2\cos\text{ec}\alpha}{3L} \sqrt{\frac{rQ}{2\mu}} \right) \qquad\qquad 5.12$$

where Q = migration period.

Worsted yarns, where a more open and low-twist structure is needed, are commonly produced as two-ply yarns. The pressure of the plies on each other generates the initial grip that can build up within the singles yarns. More recently, developments have led to increased migration, so that singles worsted yarns can be used in finer fabrics. Where there is less migration, the wrapping fibres, which occur naturally in rotor spun yarns, interact with twist to generate gripping forces. Wrapping is the sole grip mechanism in hollow-spindle spinning in which a zero-twist strand of short fibres is wrapped by a continuous filament yarn.

In woollen spun yarns, and to a greater extent in felted yarns, entanglement is the gripping mechanism. Contact pressures occur as fibres pass round one another in an irregular structure, which can be modelled as a set of overlapping helices. A similar model can be applied to air-jet textured yarns. The loops, which stick out from the yarns, are gripped by entanglement in the core of the yarns.

For more bulky yarns, whether due to the natural crimp of wool or the imposed helices or snarls of textured yarns, there is an initial stretch under low tension until the crimp is pulled out. The simplest model is to assume that there is no resistance to elongation until the fibres are taut at the shortest available path length. More detailed mechanics can be applied to the interaction of bending and twisting in the elongation of helices and snarls.

Bending resistance of yarns is a more difficult problem because of a change from one mechanism to another. When inter-fibre slip is easy, the bending stiffness is given by summing the stiffness of the individual fibres. This leads to the following expression for the yarn bending stiffness for yarns with low inter-fibre gripping, B_{yl}:

$$B_{yl} = B_f \frac{m}{m_f} \left[1 - \frac{1}{4}(1+N)\tan^2 \alpha \right] \qquad\qquad 5.13$$

where B_f is the bending stiffness of the fibre, m is the mass of a unit length of yarn, m_f is the mass of a fibre within the unit length of yarn, α is the twist angle at the yarn surface, and $N = B_f/C_f$ with C_f being the torque stiffness.

When the fibres are firmly gripped, yarns act as solid rods. The expression for the bending stiffness under this situation is

$$B_{\text{yh}} = B_{\text{f}} \left(\frac{m}{m_{\text{f}}} \right)^2 \qquad\qquad 5.14$$

While these two extremes are easy to calculate using beam bending models, the difficulty is the intermediate region, which is complicated by the fact that extension of fibre elements on the outside of a yarn can be relieved by slip along the helical path to the inside of the bend.

5.2.4 Description of yarn dimension

A yarn consists of large number of fibres in its cross-section. It is a common practice to assume that a yarn is a solid entity having the properties it should as a twist bundle of fibres. This is obviously for the convenience of analysis of fabrics that involves many yarns for their construction. In addition, although a circular shape has been popularly used to describe the yarn cross-section, notably started by Peirce (1937), the yarns in a fabric are always pressed due to the weaving tension and the pressure applied to the fabric surface. As a result, the cross-sections of yarns in most fabrics are flattened. Kemp (1958) assumed a racetrack cross-section, and Shanahan and Hearle (1978) proposed a lenticular yarn cross-section. The racetrack and the lenticular yarn cross-sections are illustrated in Fig. 5.4.

In textiles, the dimension of yarn is described by the yarn linear density, which is affected by the yarn porosity, and fibre density. The area of the cross-section of a yarn, A_x, can be described as follows, according to the definition of yarn linear density:

$$A_x = \frac{T}{f\rho} \times 10^{-5} \quad (\text{cm}^2) \qquad\qquad 5.15$$

where T is the linear density of yarn in tex, f is the fibre packing factor, and ρ is the specific density of the fibre in g/cm³.

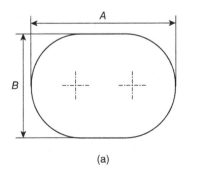 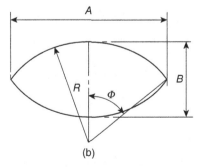

5.4 Flattened yarn cross-sections: (a) racetrack; (b) lenticular.

The simplest assumption for the yarn cross-section has been circular. For a yarn with circular cross-section, the diameter, D, can be calculated as below:

$$D = 2\sqrt{\frac{T}{\pi f \rho}} \times 10^{-5} \quad (cm) \qquad\qquad 5.16$$

For flattened cross-sectional shapes, the aspect ratio of width to height of the yarn cross-section describes the extent to which a yarn cross-section is flattened. As indicated in Fig. 5.4, the aspect ratio used for the racetrack and lenticular cross-sectional shapes is A/B.

A racetrack shape, shown in Fig. 5.4(a), is composed of two parallel lines with a half circle at each end. The area of this cross-sectional shape is, therefore,

$$A_x = B^2 \left(\frac{A}{B} - 1 + \frac{\pi}{4} \right)$$

When the aspect ratio A/B of a yarn cross-section is known, B(the same as the diameter of the half circles) can be expressed as follows after substituting equation 5.15 for A_x:

$$B = \sqrt{\frac{T \times 10^{-5}}{f \rho \left(\dfrac{A}{B} - 1 + \dfrac{\pi}{4} \right)}} \qquad\qquad 5.17$$

The lenticular shape shown in Fig. 5.4(b) is formed by integrating two face-to-face but otherwise identical arcs, where R is the radius and ϕ is half of the central angle of the arc. Subsequently, the area of the lenticular cross-sectional shape is

$$A_x = 2R^2 (\phi - \cos\phi \sin\phi)$$

and the radius of the arcs is, after substituting equation 5.15 for A_x,

$$R = \sqrt{\frac{T \times 10^{-5}}{2 f \rho (\phi - \cos\phi \sin\phi)}} \qquad\qquad 5.18$$

It is also easy to find that

$$R = \frac{A^2 + B^2}{4B}$$

$$\sin\phi = \frac{2AB}{A^2 + B^2}$$

$$A = 2R \sin\phi$$

5.3   Weave modelling

Woven fabrics are constructed by interlacing two systems of yarns perpendicular to each other; the one in the length direction of the fabric is the warp and the one that goes in the width direction of the fabric is the weft. There are many different ways of interlacing the warp and weft yarns into a fabric, and a particular plan for constructing a fabric from warp and weft yarns is known as a weave. Weaves are traditionally classified into elementary weaves, derivative weaves, combined weaves and complex weaves. In addition to these, woven fabrics can be made to have a considerable thickness from multiple sets of yarns in each of the two directions, and these are popularly termed 3D woven fabrics.

Modelling the structure of the woven fabrics is regarded as an important step towards the computerised generation of weaves. Chen and his colleagues (1988, 1996) worked on the structural modelling of woven fabrics by defining weaves as regular or irregular. A regular weave is one whose float arrangement and the step number do not change in a repeat of the weave. All other weaves are defined as irregular weaves. Many commonly used weaves are regular themselves or further developed from regular weaves. Whilst each type of irregular weave would need a distinct mathematical model to describe its construction, all regular weaves need the same mathematical model for their construction.

Figure 5.5 lists some weaves that are commonly used for making fabrics for domestic as well as technical applications.

5.3.1   Mathematical description of regular 2D fabric weaves

A model describing regular weave construction was reported by Chen *et al.* (1988, 1996), which yields the 2D binary weave matrix, W, upon the specification of the float arrangement, F_i, and the step number, S. The layout of the weave matrix is as follows:

$$W = \begin{bmatrix} w_{1,R_p} & w_{2,R_p} & \cdots & w_{R_e,R_p} \\ \cdots & \cdots & \cdots & \cdots \\ w_{1,2} & w_{2,2} & \cdots & w_{R_e,2} \\ w_{1,1} & w_{2,1} & \cdots & w_{R_e,1} \end{bmatrix} \qquad 5.19$$

Suppose that $w_{x,\,y}$ is the element of this matrix at coordinate (x, y), where $1 \leq x \leq R_e$ and $1 \leq y \leq R_p$ with R_e and R_p being the warp and weft repeat

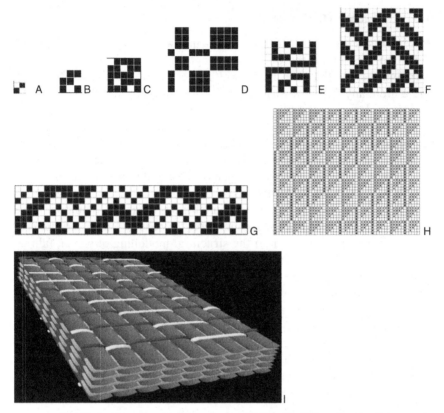

5.5 Some weaves: A plain weave; B 2/1 twill; C 5-end satin; D hopsack weave; E basket weave; F entwined weave; G see-saw weave; H 3D orthogonal fabric weave; I geometric model for the 3D orthogonal fabrics whose weave is shown in H.

respectively, then the first column of the weave matrix can be generated using the following equation:

$$w_{1,y} = \begin{cases} 1 & \text{if } i \text{ is an odd integer} \\ 0 & \text{if } i \text{ is an even integer} \end{cases} \qquad 5.20$$

where $y = \left(\sum_{j=1}^{i} F_j - F_i + 1 \right)$ to $\sum_{j=1}^{i} F_j$; and $1 \le i \le N_f$. N_f is the number of floats in the float arrangement. Then, the rest of the matrix will be assigned values as follows:

$$w_{x,z} = w_{1,y} \qquad 5.21$$

where

$$
z = \begin{cases} y+[S\times(x-1)]+R_{\mathrm{p}} & \text{if } \{y+[S\times(x-1)]\} < 1 \\ y+[S\times(x-1)] & \text{if } 1 \le \{y+[S\times(x-1)]\} \le R_{\mathrm{p}} \\ y+[S\times(x-1)]-R_{\mathrm{p}} & \text{if } \{y+[S\times(x-1)]\} > R_{\mathrm{p}} \end{cases}
$$

$$2 \le x \le R_{\mathrm{e}}$$

$$1 \le y \le R_{\mathrm{p}}$$

5.3.2 Mathematical description of 3D fabric weaves

In a conventional sense, there are three methods in weaving that can lead to the creation of 3D solid fabrics, which are characterised by structural integrity and notable dimension in the thickness direction. These are the multi-layer method, the orthogonal method, and the angle interlock method.

In order to create weaves for such fabrics, mathematical expressions of these weaves are essential. Chen and Potiyaraj (1998, 1999a, 1999b) reported on the mathematical modelling of backed cloth weaves, multi-layer fabric weaves, and orthogonal and angle interlock fabric weaves. In modelling all these 3D solid fabric weaves, the parametric approach is used to specify the features of the fabrics. For computer aided manufacture, the weave diagrams expressed in 2D form are generated using the algorithms created according to the modelling results. The modelling also leads to programs that create 3D models for such fabrics, which have been used as the geometrical models for property analyses.

Another group of 3D woven fabrics is hollow fabrics, which allow tunnels to be created in the warp, weft or diagonal direction in 3D fabrics. Chen *et al.* (2004) reported on the modelling of 3D cellular fabrics with uneven surfaces. The work involves two major issues, i.e., the description of the cross-section configuration of the fabrics, and the weave creation based on the multi-layer principle. Modelling of 3D cellular fabric with even surfaces was reported by Chen and Wang (2006). In this work, a self-opening 3D cellular fabric structure is mathematically described and the work too leads to the creation of the weave diagram.

5.4   Geometrical modelling of woven fabrics

The performance of a textile fabric is basically a function of the property of the constituent fibres/yarns, and the geometrical construction of the fibres/yarns in the fabrics. The geometry of fabrics has been studied for almost a century. Geometrical models of fabrics have led to the estimation of some structural and physical properties of the fabrics, such as the areal mass and the porosity, and the results from the modelling have been used as guidance to fabric manufacture in giving the maximum areal density of the fabrics. Geometrical modelling of textile assemblies becomes more important, as the geometrical models are believed to be the most reliable solution in providing geometrical information on textile assemblies for finite element (FE) analysis for performance simulation.

5.4.1 Peirce's plain fabric model and its derivative models

Peirce's model

Peirce's work on plain fabric geometry (1937) is regarded as the beginning of modelling woven fabric geometries. Under certain assumptions, including circular yarn cross-section, complete flexibility of yarns, incompressible yarns and arc–line–arc yarn path, a set of equations was established to describe the geometry of plain woven fabrics. The cross-section of a plain woven fabric based on Peirce's assumption is shown in Fig. 5.6.

$$D = d_e + d_p$$

$$h_e + h_p = D$$

$$c_e = \frac{l_e}{p_p} - 1$$

$$c_p = \frac{l_p}{p_e} - 1$$

$$p_p = (l_e - D\theta_e)\cos\theta_e + D\sin\theta_e$$

$$p_e = (l_p - D\theta_p)\cos\theta_p + D\sin\theta_p$$

$$h_e = (l_e - D\theta_e)\sin\theta_e + D(1 - \cos\theta_e)$$

$$h_p = (l_p - D\theta_p)\sin\theta_p + D(1 - \cos\theta_p)$$

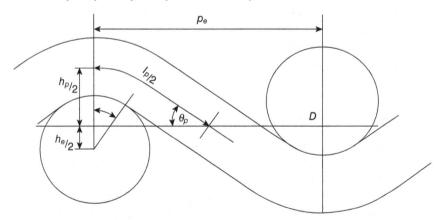

5.6 Peirce's model for plain woven fabric.

where

h_e, h_p = the modular heights of the warp and weft yarns normal to the fabric plane

c_e, c_p = the crimps of the warp and weft yarns

D = the sum of the diameters of the warp and weft yarns

d_e, d_p = the diameters of the warp and weft yarns

p_e, p_e = the thread spacing between adjacent warp and weft yarns

l_e, l_p = the modular lengths of the warp and weft yarns in one repeat

θ_e, θ_p = the weaving angles of warp and weft yarns.

Subscripts e and p in the variables above refer to warp (ends) and weft (picks) respectively.

There are 13 variables in these eight equations. Therefore, with five variables known, these simultaneous equations can be solved for definite fabric geometry. Ai (2003) presented an algorithm to calculate the geometry assuming that five variables, p_e, p_p, d_e, d_p and one of c_e and c_p are specified.

As mentioned earlier, Kemp's racetrack model (1958) and Shanahan and Hearle's lenticular model (1978) are extensions to Peirce's model, in that the circle (arc) and straight line elements are adopted although the yarn flattening is satisfactorily addressed. For this reason, the two latter models are regarded as derivative models from Peirce's.

Derivative models

Kemp's fabric model based on the racetrack yarn cross-section is shown in Fig. 5.7, where the section of the model between S_1 and S_2 is the same as in the Peirce model with S_1 and S_2 going through the origins of the half circles in the two adjacent yarns. In this model, A_e, A_p and B_e, B_p are the widths and heights of the cross-section of warp and weft yarns respectively. The variables h_e, h_p, c_e, c_p, p_e, p_p, l_e, l_p, θ_e and θ_p have the same definition as for the Peirce's model. The variables p_e' and p_p' are the distances between S_1 and S_2 in the fabric cross-sections along warp and weft respectively, and l_e' and l_p' are the yarn lengths between S_1 and S_2. Variables c_e' and c_p' are the warp and weft crimp between S_1 and S_2 in the due direction.

The following relations can be achieved from the racetrack model described in Fig. 5.7.

$$p_e' = p_e - (A_e - B_e) = (l_p - D'\theta_p)\cos\theta_p + D'\sin\theta_p$$

$$p_p' = p_p - (A_p - B_p) = (l_e - D'\theta_e)\cos\theta_e + D'\sin\theta_e$$

$$h_e = (l_e - D'\theta_e)\sin\theta_e + D'(1 - \cos\theta_e)$$

$$h_p = (l_p - D'\theta_p)\sin\theta_p + D'(1 - \cos\theta_p)$$

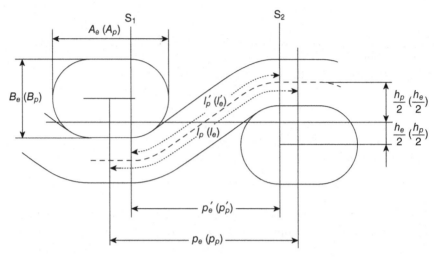

5.7 Kemp's racetrack fabric model.

$$l'_e = l_e - (A_p - B_p)$$

$$l'_p = l_p - (A_e - B_e)$$

$$c'_e = \frac{l'_e - p'_p}{p'_p} = \frac{c_e p_p}{p_p - (A_p - B_p)}$$

$$c'_p = \frac{l'_p - p'_e}{p'_e} = \frac{c_p p_e}{p_e - (A_e - B_e)}$$

$$D' = B_e + B_p$$

Shanahan and Hearle (1978) constructed a woven fabric model with lenticular yarn cross-section. This fabric model is shown in Fig. 5.8, where the composition of the yarn path is the same as that in Peirce model. The following relationships can be established from Shanahan and Hearle's fabric model:

$$h_e + h_p = B_e + B_p$$

$$c_e = \frac{l_e}{p_p} - 1$$

$$c_p = \frac{l_p}{p_e} - 1$$

$$p_p = (l_e - D_e \theta_e) \cos \theta_e + D_e \sin \theta_e$$

$$P_e = (l_p - D_p \theta_p) \cos \theta_p + D_p \sin \theta_p$$

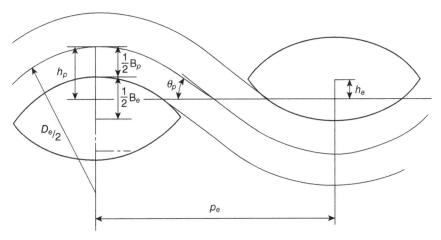

5.8 Shanahan and Hearle's lenticular fabric model.

$$h_e = (l_e - D_e\theta_e)\sin\theta_e + D_e(1 - \cos\theta_e)$$

$$h_p = (l_p - D_p\theta_p)\sin\theta_p + D_p(1 - \cos\theta_p)$$

$$D_e = 2R_e + B_p$$

$$D_p = 2R_p + B_e$$

where h_e, h_p, p_e, p_p, l_e, l_p, c_e, c_p, θ_e and θ_p have the same definitions as those used in Peirce's model. In addition, B_e and B_p are the heights of the cross-sections of warp and weft yarns, D_e and D_p are the diameters of the arcs of the warp and weft yarn paths, and R_e and R_p are the radii of arcs for lenticular shapes for warp and weft yarns.

5.4.2   Relationships between the fabric parameters

Fabric jam

In each case, the models can be used to determine the maximal warp and weft densities in the fabric using the jammed fabric theory. According to the Peirce model, the straight-line segment in the warp yarn path decreases when the weft density increases. When a fabric reaches its maximum weft density, the straight-line segment in the warp yarn path will disappear. Figure 5.9(a) shows that that yarn path is formed by two arcs tangentially connected together for a jammed situation.

For a jammed fabric geometry, according to Peirce's assumptions, there must be $p_e \geq d_e$ or $p_p \geq d_p$ for fabrics with circular yarns, and $p_e \geq A_e$ or $p_p \geq A_p$ for flattened yarns. In a given fabric, the warp and weft spacings p_e and p_p cannot both be set to be equal to the diameter or width of the

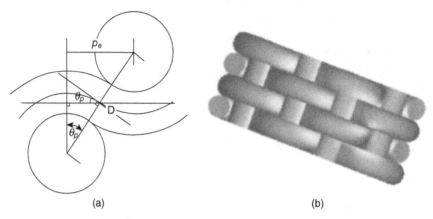

(a)   (b)

5.9 Jammed fabric geometry: (a) model; (b) simulated fabric.

yarn at the same time in the due direction. For example, when the warp yarn spacing p_e assumes the value of the diameter of the weft yarn, the weft spacing in the fabric will need to have a value which must be smaller than the diameter of warp yarn. The two spacing values limit each other according to the yarn geometrical shape.

Modular heights and yarn spacings

In the Peirce model, when the warp and weft yarns are selected, $h_e + h_p = D$, which is a constant. It is obvious that if h_e is increased, h_p will decrease; and vice versa. When h_e becomes $h_{e\,max} = D$, h_p becomes $h_{p\,min} = 0$. At this extreme situation, the warp ends will be straight in the fabric, and $p_e = D$.

From Fig. 5.6, it can be seen that θ_p and h_p change in the same direction, and θ_e and h_e change in the same direction. The nature of the relationship between p_e and θ_p, and between p_e and θ_e, can be found from the following exercise.

From Peirce's model equations, we have

$$\frac{\partial p_p}{\partial \theta_e} = -(l_e - D\theta_e)\sin\theta_e$$

$$\frac{\partial p_e}{\partial \theta_p} = -(l_p - D\theta_p)\sin\theta_p$$

By definition, the minimum length of the straight line segment in the yarn path is 0, i.e.,

$$l_e - D\theta_e \geq 0$$

$$l_p - D\theta_p \geq 0$$

In a woven fabric, the values of θ_e and θ_p are always within the range [0, $\pi/2$]. Therefore,

$$\frac{\partial p_p}{\partial \theta_e} = -(l_e - D\theta_e)\sin\theta_e \le 0$$

$$\frac{\partial p_e}{\partial \theta_p} = -(l_p - D\theta_p)\sin\theta_p \le 0$$

This indicates that p_e and θ_p and θ_e change in the inverse directions respectively. For example for any given p_e, when θ_p increases p_e will decrease.

The relationships of these parameters are summarised below:

$$p_p \uparrow \rightarrow \theta_e \downarrow \rightarrow h_e \downarrow, \text{ therefore } p_p \uparrow \rightarrow h_e \downarrow$$

$$p_p \downarrow \rightarrow \theta_e \uparrow \rightarrow h_e \uparrow, \text{ therefore } p_p \downarrow \rightarrow h_e \uparrow$$

and similarly,

$$p_e \uparrow \rightarrow \theta_p \downarrow \rightarrow h_p \downarrow, \text{ therefore } p_e \uparrow \rightarrow h_p \downarrow$$

$$p_e \downarrow \rightarrow \theta_p \uparrow \rightarrow h_p \uparrow, \text{ therefore } p_e \downarrow \rightarrow h_p \uparrow$$

Therefore, for a given p_p, when h_p reaches its maximum value $h_{p\,max}$, p_e will reach the minimum value $p_{p\,min}$; and for a given p_e, when h_e reaches its maximum value $h_{e\,max}$, p_p will reach the minimum value $p_{p\,min}$.

Yarn cross-section

Regular cross-sectional shapes, such as circular, lenticular, racetrack and super-ellipse, have been used as idealised cross-sectional shapes for yarns. However, in most cases, the yarn cross-sections in a fabric may take a much more complicated and random cross-sectional shape. This section explains a method for creating irregular yarn cross-sectional shapes.

The 2D polar coordinate system is used to construct irregular yarn cross-sections. The number of discrete nodes is adjustable between three and 20. Figure 5.10 shows an example of creating an irregular yarn cross-section with 12 nodes. The discrete nodes for warp yarn and weft yarn can be expressed as $[\theta_{i,warp}, R_{i,warp}]$ and $[\theta_{i,weft}, R_{i,0}]$ respectively. Then the piecewise quadratic B-Splines are used to create the function of cross-sectional shape of the yarn. The corresponding Cartesian representation for the nodes is as follows.

For warp:

$$\begin{cases} x_{i,warp} = x_{c,warp} + R_{i,warp}\cos\theta_{i,warp} \\ y_{i,warp} = y_{c,warp} \\ z_{i,warp} = z_{c,warp} + R_{i,warp}\sin\theta_{i,warp} \end{cases} \qquad i = 0, 1, \ldots, N \qquad 5.22$$

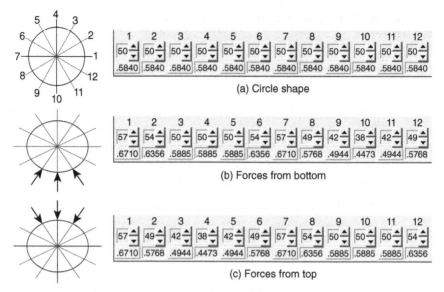

5.10 Irregular cross-section expressed by polar coordinates: (a) circle shape; (b) forces from bottom; (c) forces from top.

For weft:

$$\begin{cases} x_{i,\text{weft}} = x_{c,\text{weft}} \\ y_{i,\text{weft}} = y_{c,\text{weft}} + R_{i,\text{weft}} \cos\theta_{i,\text{weft}} \qquad i = 0, 1, \ldots, N \\ z_{i,\text{weft}} = z_{c,\text{weft}} + R_{i,\text{weft}} \sin\theta_{i,\text{weft}} \end{cases}$$
5.23

where θ_i and R_i are the angle and radius of the ith node, and (x_c, y_c, z_c) are the coordinates of the centreline of the yarn path. Using these nodes as the control points, a spline curve will be created to represent the cross-sectional shape of the yarn.

Figure 5.10(a) is a circular shape created by our irregular generator of cross-sectional shape, (b) simulates the shape when the yarn is deformed from the bottom, and (c) simulates a deformed shape with force coming from the top. The principle in creating irregular yarn cross-sections is that the yarn cross-sectional area is kept the same. In Fig. 5.10, the cross-sections in situations (a), (b) and (c) have the same area.

Relationship between yarn dimension and yarn linear density

Yarn count is popularly expressed in linear density. Equation 5.15 reveals the relationship between the yarn linear density and the yarn cross-sectional area.

To simulate irregular yarn cross-sectional shape, we use parameter r_i to describe the radii of the nodes to the yarn centre in the cross-section of a yarn model. The relevant cross-sectional area of the model can be approximately expressed as

$$A_x = \frac{\sin\theta}{2}\sum_{i=0}^{N-1} r_i r_{i+1} \qquad\qquad 5.24$$

where N is the number of nodes in a yarn cross-section, and θ is the angle between two adjacent radii. Substituting equation 5.15 into 5.24 leads to equation 5.25 for the determination of r_{i+1}:

$$T = \left(\frac{f\rho\sin\theta}{2}\times10^5\right)\sum_{i=0}^{N-1} r_i r_{i+1} \qquad\qquad 5.25$$

5.4.3 Geometrical modelling of 3D fabrics

Based on similar principles, different types of 3D fabrics can be geometrically modelled. Typical types of 3D solid fabrics include multi-layer fabrics, orthogonal fabrics and angle-interlock fabrics. 3D cellular fabrics are also made with flat surfaces and with uneven surfaces. Packages for creating 3D fabric geometries include WiseTex (Lomov *et al.*, 2001), TexGen (Robitaille *et al.*, 2003) and Weave Engineer (Chen, 2008). It is not intended to discuss the methods used for creating geometric models for 3D fabrics here. The interested reader may follow the literature mentioned above.

5.5   Finite element (FE) modelling of woven fabrics

Based on the results from weave modelling and fabric geometric modelling, solid models for different types of woven fabrics are created. For example, a program named *UniverWeave* was developed to create woven fabric geometrical models efficiently (Nazarboland *et al.*, 2008). The defined geometry can be picked up by major FE software packages to carry out FE simulations for material and component analysis.

5.5.1 FE modelling of filtration through fabrics

Since the geometry and porosity of the fabric filter are determined by the weave pattern and the various parameters of yarns constituting the fabric (Backer, 1948), it is important to optimise its structure to achieve the most efficient filtration. The past and current practice is to rely on the practical skill and experience of the fabric designers and empirical trials. Readily

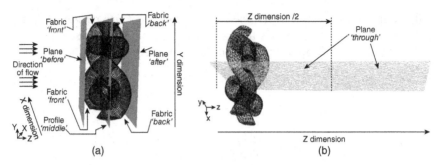

Direction of flow
Fabric 'front'
Fabric 'back'
Plane 'before'
Plane 'after'
Fabric 'front'
Fabric 'back'
Profile 'middle'
Y dimension
X dimension
Z dimension /2
Plane 'through'
Z dimension

(a)   (b)

5.11 Fabric in relation to planes *before, middle, after* and *through.*

available computational power provides an opportunity to develop computer-aided design (CAD) procedures. CAD software would enable predictions of filter performance to be made, leading to improved filters and reduced cost of trials. In the first stages of developing CAD programmes, it is necessary to produce good models of fabric structure and then predict the flow through the fabrics.

Among the numerous outputs from the analysis, the fluid pressure, fluid velocity and shear stress on the fabric are used as the performance indices (Nazarboland *et al.*, 2008). Figure 5.11 shows the positions where the data were extracted from the FE model in relation to the fabric and the direction of flow.

In each analysis, three different fabric models (all plain) are used. Every fabric model is specified using warp yarn linear density (t_1), weft yarn linear density (t_2), warp density (d_1), weft density (d_2), warp crimp (c_1), cross-sectional shape of the yarns and width to height ratio (WHR) of the yarn cross-section. Warp crimp is then a dependent parameter. The inlet pressure for all cases is 3 bars and the operating pressure 1 bar. Density and viscosity of the fluid were assumed to be constant, corresponding to the isothermal approach. Liquid water was used as the Newtonian fluid (with density of 998.2 kg/m^3 and viscosity of 10^{-3} kg/m.s). As an example, when the cross-sections of the yarns in the filter fabrics take circular, racetrack and lenticular shapes, Fig. 5.12 summarises the effect of yarn cross-sectional shape on fluid pressure on the front and back surfaces.

In this experiment three rather ideal yarn cross-sectional shapes, namely circular (FS4), racetrack (FS5) and lenticular (FS6), are investigated. From circular to racetrack to lenticular, the yarn width increases as its height decreases. By changing yarn cross-sectional shapes from circular to racetrack to lenticular, the fluid pressure is increased on the front face of the fabric due to higher flow resistance and drops dramatically at the back face.

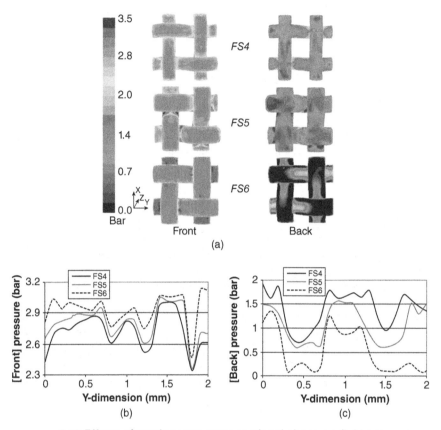

5.12 Effects of varying yarn cross-sectional shape on fluid pressure.

5.5.2   FE modelling of ballistic impact on fabrics

FE modelling of ballistic impact on fabrics is another field that is much needed in order to understand the strain/stress distribution in each of the fabric layers and among all fabric layers. Results from such work provide guidance for body armour engineering. Modelling of ballistic fabrics is carried out on two levels, for a single layer of fabric and for layered panels of fabrics.

Figure 5.13 shows that the modelled exit velocities agree well with the measured exit velocities, with the correlation coefficient being 0.9939, according to research at the University of Manchester (Sun and Chen, 2010). With the models validated by the experimental data, a series of FE simulations was carried out. Figure 5.14 reveals the ballistic impact process, with the projectile impact velocity $v_0 = 494.217$ m/s.

It is clear in Fig. 5.14 that the distribution of stress caused by the ballistic impact on the fabric is mainly along the warp and weft yarns in the fabric

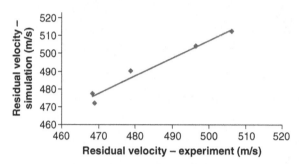

5.13 Agreement between the model and experiment.

5.14 Stress distribution on an impacted fabric at (a) $t = 0$ μs, (b) $t = 0.75$ μs, (c) $t = 1.37$ μs, (d) $t = 5.25$ μs, (e) $t = 6.0$ μs, and (f) $t = 8.12$ μs.

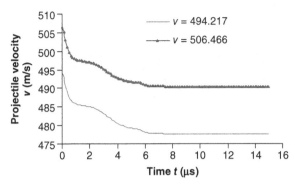

5.15 Modelled project velocity change due to impact.

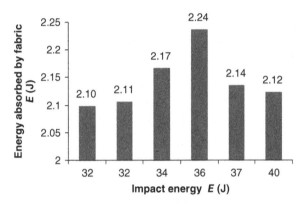

5.16 Energy absorption vs. impact energy.

before the projectile penetrates the fabric. This implies that the currently used plain woven fabric may not be the most efficient construction for ballistic applications, because such constructed fabric is unable to mobilise more areas of fabric to absorb the impact energy. To improve the fabric's ability to absorb impact energy, fabrics with better yarn grip have been designed and manufactured. The test results confirmed the superiority of such fabrics.

The models also give information that enhances understanding of the impact process. Figure 5.15 shows the change of the projectile velocity due to the impact on the fabric. The curves for both cases demonstrate similar trends. They indicate that the initial impact causes a sharp reduction of projectile velocity before the fibres start to break at about 1 µs into the impact. Projectile going through the fabric further reduces its velocity. The penetration of the projectile for both cases seems to have taken place at around 8 µs.

Figure 5.16 reveals that for a given assembly of ballistic fabrics there exists an impact energy that relates to the maximum energy absorption of

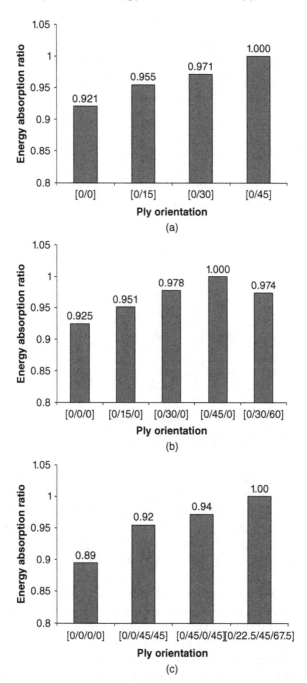

5.17 Effect of fabric orientation on energy absorption: (a) 2-layer assembly; (b) 3-layer assembly; (c) 4-layer assembly.

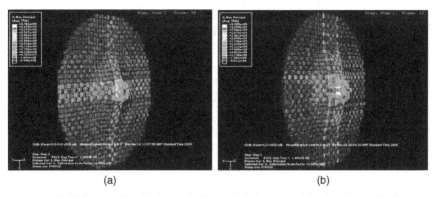

(a)   (b)

5.18 Stress distribution of 4-layer fabric assembly: (a) aligned 4-layer; (b) angled 4-layer.

the assembly of the fabrics. The impact energy is only a quarter of the energy applied to the fabric assembly, due to the geometrical symmetry of the projectile. Under this circumstance, the fabric assembly absorbs most of the energy at 36 J. This provides support for the use of V_{50} in ballistic testing.

The results from simulating single-layer fabrics give valuable information for fabric design so that the fabric is more absorbent to impact energy. It is also important to consider how the fabric is used in the fabric assembly. Obviously, different fabrics/materials can be used at different positions in the assembly because the impact action and reaction are different in different layers. Even if the same fabric is used, the fabric orientation in the assembly will also matter, as fabric orientation has a direct influence on the strain/stress distribution. Figure 5.17(a), (b) and (c) show that fabric layers with different orientation angle lead to better impact energy absorption.

Figure 5.18 demonstrates that for up to four layers of fabrics, a more evenly aligned fabric assembly absorbs most of the impact energy, and the aligned fabric assemblies are related to the least energy absorption. Experimental results show the same trend.

5.6   Future development of textile modelling

Modelling of textile materials and structures will always be an important issue. Close-to-reality models are essential for us to understand the performance and behaviour of textiles used either independently as a product or as constituent parts of a product. In geometrical terms, textiles always have irregular shapes compared to other types of materials, and due to the composition of the textile product the textiles are always inhomogeneous and anisotropic. Almost without exception, textiles are flexible, which has imposed complexity into the mathematical and mechanical descriptions. Despite all of these difficulties, textiles as special and non-replaceable types of materials are playing an increasingly important role in today's world, not only in traditional areas such as clothing, bedding and domestic decoration, but in many new areas including textile composites, medical scaffolds, geo-textiles for dam and road construction, and ballistic body armour protecting soldiers' lives. In all these applications, the capability of engineering textiles with more accuracy is essential. Therefore, it is reasonable to believe that textile modelling will continue to meet the challenges arising from the various applications, and that new techniques will be developed for modelling textiles with high accuracy.

5.7   Acknowledgements

The author appreciates the important contributions from Xuegong Ai, Pranut Potiyaraj, Yingliang Ma, Yong Jiang, Ali Nazarboland, Danmei Sun and Ying Wang as research students and assistants over the years.

5.8   References and further reading

Ai, X. (2003), Geometrical modelling of woven and knitted fabrics for technical applications, MPhil Thesis, UMIST, Manchester, UK

Backer, S. (1948), The relationship between the structural geometry of textile and its physical properties, I: Literature review, *Text. Res. J.*, 18, 650

Bogdanovich, A.E. and Pastore, C.M. (1996), *Mechanics of Textile and Laminated Composites*, Chapman & Hall, London

Chen, X. (2008), CAD/CAM for 3D fabrics for conventional looms, *Proceedings to the 1st World Conference on 3D Fabrics and Their Applications*, Manchester, UK

Chen, X., editor (2009), *Modeling and Predicting Textile Behaviour*, Woodhead Publishing, Cambridge, UK

Chen, X. and Potiyaraj, P. (1998), CAD/CAM of complex woven fabrics, Part 1 Backed cloths, *J. Text. Inst.*, 89, Part 1, No. 3, 532

Chen, X. and Potiyaraj, P. (1999a), CAD/CAM of complex woven fabrics, Part 2 Multi-layer fabrics, *J. Text. Inst.*, 90, Part 1 No. 1, 73

Chen, X. and Potiyaraj, P. (1999b), CAD/CAM of the orthogonal and angle-interlock woven structures for industrial applications, *Text. Res. J.*, 69, No. 9, 648

Chen, X. and Wang, H. (2006), Modelling and computer aided design of 3D hollow woven fabrics, *J. Text. Inst.*, 97, No. 1, 79

Chen, X., Li, M.-E. and Liu, Z. (1988), Mathematical models for woven fabric weaves and their application to fabric CAD, *Journal of Textile Research*, 9

Chen, X., Knox, R.T., McKenna, D.F. and Mather, R.R. (1996), Automatic generation of weaves for the CAM of 2D and 3D woven textile structures, *J. Text. Inst.*, 87, Part 1, No. 2, 356

Chen, X., Ma, Y. and Zhang, H. (2004), CAD/CAM for cellular woven structures, *J. Text. Inst.*, 95, Nos 1–6, 229

Chou, T.-W. and Ko, F.K., editors (1989), *Textile Structural Composites*, Elsevier, Amsterdam, Netherlands

Grosberg P. and Iype C. (1999), *Yarn Production: theoretical aspects*, Woodhead Publishing, Cambridge, UK

Hearle J W S (1965), Theoretical analysis of the mechanics of staple fibre yarns, *Textile Res. J.*, 35, 1060

Hearle, J.W.S., Grosberg, P. and Backer, S. (1969), *Structural Mechanics of Fibers, Yarns and Fabrics*, Wiley-Interscience, New York

Hearle, J.W.S., Thwaites, J.J. and Amirbayat, J., editors (1980), *Mechanics of Flexible Fibre Assemblies*, Sijthoff and Noordhoff, Alphen aan der Rijn, Netherlands

Hearle, J.W.S., Hollick, L. and Wilson, D.K. (2001), *Yarn Texture Technology*, Woodhead Publishing, Cambridge, UK

Hu, J. (2004), *Structure and Mechanics of Woven Fabrics*, Woodhead Publishing, Cambridge, UK

Jiang, Y. and Chen, X. (2005), Geometric and algebraic algorithms and methods for modelling yarns in woven fabrics, *J. Text. Inst.*, 96, No. 4, 237

Jiang, Y. and Chen, X. (2006), Asymptotic iterative approximation of intellectualized periodic interpolating spline and its application, *Journal of Information and Computing Science*, 1, No. 1, 47

Kemp, A. (1958), An extension of Peirce cloth geometry to the treatment of non-circular threads, *J. Text. Inst.*, 49, T44

Lawrence C.A. (2003), *Fundamentals of Spun Yarn Technology*, Woodhead Publishing, Cambridge, UK

Lomov, S.V., Huysmans, G., Luo, Y., Parnas, R.S., Prodromou, A., Verpoest, I. and Phelan, F.R. (2001), Textile composites: modelling strategies, *Composites Part A: Applied Science and Manufacturing*, 32, No. 10, 1379

Long, A.C., editor (2005), *Design and Manufacture of Textile Composites*, Woodhead Publishing, Cambridge, UK

Lord P.R. (2003), *Handbook of Yarn Production*, Woodhead Publishing, Cambridge, UK

Miravete, A., editor (1999), *3-D Textile Reinforcements in Composite Materials*, Woodhead Publishing, Cambridge, UK

Morton, W.E. and Hearle, J.W.S. (2008), *Physical Properties of Textile Fibres*, 4th edition, Woodhead Publishing, Cambridge, UK

Nazarboland, M.A., Chen, X., Hearle, J.W.S., Lydon, R. and Moss, M. (2008), Modelling and simulation of filtration through woven media, *International Journal of Clothing Science and Technology*, 20, No. 3, 150–260

Peirce, F.T. (1937), The geometry of cloth structure, *J. Text. Inst.*, 28, T45

Postle, R., Carnaby, G.A. and de Jong, S. (1988), *The Mechanics of Wool Structures*, Ellis Horwood, Chichester, UK

Robitaille, F., Long, A.C., Jones, I.A. and Rudd C.D. (2003), Automatically generated geometric descriptions of textile and composite unit cells, *Composites Part A: Applied Science and Manufacturing*, 34, No. 4, 303

Shanahan, W.J. and Hearle, J.W.S. (1978), An energy method for calculations in fabric mechanics, part II: examples of application of the method to woven fabrics, *J. Text. Inst.*, 69, No. 4, 92

Sun, D. and Chen, X. (2010), Development of improved body armour, Interim progress report – 6, MoD Contract RT/COM/5/030

6

Digital technology and modeling for fabric structures

G. BACIU, E. ZHENG and J. HU,
The Hong Kong Polytechnic University, Hong Kong

Abstract: Garment texture and appearance are largely determined by the fabric geometry and yarn interlacing structure. The possibility of determining the yarn interlacing and weave parameters to obtain the desired realistic appearance and properties of fabric is still an open challenge in modern textile research. The purpose of this chapter is to reveal the mathematical description for fabric structure. Techniques to describe weave patterns are proposed based on the yarn interlacing rules, i.e. fabric structures. Definitions and theoretical models for identifying pattern orientations in weave pattern comparison and design are discussed and examined in this chapter, which would improve the efficiency and accuracy of pattern classification and matching in the textile design industry.

Key words: weave pattern, fabric structure, pattern description, digital technology.

6.1   Introduction

Woven fabric pattern generation and rendering are key components in intelligent CAD/CAM assisted design for the textile manufacturing industry (TMI). The inverse problem, that is determining the yarn and weave parameters from a given woven fabric sample in order to obtain the desired fabric properties, is one of the most challenging problems in the TMI.

Fabric geometry structure is the fundamental organization of the yarn network. Fabric texture appearance is very closely related to the geometrical structure of fiber and yarn. Fabric texture is ultimately determined by fabric weave pattern. With the development of image processing, computer graphics and Internet communication technology, the TMI has raised higher demands on weave pattern collection management and intelligent weave pattern description. In this study, we focus on fabric geometry structure in two dimensions. We look at the interlacing rules and their representation, the weave pattern. We give some experimental results of our work on fabric 2D weave pattern description in terms of main characteristic perception, indexing and classification in the fabric design domain and the information technology domain. We also point out future research directions in this area.

122

6.2 Background on woven fabric structure

Woven fabric geometry research began roughly in the 1930s as research in textile materials evolved from one of the first technical schools in Reutlingen, School of Weaving, established in 1855. Since then this area has developed into a topic of intense research.

More recently, the need for rendering cloth and fabric arises frequently in computer graphics-generated scenes. Typical cases include indoor scenes with furnishings, like curtains, sofas, and scenarios with virtual characters. The work on fabric geometry and weave pattern is relevant in two fields: computer graphics/image processing and textile engineering.

In the past, the goals of the two communities have been different from each other; computer graphics focuses on the visual appearance of virtual cloth, whereas textile engineering focuses on the 'exact' physical structure of real fabrics. Currently, however, with the development of computer graphics and Internet communication technology, the TMI puts much emphasis on CAD/CAM technology based on exact woven structure. Virtual weaving online attracts a lot of attention in the current TMI. Researchers and industry R&D groups are putting much effort into modeling not only visual appearance but also realistic and exact fabric geometric structure.

Woven fabrics exhibit micro-, milli- and macro-geometric details. Microgeometry refers to the fibers that constitute the threads, milli-geometry refers to the interweaving of threads, i.e. fabric weave pattern, and macrogeometry refers to colorways of fabric, also called the color effects map in textile design. Among these three scales, the milli-geometry structure is very important for fabric CAD technology as it provides the transition in visual texture from coarse to rich.

Weave pattern design and fabric designs are conducted on a CAD/CAM platform including generating, editing, modifying and simulating patterns and fabric on-loom or off-loom. These 2D graphics, i.e. the weave pattern diagram (also referred to as the point map diagram), and 3D graphics, i.e. fabric simulation, are very relevant to computer graphics technologies. The development of computer graphics, simply referred to as CG, has made computers easier to interact with, and better for understanding and interpreting many types of fabric data. Developments in computer graphics have had a profound impact on fabric manufacturing and have revolutionized the business model in the textiles and clothing industry.

Weave pattern is the two-dimensional representation of the threedimensional interwoven structure of the yarns within a fabric. Weave pattern is the basic framework of fabric. Fabric geometry properties depend mainly on the design of the weave pattern. We can interpret fabric geometry structure as a 3D description and weave pattern as a 2D diagram of a fabric structure. The main issues related to fabric geometry structure include fabric simulation with texture mapping technology, yarn modeling and positioning, and fiber distribution modeling. The main issues related to 2D weave pattern are fabric weave pattern recognition, indexing, classification, automatic computer generation and matching. In the following sections, we will first review the works on fabric geometry structure models and focus on the essence of fabric geometry representation, i.e. weave pattern description.

6.3   Fabric geometry structure models

The ubiquity and richness of patterns is highlighted in the series of three articles by Glassner [Glassner, 2002, p. 108; Glassner, 2003a, p. 77; Glassner, 2003b, p. 84]. There are two ways to represent a fabric structure [Zhong *et al.*, 2001, p. 13; Daubert *et al.*, 2001; Daubert and Seidel, 2002]. One uses basic yarn shape modeling and the other is based on image reconstruction through image analysis technologies. The simulation technology can be considered as 2D and 3D representation of the fabric structure, at least on the yarn level. In commercial CAD/CAM fabric software, 2D simulation is widely used based on weave pattern structure in a 2D image as it is a simple and efficient means of obtaining real size visualization.

Different fabric models have been proposed in the literature. The most easily understood approach, and the one usually used in the computer graphics community, is to treat the fabric as an elastic sheet. Some models are explicitly continuous on surface representation [Terzopoulos *et al.*, 1987, p. 205]. A discrete approximation to some continuous surface has also been adopted [Baraff and Witkin, 1998]. Focus has also been given to limiting the amount of stretching fabric either by a strain-limiting iterative process [Provot, 1995; Bridson *et al.*, 2002] or by a constraint satisfaction phase [Goldenthal *et al.*, 2007]. While simulation speeds are relatively fast, there is in general a problem of mapping physical cloth properties to the parameter space of the elastic model. One way to solve this problem is that range scan data of static fabric configurations can be used to estimate the elastic parameters [Jojic and Huang, 1997]. The literature emphasizes the structural properties and the related mechanical performances. However, from the point of the TMI, there is little need to apply this kind of simulation, especially in the clothing industry or the home textiles industry. The reason is that people still want to do some real tests of these kinds of mechanical properties as the fabric sample contains more information, such as hand feeling, and the real test is more reliable.

With the development of computer graphics technology and Internet communication improvement, virtual design and virtual manufacturing have become powerful and quick communication tools due to the increasing needs to provide efficiency and environmental protection. As the fabric can be viewed from different distances, a level of detail representation is created for the weave pattern from the color maps. Distant viewing of a fabric weave pattern is used to define the properties of the fabric that occur at a pixel level and the illumination model is evaluated accordingly. For close-up viewing, the constitution of the cloth at a particular pixel (i.e. single thread, both warp and weft or gap) is identified using the weave pattern representation and the illumination is evaluated accordingly. Thus, fabric details, such as weave pattern, yarn materials and yarn shape, need to be addressed in the VR technologies. Several models have attempted to address the fact that cloth is composed of a discrete set of yarns. Geometric modeling of yarns arguably began with Peirce, who derived a set of parameters and equations for modeling the crossing of yarns in a woven fabric as inextensible curves [Peirce, 1937]. Kawabata *et al.* proposed a beam and truss model for yarn crossings in a woven fabric, as well as a system for measuring the physical force curves resulting from stretch, shear and bend motions of cloth [Kawabata *et al.*, 1973, p. 21]. Variations of the beam-and-truss model have been used in the textile community to simulate the behavior of plain-weave fabrics like Kevlar [King *et al.*, 2005, p. 3876; Zeng *et al.*, 2006, p. 1309]. Woven fabric was modeled as a particle system, where a particle ideally represented a yarn crossing [Breen *et al.*, 1994, p. 663; Eberhardt *et al.*, 1996, p. 52]. These models addressed the fabric details at the yarn level and also gave some specific consideration to fabric properties, e.g. draping. In the computer game community and garment VR demo on a model, these kinds of models have attracted more and more interest and application needs.

According to the yarn shape and its crossing path, woven yarn crossings have also been modeled as a pair of curves [Warren, 1990, p. 1309]. In particular, Nadler *et al.* employed a two-scale model, treating cloth at the high level as a continuous sheet, and at the fine level as a collection of pairs of orthogonal curves in contact with each other, with feedback from the fine scale driving the simulation of the large scale [Nadler *et al.*, 2006, p. 206]. Yarns have also been modeled as splines, with Rémion developing the basic equations for using splines and using springs between the control points to preserve length [Rémion *et al.*, 1999]. Jiang and Chen used a spline-based yarn model to generate plausible static woven fabric configurations [Jiang and Chen, 2005, p. 237]. Similar to this, work has also been done in computer graphics on modeling and simulating thin flexible rods [Pai, 2002; Bertails *et al.*, 2006, p. 1180; Spillmann and Teschner, 2008; Theetten *et al.*, 2007], although the simulated rods are typically much shorter than the spline curves used in fabric. It is understandable that these models in the computer graphics community can be extended to fabric geometry structure description.

6.4   Fabric weave pattern

In order to address the essence of the fabric yarn interwoven structure, the weaving rule is predominantly presented as a weave point map. In textile design, we normally classify these patterns according to their interlacing rule in three categories: plain weave, twill weave and satin weave. The taxonomy we use here is illustrated in Fig. 6.1.

6.5   Description and classification of regular patterns

Fabric structure is generated from the interlacing rule of warp and weft, the grid plots with black and white cells as shown in Fig. 6.2. Here, we use black cells to indicate warp points and white cells to indicate weft points (yarn draw-downs, with black cells where the warp is on top and white cells where the weft is on top). We use a matrix Ψ to describe the weave pattern as follows:

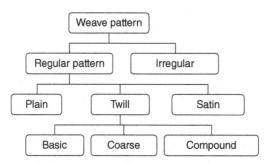

6.1 Taxonomy of fabric weave pattern.

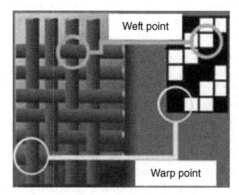

6.2 Fabric structure and weave pattern.

$$\Psi = \begin{pmatrix} a_{11} & \cdots & a_{1n} \\ \vdots & \ddots & \vdots \\ a_{m1} & \cdots & a_{mn} \end{pmatrix} \qquad \qquad 6.1$$

where $a_{ij} = \{0, 1\}$; $i = 1, 2, \ldots, m$; $j = 1, 2, \ldots, n$; a_{ij} taking the value 0 means weft point and 1 means warp point. We use each column to denote the warp interlacing rule in the fabric: $\{a_1, \ldots a_j, \ldots a_n\}$. In this way, the interlacing rule for each warp can be described by the following formula:

$$a_j = \frac{c_1 c_2 \ldots c_x \ldots c_p}{d_1 d_2 \ldots d_x \ldots d_p} \qquad \qquad 6.2$$

where c_x means the warp float points and d_y means the weft float points; $(p + q)$ is the number of interlacing between the warp and weft float points.

Based on the above definitions, we can describe the three types of regular weave patterns: regular plain weave pattern (RPWP), regular twill weave pattern (RTWP) and regular satin weave pattern (RSWP). The value of a_{ij} can be deduced from the index of the previous warp point position (subscript position) in the matrix. We use steps in warp and weft to describe the warp point distribution rule, s_j meaning a step in the warp direction and s_w meaning a step in the weft direction. For these regular patterns, there is a prerequisite condition

$$m = n = \sum_{j=1}^{p} c_j + \sum_{j=1}^{q} d_j \qquad \qquad 6.3$$

For RPWP, this can be expressed as

$$a_1 = \frac{1}{1}, (c_1 = 1, d_1 = 1, p = q = 1), s_j = s_w = \pm 1 \qquad \qquad 6.4$$

For RTWP, it can be subdivided into three categories. The first one is basic twill, which can be described as

$$a_1 = \frac{c_1}{d_1}, (c_1 + d_1 \geq 3, p = q = 1), s_j = s_w = \pm 1 \qquad \qquad 6.5$$

The second one is coarse twill, which can be given by

$$a_1 = \frac{c_1}{d_1}, (c_1 + d_1 \geq 4, c_1 \neq 1, d_1 \neq 1, p = q = 1), s_j = s_w = \pm 1 \qquad \qquad 6.6$$

When $c_1 > d_1$, technicians call this a warp effect coarse twill; when $c_1 < d_1$, technicians refer to it as a weft effect coarse twill; $c_1 = d_1$ is double face coarse twill. The third type is compound twill, that is

$$a_1 = \sum (c_1 + c_2 + \ldots + c_p + d_1 + d_2 \ldots + d_q) \geq 5, m = n \geq 2, s_j = s_w = \pm 1 \qquad \qquad 6.7$$

For RSWP, there are also two types, warp effect:

$$a_j = \frac{c_1}{1}, c_1 \geq 4 \, (c_1 \neq 5), \, p = q = 1 \qquad\qquad 6.8$$

where $1 < s_j < c_j$, $(c_j + 1)$ and s_j are relatively prime, and weft effect:

$$a_j = \frac{1}{d_j - 1}, d_1 \geq 5 \, (c_1 \neq 6), \, p = q = 1 \qquad\qquad 6.9$$

Here, $1 < s_w < (d_j - 1)$, and d_j and s_w are relatively prime.

6.6   Description and classification of irregular patterns

For many fabric weave patterns, the step in warp or weft is not a fixed number. In this case, we use smoothness and connectivity to describe and classify these irregular patterns. The smoothness is measured by the horizontal, vertical and diagonal transitions, and the connectivity is measured by the number of black (warp point, nominal value equals 1) and white (weft point, nominal value equals 0) clusters.

Before doing the calculations of smoothness and connectivity, we use skill to differentiate satin weave from plain and twill weave categories. As shown in Fig. 6.3, the center interlacing point of interest has two kinds of connections with its neighbors: touching each other by 45° (i.e., $(i + 1, j \pm 1)$ or $(i - 1, j \pm 1)$) or by 90° (i.e., $(i, j \pm 1)$ or $(i \pm 1, j)$). Touching by 90° is known as four-connectivity, and 90°/45° touching is known as eight-connectivity. Satin weaves may or may not have 90° touch with the center. But for 45° touch with the center, it is not common to have this kind of connection. For reasons of simplicity, we take this as a criterion to discriminate satin weaves from plain and twill weaves. (If there are exceptions, the user can adopt an enforcement rule to take this situation into consideration.)

The smoothness of the interest neighborhood around pixel $\tau_{i,j}$ is measured by the total number of horizontal, vertical, diagonal and anti-diagonal

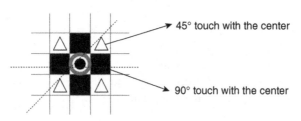

6.3 45° and 90° connection.

transitions: E_1 (horizontal), E_2 (vertical), E_3 (diagonal) and E_4 (anti-diagonal). $\tau_{i,j}$ is the value of the ith weft and jth warp in the weave pattern. In order to describe the distributions of the four directions of transitions, we use variance of the four-direction transitions array $\{E_1, E_2, E_3, E_4\}$:

$$v'^2 = \frac{1}{n} \sum_{i=1}^{n} (E_\varphi - E_i)^2 \qquad\qquad 6.10$$

$$E_1(i,j) = \sum_{\mu=-1}^{1} \sum_{v=-1}^{1} \xi(\{\tau_{i+\mu,j+v} \neq \tau_{i+\mu,j+v}\}) \qquad\qquad 6.11$$

$$E_2(i,j) = \sum_{\mu=-1}^{1} \sum_{v=-1}^{1} \xi(\{\tau_{i+\mu,j+v} \neq \tau_{i+\mu,j+v}\}) \qquad\qquad 6.12$$

$$E_3(i,j) = \sum_{\mu=-1}^{0} \sum_{v=-1}^{0} \xi(\{\tau_{i+\mu,j+v} \neq \tau_{i+\mu+1,j+v+1}\}) \qquad\qquad 6.13$$

$$E_4(i,j) = \sum_{\mu=0}^{1} \sum_{v=-1}^{0} \xi(\{\tau_{i+\mu,j+v} \neq \tau_{i+\mu-1,j+v+1}\}) \qquad\qquad 6.14$$

where $\xi(\cdot)$ is an indicator function taking values in $\{0, 1\}$. An example of broken twill weave is given in Fig. 6.4 to illustrate the calculation of transitions in four directions. Given a reference value $v'^2 = \varpi$ as a plain weave category for differentiation of plain weave and twill weave, if $v'^2 \in (\varpi - \delta, \varpi + \delta)$ the pattern can be classified into the plain weave category, otherwise it is a twill weave pattern. Here the increment $|\pm\delta|$ is the threshold value for the judgment.

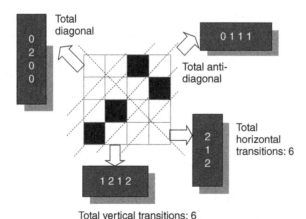

6.4 Illustrations of transitions calculation.

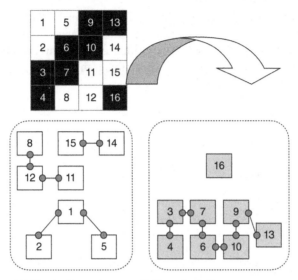

6.5 Illustration of connectivity calculation.

After the calculation of smoothness, we also use connectivity to differentiate plain weave and twill weave patterns. As illustrated in Fig. 6.5, the connectivity is measured by the number of the black and white clusters. In order to make a differentiation between plain weave and twill weave, we can use the 90° connection criterion shown in Fig. 6.3 to describe the connectivity. Figure 6.5 shows an example of 4 × 4 pattern with eight white points forming two clusters and with eight black points forming two clusters. The number of clusters can be automatically identified by traversing the graph using depth-first search strategy. Here we use a stack-based search algorithm to calculate the connectivity of 90°.

Suppose that there are Γ interlacing points in total and the final value of ε indicates the number of clusters. Let $v(k)$ store the value of the kth interlacing point and v_0 be the interlacing value of interest. Set up an empty stack and a P-element array $L[\cdot]$ for storing the index of the cluster that the interlacing point belongs to; set $L[k] = 0$ for all $k = 1, 2, \ldots, P$; $i = 1$; $\varepsilon = 0$.

Step I: if $L[i] \neq 0$ or $v[i] \neq v_0$, go to the final step (Step V).
Step II, $\varepsilon = \varepsilon + 1$; push node i into the stack.
Step III, if the stack is empty, go to Step V.
Step IV, $L[k] = \varepsilon$. Find all pixels connected with k. For each connected interlacing point j, if $L[j] = 0$ (i.e. it has not been visited or pushed into the stack), let $L[j] = -1$, push node j into the stack (by the definition of connected, $v[j] = v_0$). Go back to Step III.
Step V, $i = i + 1$; if $i > \Gamma$, stop, otherwise go to Step I.

The number of clusters is used to supplement the judgment criterion between plain pattern and twill pattern.

6.7   Weave pattern and fabric geometry surface appearance

From the point of view of vision perception, there is not always consistency between the weave pattern diagram and the appearance of the real fabric structural surface. One could get an accurate description of the weave pattern diagram, but for the real sample's pattern, it may become very different in appearance from its weave pattern diagram. The major reason for this is largely attributed to the color effects and interactive force between the yarns in the fabrics which result in yarn shifting, skewness and even deformation. From this viewpoint, the perceived visual appearance of the fabric image is determined not only by the weaving structure but also by the yarn properties.

6.8   Experimental pattern analysis

We chose 14 different weave patterns to investigate the ideal weaving pattern and the actual scanned fabric pattern. The patterns are shown in Fig. 6.6. The ideal patterns could be generated on a computer. The real fabrics need to be manufactured on the weaving machine. We used the same warp and weft yarns with the same physical structures: yarn count, number of twists, twist direction and composition. In this way, we could guarantee all the real samples were under the same comparative prerequisite conditions. The relevant weaving and structural details of all the samples are given in Table 6.1.

Table 6.1 Fabric composites and manufacturing parameters

Items	Specific details	Notes
Weaving machine	Dornier rapier loom, Staubli F69680 head	Made in Germany and France
Width	168.0 cm	Excluding selvage
Density	120 ends/inch × 76 picks/inch	–
Content	100% polyester	–
Yarn count	100D × 100D	Low twist, 5–6 twists/cm
Color effect	Warp: white Weft: dark green	–

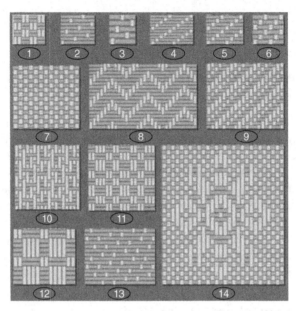

6.6 Fabric ideal weave pattern schematic diagram.

6.9   Methodology

In information theory, for a discrete source X, the self-information of symbol x_i which occurs with probability $P(x_i)$ is defined as $I(x_i) = -\log p(x_i)$. The uncertainty or entropy of the source is defined as

$$H(X) = E(I(X)) = \sum_i p(x_i) \log p(x_i) \qquad \qquad 6.15$$

$H(X)$ is the average information per source output. It is related to the distribution of source probability. Indeed, the weave pattern diagram can be regarded as an image of black and white point distribution. The distribution of different brightness will produce the weave pattern. Each weave pattern diagram will have different entropy. It is possible and efficient to describe the weave pattern diagram by its entropy.

For an image, the size is $M \times N$ and function $f(x, y)$ denotes the brightness of pixels. If $f(x, y) = \{k_1, k_2, \ldots, k_L\}$ and the probability of each brightness in the image is $P_f = \{p_1, p_2, \ldots, p_L\}$, the information entropy for the weave pattern image can be given as

$$H_L(p_1, p_2, \cdots, p_L) = \sum_{i=1}^{L} p_i \log_2 p_i, i = 1, 2, \cdots, L \qquad 6.16$$

For the weave pattern diagram, the interlacing information includes only two situations, i.e. weft point denoted as white ($f(x, y) = 0$) and warp point denoted as black ($f(x, y) = 1$). In this case, Equation 6.16 becomes:

$$H_2(p_0, p_1) = -p_0\log_2 p_0 - p_1\log_2 p_1 \qquad 6.17$$

where

$$p_1 = \sum_M \sum_N \frac{f(x, y)}{M \times N}, \ p_0 = 1 - p_1 \qquad 6.18$$

Since entropy has a physical interpretation, the calculation of a pattern's entropy involves the sense of the microscopic essence of the pattern complexity based on the probability of weft or warp yarn points appearing. However, for a specific weave pattern description, entropy cannot provide the distribution information for the warp and weft float arrangement and distance. For example, theoretically, ideal patterns 1 and 7 have the same entropy value, but they have different spatial distributions. Here, we use fast Fourier transform (FFT) to extract the information for the spatial distribution of yarn interlacing points.

If a function $f(x)$ in the time or spatial domain is known, then $F(u)$, the Fourier transform of $f(x)$, can be defined as

$$F(u) = \int_{-\infty}^{+\infty} f(x)\exp(-j2\pi ux)dx \qquad 6.19$$

where $j = \sqrt{-1}$. The function $f(x)$ may be reconstructed by the inverse Fourier transform:

$$f(x) = \int_{-\infty}^{+\infty} F(u)\exp(j2\pi ux)du \qquad 6.20$$

where $f(x)$ and $F(u)$ are referred to as the Fourier transform pair. The Fourier transform of a real function is generally complex, with real and imaginary components $I(u)$ and $J(u)$, so the magnitude $M(u)$ can be calculated by

$$M(u) = |F(u)| = \sqrt{I^2(u) + J^2(u)} \qquad 6.21$$

The square of the magnitude function is commonly referred to as the power spectrum $P(u)$ of $f(x)$:

$$P(u) = |M(u)|^2 = I^2(u) + J^2(u) \qquad 6.22$$

Similarly, for the discrete case, the corresponding pair of the two-dimensional Fourier transform of a pattern image is

$$F(u, v) = \frac{1}{MN}\sum_{x=0}^{M-1}\sum_{y=0}^{N-1} f(x, y)\exp\left(-j2\pi\left(\frac{ux}{M} + \frac{vy}{N}\right)\right) \qquad 6.23$$

for $u = 0, 1, 2, \ldots, M - 1$, and $v = 0, 1, 2, \ldots, N - 1$, and

$$f(x, y) = \frac{1}{MN} \sum_{x=0}^{M-1} \sum_{y=0}^{N-1} F(u, v) \exp\left(j2\pi \left(\frac{ux}{M} + \frac{vy}{N} \right) \right)$$

6.24

for $x = 0, 1, 2, \ldots, M - 1$, and $y = 0, 1, 2, \ldots, N - 1$.

This can reduce N^2 operations in the discrete Fourier transform (DFT) into $N \log_2 N$ operations in the FFT. The basic idea of the FFT is that a DFT of an N data set is the sum of the DFTs of two subsets: even-numbered points and odd-numbered points. The data set can be recursively split into even and odd until the length equals 1.

We can filter the Fourier transform so as to select those frequency components deemed to be of interest to a particular application, e.g. fabric density calculation and even pattern recognition [Xu, 1996, p. 496]. Alternatively, it is convenient to collect the magnitude transform data in achieve a reduced set of measurements. First, the FFT data can be normalized by the sum of the squared values of each magnitude component instead of histogram equalization, so that the magnitude data is invariant to linear shifts in illumination to obtain normalized Fourier coefficients NF as

$$NF_{u,v} = \frac{|F_{u,v}|}{\sqrt{\displaystyle\sum_{(u \neq 0) \wedge (v \neq 0)} |F_{u,v}|^2}}$$

6.25

As we know, histogram equalization can also provide such invariance, but it is more complicated than using Equation 6.25. Then the spectral data can be described by the entropy, $F_{u,v}h$, as

$$F_{u,v}h = \sum_{u=1}^{M} \sum_{v=1}^{N} NF_{u,v} \log_2 (NF_{u,v})$$

6.26

or by their energy, $F_{u,v}e$, defined as

$$F_{u,v}e = \sum_{u=1}^{M} \sum_{v=1}^{N} (NF_{u,v})^2$$

6.27

Another measure is their inertia, $F_{u,v}i$, as,

$$F_{u,v}i = \sum_{u=1}^{M} \sum_{v=1}^{N} (u - v)^2 NF_{u,v}$$

6.28

Clearly, the entropy, inertia and energy are relatively immune to rotation, since order is not important in their calculation. By Fourier analysis, the measures are inherently position-invariant and can extract the information on the spatial arrangement of yarns inside the image.

6.10   Results and discussion

We assume that the ideal weave pattern is composed of two parts, i.e. the ideal simulation weave pattern (ISWP) and the diagram of weave pattern (DOWP). The former may include color and yarn edge or configuration information. To some extent, it is a simple simulation of a real fabric without consideration of the interactive force between the yarns and the relevant distortion or deformation, etc. The latter only consists of white and black points in a mesh. Usually, a white point corresponds to the descending motion of the weft yarn and a black point corresponds to the ascending motion of the warp yarn. These motions are controlled by the heald wire. In order to validate the algorithm proposed, we investigated the following three types of patterns: diagram of weave pattern (DOWP), ideal simulation weave pattern (ISWP), and real samples weave pattern (RSWP).

In this chapter, we first discuss the classification of 14 samples to bring into alignment the basic design principles of woven fabric, that is, plain, twill, satin and jacquard effect fabrics. The details of the classification are as follows:

- Plain weave effect pattern (PWEP), including weave patterns 1 (basket plain weave), 7 (basic plain weave), 11 (variant plain weave) and 12 (variant plain weave)
- Twill weave effect pattern (TWEP), including 4 (1/4 right twill), 5 (1/3 right twill), 6 (1/2 right twill), 8 (variant sawtooth twill) and 9 (compound right twill)
- Satin weave effect pattern (SWEP), including 2 (8-3 weft effect satin weave), 3 (5-3 weft effect satin weave) and 13 (two-basic compound satin weft effect weave)
- Jacquard figured effect pattern (JFEP), including 10 (birds-eye figured weave) and 14 (jacquard flower weave).

The four types of fabric ISWP are illustrated in Fig. 6.6 and the corresponding RSWP and their DOWP's FFT entropy, energy and inertia are calculated respectively as shown in Table 6.2. All numbers in the weave pattern diagrams and real sample weave pattern were consistent with the serial numbers in Fig. 6.6.

6.10.1 Comparisons

There is no standard method to subjectively evaluate a pattern's complexity and its content in textile design. Different people have different preferences when defining the complexity of DOWP. Actually, the pattern complexity should be defined by using different viewpoints in different

Table 6.2 Calculation results of $F_{u,v}h$, $F_{u,v}e$ and $F_{u,v}i$ for DOWP, ISWP and RSWP

(a) $F_{u,v}h$, $F_{u,v}e$ and $F_{u,v}i$ for DOWP

DOWP type	No.	$F_{u,v}h$	$F_{u,v}e$	$F_{u,v}i$
PWEP	1	269.99	1.9328	397240
	7	222.70	2.8022	299810
	11	501.05	1.9289	1978200
	12	426.33	1.7491	1331900
TWEP	4	375.81	4.3873	872970
	5	303.48	3.6222	439670
	6	283.65	2.8022	506170
	8	635.17	2.7963	19639000
	9	382.13	1.933	1105800
SWEP	2	377.06	6.4339	579650
	3	358.49	4.3746	844390
	13	539.50	5.7522	2433000
JFEP	10	378.22	1.5673	458080
	14	1293.30	1.6939	15014000

(b) $F_{u,v}h$, $F_{u,v}e$ and $F_{u,v}i$ for ISWP

ISWP type	No.	$F_{u,v}h$	$F_{u,v}e$	$F_{u,v}i$
PWEP	1	10.117	54.653	600.08
	7	1.2037	60.083	214.4
	11	115.48	53.539	17171
	12	67.301	53.347	2914.7
TWEP	4	37.133	61.074	3214.7
	5	37.807	61.828	1582.3
	6	18.902	57.466	490.03
	8	231.04	55.697	221850
	9	44.296	57.636	2271.8
SWEP	2	57.546	62.724	5339.6
	3	57.903	63.436	3691.9
	13	133.94	61.34	23927
JFEP	10	64.674	56.001	3822.3
	14	450.19	62.379	266410

design procedures. For designers, DOWP is the basic foundation of fabric structure. Since a fabric includes information on color and yarn characteristics except for the yarn interlacing rule, ISWP and RSWP are just part of the comprehensive information.

The complexity of a DOWP refers to the time spent to reproduce it from the viewpoint of a designer. Based on this commonsense application, we invited 20 designers to draw the 14 DOWPs by clicking the check in a mesh

Table 6.2 Continued

(c) $F_{u,v}h$, $F_{u,v}e$ and $F_{u,v}i$ for RSWP

RSWP type	No.	$F_{u,v}h$	$F_{u,v}e$	$F_{u,v}i$
PWEP	1	2327.7	6.2704	20297000
	7	2735.8	8.1474	26889000
	11	2554.7	5.6064	21419000
	12	2417.0	5.3701	20926000
TWEP	4	3498.8	4.7637	35629000
	5	3326.3	5.0368	33735000
	6	2844.8	6.0753	28553000
	8	3047.3	4.3247	26383000
	9	2623.3	6.9469	25432000
SWEP	2	3973.6	5.6904	37264000
	3	3644.3	4.9177	36683000
	13	4528.6	5.997	46754000
JFEP	10	2657.1	6.1669	22583000
	14	3320.9	6.1247	27626000

to reproduce them. Generally, it is necessary to choose the weft face of DOWP to eliminate influences caused by the unnecessary clicking time among different patterns. We recorded the time spent on every DOWP for each designer and sorted them in order according to the time to decide the pattern diagram complexity. The more complex a DOWP was, the longer the drawing time duration. After sorting into ascending order for each designer's DOWP complexity dataset (there were 14 elements corresponding to each DOWP number), we obtained a matrix of 20 rows by 14 columns. In each column, we chose the most frequently appearing DOWP number and obtained a standard DOWP complexity order. For PWEP, the order is [7, 1, 11, 12]; for TWEP, SWEP and JFEP, the corresponding order is [6, 5, 4, 9, 8], [3, 2, 13] and [10, 14] respectively. After this, we did another experiment to set the pattern complexity order. We asked designers to set the complexity order according to their textile design experience and found that most of the designers made the same aforementioned order for the four types of DOWP. We call this the human perception order in the following discussion for comparison. For ISWP and RSWP, we can take the DOWP comparison order as the reference. We need to note that the yarn color effects might make the reference of limited use in ISWP. With regards to RSWP, so many factors will influence the final appearance, such as internal force interaction, yarn float and shift, fabric density, yarn parameter, compositions, weaving machine parameter, etc. Since we use the scanning method to capture the image of the real fabric appearance, the weave

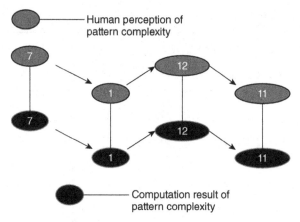

6.7 Comparison between computation result for DOWP and human perception for PWEP.

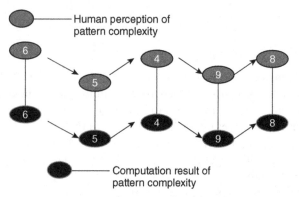

6.8 Comparison between computation result for DOWP and human perception for TWEP.

pattern is just the underlying information and there might be some discrepancy between DOWP calculation and the RWEP calculation.

From Table 6.2(a), we can make a comparison for the four types of pattern, as shown in Figs 6.7 to 6.12. There was a good agreement between pattern computation results and human perception order for DOWP's $F_{u,v}h$ computation results. Specifically, from Figs 6.7, 6.8 and 6.9, all the complexity orders were identical with each other, so they could be used in pattern subdivision and matching for PWEP, TWEP, SWEP and JFEP in DOWP.

Since the DOWP calculated order was in good agreement with the comparison principle, as discussed before, we could choose the DOWP $F_{u,v}h$ computation order as a comparison reference to ISWP and RSWP. The

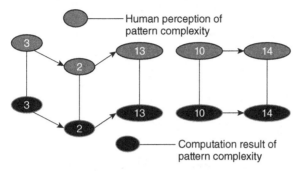

6.9 Comparison between computation result for DOWP and human perception for SWEP and JFEP.

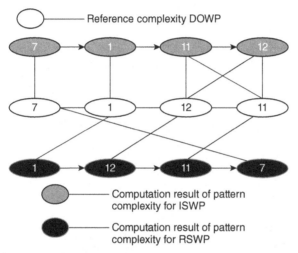

6.10 Comparing computation results of ISWP and RSWP with DOWP for PWEP.

comparison orders are shown in Figs 6.10 to 6.12. Because there were some differences in the pattern complexity calculation order, we needed to evaluate the extent of discrepancy or misplacement between the computation results and the reference standard. If the position of the pattern to compare had the same relative position in the DOWP, its misplacement extent value was defined as 0. Otherwise, the relative position difference value had to be calculated according to the element position difference from the corresponding one in the DOWP reference. For example, in Fig. 6.10, the four PWEP misplacement extent values were 0, 0, 1 and 1 respectively in ISWP. The corresponding four RSWP misplacement extent values were 1, 1, 1 and 3. Obviously, the sum of the misplacement extent values could be used to evaluate the effectiveness of the computation results.

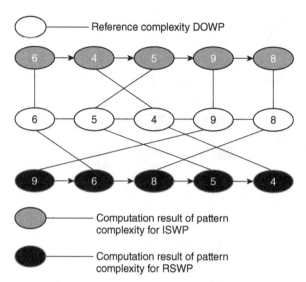

6.11 Comparing computation results of ISWP and RSWP with DOWP for TWEP.

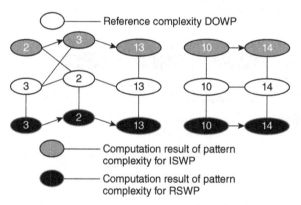

6.12 Comparing computation results of ISWP and RSWP with DOWP for SWEP and JFEP.

From Figs 6.10 to 6.12, it is seen that there were three pairs of misplacements for ISWP. Theoretically, only the color effect would affect the computation results compared to the calculation results of DOWP. Since we used only one set of warp and one set of weft, the influence of color effects could be reduced as much as possible. The problems were probably caused by the ISWP generation method. We selected one repetition of the ISWP image in a textile design CAD software interface, since the whole simulation pattern, composed of the repetition of many pattern units, was difficult

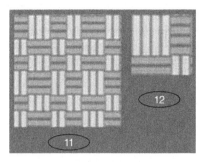

6.13 ISWP image comparison of no. 11 and no. 12.

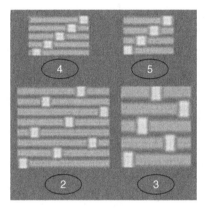

6.14 ISWP image comparison of nos 4 and 5 and nos 2 and 3.

for calculation. When we wanted to choose a more accurate boundary of ISWP, we zoomed in and captured a pattern repetition. Clearly, there were no essential changes of the pattern for human perception, but different times of zoom in or out may have a great influence on the computation results, just like the misplacement of 11 and 12, 4 and 5, and 2 and 3, as shown in Figs 6.13 and 6.14. Table 6.2(b) shows that the calculated difference was quite small for each pair, but it caused pattern complexity order mixture. From the explanation given by the algorithm, this operation would result in a different distribution of pixels in the grayscale, which made the lack of comparison of the computation results inevitable. So our computation will be effective under the same generating and capturing image situations.

In terms of the RSWP calculation results, the scanning reflected pattern of a real object showed much discrepancy from its underlying weaving structure. In Figs 6.10, 6.11 and 6.12, the misplacement extent was quite significant except for two samples of JFEP.

Table 6.3 Pattern complexity order of $F_{u,v}e$ calculation results

Pattern type	DOWP order	ISWP order	RSWP order
PWEP	7	7	7
	1	1	1
	11	11	11
	12	12	12
TWEP	4	5	9
	5	4	6
	6	9	5
	8	6	4
	9	8	8
SWEP	2	3	13
	13	2	2
	3	13	3
JFEP	14	14	10
	10	10	14

According to the $F_{u,v}e$ calculation results of DOWP (Table 6.2a), ISWP (Table 6.2b) and RSWP (Table 6.2c), the higher the FFT energy was, the more stable the pattern was. In this way, the basic plain weave would have a higher value than variant or compound plain weave effect patterns. From the viewpoint of fabric weaving, the interlacing of the basic plain weave was greatest compared with its relatives. This method could be used as an evaluation measure to compare the interlacing times at this point. However, this energy method was not stable enough to obtain a consistent result as an $F_{u,v}h$ computation result. Based on theoretical computation results, calculated pattern complexity orders were obtained as shown in Table 6.3. This could provide only a rough complexity explanation for PWEP, TWEP and SWEP.

The matching condition extent of the $F_{u,v}i$ computation result was between the $F_{u,v}h$ method and the $F_{u,v}e$ method as shown in Table 6.4. On the whole, it could provide a better pattern complexity description for the four types of patterns in terms of texture description. Here, the texture represents the whole appearance or motif perceived by a human. By directly comparing the computational value of RSWP $F_{u,v}i$ in Table 6.2(c), the $F_{u,v}i$ method could discriminate relatively satisfactorily among simple PWEP, TWEP and SWEP: PWEP had the lowest calculated value, TWEP had the middle value, and SWEP had the highest computational value. This matched the understanding of the designer's pattern complexity. In the real scanned image, there were a lot of factors which may be regarded as noise affecting the final calculation results. As shown in Fig. 6.15, the edges of the warp yarn were quite ambiguous in sample 10. According to human perception, sample 7 had a much finer texture than sample 1, although from the textile design

6.15 Real scanned fabric image of no. 1 and no.7.

Table 6.4 Pattern complexity order of $F_{u,v}i$ calculation results

Pattern type	DOWP order	ISWP order	RSWP order
PWEP	7	7	1
	1	1	12
	12	12	11
	11	11	7
TWEP	5	6	9
	6	5	8
	4	9	6
	9	4	5
	8	8	4
SWEP	2	3	3
	3	2	2
	13	13	13
JFEP	10	10	10
	14	14	14

viewpoint the former pattern was more complex than the latter. By theoretical analysis of the algorithm, the finer the texture was, the higher the calculated value was. This method could be used to describe the refined extent of a sample. However, this description will be subject to noise. That might be the reason for some discrepancy between the theoretical calculation and human visual perception in Table 6.2. To handle a large number of patterns we need a large set of measurements (larger than the three given here) to discriminate better between different real scanned fabric images. Since this problem refers to the color effects of the fabric, we will not discuss it here in terms of a pattern topic.

6.10.2 Findings

By using the Fourier transform and information complexity calculation, we proposed three methods to describe relative pattern complexity. In this way, we could use quantitative methods to represent the fabric design pattern and spatial distribution information for real scanned samples.

Among the three calculation methods, $F_{u,v}i$ and $F_{u,v}e$ were more suitable for describing the real sample's refined extent of texture; $F_{u,v}h$ was very appropriate to represent the weave pattern diagram complexity which kept good consistency with the experiments and designer visual perception. The $F_{u,v}h$ method was efficient at discriminating the four types of DOWP: PWEP, TWEP, SWEP and JFEP.

Theoretically, the four types of ISWP complexity could also be represented by this method. However, the calculations should be based on the same picture generating method in order to avoid some unnecessary mixing of similar patterns. Since both ISWP and RSWP were just a part of the information extracted to the corresponding images, this method might be subject to all kinds of noise influences in addition to color for ISWP and RSWP.

A more detailed study should be conducted based on the specific characteristics of different textile samples to minimize or even eliminate the noise. In conclusion, our methods provide a promising direction to sort the weave patterns based on their complexity from the viewpoint of the fabric designer, especially in resorting the weave pattern diagram for plain effects patterns, twill effects patterns and satin effects patterns for facilitating fabric CAD technology.

6.11 Acknowledgment

We acknowledge the partial support of Hong Kong RGC GRF grants PolyU 5335/06E, PolyU 5101/07E and PolyU 5101/08E.

6.12 References

Baraff D and Witkin A (1998), 'Large steps in cloth simulation', in *Proc. SIGGRAPH '98*, ACM Press/ACM SIGGRAPH, 43–54.

Bertails F, Audoly B, Cani M P, Querleux B, Leroy F and Lévêque J-L (2006), 'Super-helices for predicting the dynamics of natural hair', *ACM Trans. Graph.*, 25, 1180–1187.

Breen D, House D and Wozn M (1994), 'A particle-based model for simulating the draping behavior of woven cloth', *Text. Res. J.*, 64, 663–685.

Bridson R, Fedkiw R and Anderson J (2002), 'Robust treatment of collisions, contact and friction for cloth animation', in *Proc. SIGGRAPH '02*, ACM Press/ACM SIGGRAPH, 594–603.

Daubert K and Seidel H P (2002), 'Hardware-based volumetric knit-wear', *EUROGRAPHICS: Computer Graphics Forum*, 21, 575–583.

Daubert K, Lensch H P A, Heindrich W and Seidel H P (2001), 'Effecient cloth modeling and rendering', in *Rendering Techniques 2001: Proc. 12th Eurographics Workshop on Rendering*, 63–70.

Eberhardt B, Weber A and Strasser W (1996), 'A fast, flexible, particle-system model for cloth draping', *IEEE Computer Graphics and Applications*, 16, 52–59.

Glassner A (2002), 'Digital weaving, part 1', *IEEE Computer Graphics and Applications*, 22 (6), 108–118.

Glassner A (2003a), 'Digital weaving, part 2', *IEEE Computer Graphics and Applications*, 23 (1), 77–90.

Glassner A (2003b), 'Digital weaving, part 3', *IEEE Computer Graphics and Applications*, 23 (2), 80–89.

Goldenthal R, Harmon D, Fattal R, Bercovier M and Grinspun E (2007), 'Efficient simulation of inextensible cloth', in *Proc. SIGGRAPH '07*, 281–290.

Jiang Y and Chen X (2005), 'Geometric and algebraic algorithms for modelling yarn in woven fabrics', *J. Text. Inst.*, 96, 237–245.

Jojic N and Huang T (1997), 'Estimating cloth draping parameters from range data', *Int. Workshop on Synthetic-Natural Hybrid Coding and Three Dimensional Imaging*, 73–76.

Kawabata S, Niwa M and Kawai H (1973), 'The finite deformation theory of plain-weave fabrics, part I: The biaxial deformation theory', *J. Text. Inst.*, 64, 21–46.

King M, Jearanaisila Wong P and Scorate S (2005), 'A continuum constitutive model for the mechanical behavior of woven fabrics', *Int. J. Sol. Struct.*, 42, 3876–3896.

Nadler B, Papadopoulos P and Steigmann D J (2006), 'Multiscale constitutive modeling and numerical simulation of fabric material', *Int. J. Sol. Struct.*, 43, 206–221.

Pai D (2002), 'Strands: Interactive simulation of thin solids using Cosserat models', in *Proc. Eurographics*, 347–352.

Peirce F (1937), 'The geometry of cloth structure', *J. Text. Inst.*, 28, T45–T97.

Provot X (1995), 'Deformation constraints in a mass-spring model to describe rigid cloth behavior', in *Proc. Graphics Interface '95*, 147–154.

Rémion Y, Nourrit J M and Gillard D (1999), 'Dynamic animation of spline like objects', in *Proc. WSCG'99*, 426–432.

Spillmann J and Teschner M (2008), 'An adaptive contact model for the robust simulation of knots', in *Proc. Eurographics '08*, 1–10.

Terzopoulos D, Platt J, Barr A and Flerscher K (1987), 'Elastically deformable models', *Computer Graphics*, 21, 205–214.

Theetten A, Grisoni L, Duriez C and Merlhiot X (2007), 'Quasi-dynamic splines', in *Proc. ACM Symposium on Solid and Physical Modeling '07*, 409–414.

Warren W (1990), 'The elastic properties of woven polymeric fabric', *Polym. Eng. Sci.*, 30, 1309–1313.

Xu B (1996), 'Identifying fabric structures with fast fourier transform techniques', *Text. Res. J.*, 66, 496–506.

Zeng X, Tan V B C and Shin V P W (2006), 'Modelling inter-yarn friction in woven fabric armor', *Int. J. Num. Meth. Eng.*, 66, 1309–1330.

Zhong H, Xu Y, Guo B and Shum H (2001), 'Realistic and efficient rendering of free-form knitwear', *J. Visualization and Computer Animation*, 12, 13–22.

7

Modeling ballistic impact on textile materials

M. S. RISBY, National Defence University, Malaysia and
A. M. S. HAMOUDA, Qatar University, Qatar

Abstract: Modeling the damage response of materials to ballistic impact
has received a great deal of attention, with particular focus on its
applications in the defense and aerospace industries. With the constant
advances in material science leading to high performance applications of
damage response technology, predicting the ballistic resistance and
behavior of protective materials subjected to impacts is the subject of
much research. This chapter reviews the various techniques used to
model the response of textile materials to ballistic impact.

Key words: modeling, ballistic impact, textiles.

This chapter is a revised and updated version of Chapter 4, 'Modelling
ballistic impact', by A. M. S. Hamouda and R. M. Sohaimi, originally published in *Lightweight Ballistic Composites: Military and Law-enforcement
Applications*, ed. A Bhatnagar, Woodhead Publishing Limited, 2006, ISBN
978-1-85573-941-3.

7.1 Introduction

Modeling the damage response of materials to ballistic impact has received
a great deal of attention, with particular focus on its applications in the
defense and aerospace industries. With the constant advances in material
science leading to high performance applications of damage response technology, predicting the ballistic resistance and behavior of protective materials subjected to impacts is the subject of much experimental, analytical and
numerical research.

Ballistic experiments help to provide a greater understanding of the
complexity of penetration mechanics; this is turn allows the identification
of important parameters which define the perforation and damage phenomena of textile based protective materials such as soft body armors. The
complexity of the ballistic problems caused by key influential factors such
as relative stiffness and mass of the materials, relative velocity, projectile
nose shapes, contact points, material characteristics, dimensions and boundary conditions, etc., increases when textile-based armor materials are
involved. These textile materials normally possess orthotropic-like properties and specific failure modes that may occur during the impact event

146

(Cismasiu *et al.*, 2005). The process of designing protective materials requires a huge number of experimental tests, which can be both time-consuming and expensive (Cismasiu *et al.*, 2005; Franzen *et al.*, 1997). Using the latest developments in understanding failure mechanisms and mechanics of textile-based fabrics and composites, along with advanced orthotropic and anisotropic-based material models, computer-based simulations have been carried out, which have the potential to reduce the need for numerous experimental tests (Walker, 1999; Lee *et al.*, 2001; Barauskas and Abraitiene, 2007).

At present, three common methods are used to compute the interaction between impactor and target: empirical, analytical and numerical. After extensive research into the technique, and with the development of high performance computing, computer-based simulations have become both cost-effective and viable, thus enabling scientists to reduce the need for physical experimentation (Parga-Landa and Hernandez-Olivares, 1995). Complex iterations required for the optimization process of the above-mentioned parameters involved in ballistic impact phenomena can now be performed without any glitches. However, any results gained through computer simulations should be treated with caution and should be continually verified by actual experimental tests.

Ballistic researchers typically employ the empirical method to explain relations between several physical test parameters, which can be used to define the interaction between impactor and target, including their material properties, geometry and the speed of impact. This method depends on the analysis of data obtained through experimental work (i.e. penetration depth, ballistic limit V_{50}, absorbed energy, etc.) in order to determine the damage response and to obtain the constitutive relations and failure criterion (Parga-Landa and Hernandez-Olivares, 1995; Tabiei and Nilakantan, 2008). Further processes include curve fitting, nonlinear regression analysis of experimental data and the use of statistical distributions, usually with large data points, which are analyzed using commercial mathematical software. The output is then established into certain parametric equations that relate to the selected parameters studied during the experiment. The selection of the correct empirical approach (i.e. Taguchi, Response Surface Methodology, Full Factorial Method, etc.) for the experimental design is also important to ensure that the model which is developed is accurate, as reported by Park *et al.* (2005). This method is practical when there are only limited numbers of variables to correlate with the obtained results. The disadvantage of this technique is that the accuracy of the developed model will depend on the accuracy and completeness of the obtained results. Limited test experiments (or lack of replication) may also result in invalid conclusions (Cunniff, 1992; Kirkwood *et al.*, 2004; Van Gorp *et al.*, 1993).

Analytical methods are typically based on general continuum mechanics equations (i.e. the law of conservation and the law of momentum and energy) in which the kinematics and the mechanical response of materials are modeled as a continuous mass and not as discrete entities. With relation to ballistic impact, a solid understanding of the penetration mechanism during the armor penetration process is essential in determining the correct parameters (Parga-Landa and Hernandez-Olivares, 1995). The governing equations of the impactor–target interaction during the impact event will take these parameters into account. Although the analytical approach provides a significant reduction in computing time, it also reduces the accuracy of such a model when compared with results obtained from a numerical simulation (Tabiei and Nilakantan, 2008).

These analytical approaches are useful in addressing simple physical phenomena, but become increasingly complicated with phenomena which are more complex and which involve many variables (Parga-Landa and Hernandez-Olivares, 1995; Tabiei and Nilakantan, 2008). Examples of an analytical model using these approaches with respect to certain impact dynamics of fabric materials have been developed by some researchers (Gu, 2003; Chocron-Benloulo et al., 1997a; Billon and Robinson, 2001). The principal fabric's yarn interaction or contact with the projectile is the primary focus in these models, while the energy absorbed by other yarns and the kinetic energy of the fabric in the damaged section are not taken into consideration. Gu (2003) also presented an analytical model to calculate the decrease of kinetic energy and residual velocity of a projectile penetrating targets composed of multilayered planar plain-woven fabrics. Due to fibers (in the fabrics) being subjected to a high strain rate during the projectile perforation process (at high velocity), the strain rates of the two kinds of fibers, Twaron and Kuralon, were reported to be 1.0×10^{-2} to 1.5×10^3 s^{-1}, respectively. These strain rates are used to calculate the residual velocity of a projectile. It is noted in most published works that the mechanical properties of fibers at high strain rate should be adopted in modeling rate-sensitivity materials (Parga-Landa and Hernandez-Olivares, 1995; Tabiei and Nilakantan, 2008; Chocron-Benloulo et al., 1997a; Mamivand and Liaghat, 2010; Ivanov and Tabiei, 2004; Phoenix and Porwal, 2003; Bilion and Robinson, 2001).

Numerical methods are based on finite element or finite difference approaches, which focus on finding approximate solutions to partial differential equations (PDE) and also to integral equations. In general, the governing equations for the ballistic impact of a material behave nonlinearly; numerical modeling tools can help to provide a more precise virtual representation of the damages to the material or structure and its responses. With the advancements in pre-processing tools with solid modeling capabilities, a simple fabric structure and projectiles as shown in Figs 7.1 and 7.2

LS-DYNA keyword deck by LS-PRE

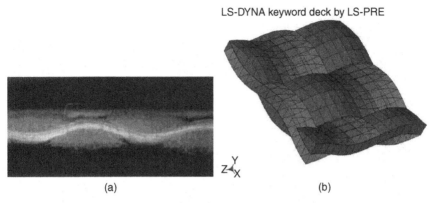

(a)   (b)

7.1 (a) Cross-section of Twaron plain woven fabric; (b) a three-dimensional fabric of Twaron plain woven model in LS-DYNA (Talebi, 2006).

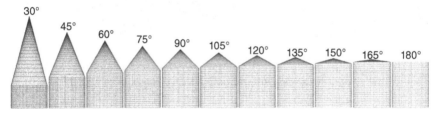

7.2 Cross-sectional view of the modeled projectiles with different nose angles (Talebi *et al.*, 2009).

can be converted into a two- or three-dimensional representation. Ballistic scientists are now able to mimic the actual ballistic experimentation in their respective studies and predict the outcomes. Complex objects such as yarns, fabric plies, projectiles and projectile–textile interaction can be modeled explicitly and results can be simulated to enable a better understanding of the impact process (as shown in Fig. 7.3).

Every high velocity impact numerical model is built using hydrocodes. Hydrocodes follow the common numerical methods often used in engineering analysis. However, in high velocity impact events certain options may need to be specified. The accuracy of the hydrocode is hugely reliant upon definite constitutive equations used to correspond to the action of each individual material (Zaeraá and Sánchez-Gálvez, 1998). The disadvantage of this approach is the long period of computing time required for a specific simulation process. This approach is mostly adopted into commercial packages, such as ABAQUS, ANSYS, LSDYNA, etc., to conduct the analysis or simulation.

In this chapter, the primary focus will be on numerical aspects of ballistic modeling which widely utilize computer technology. The next section of this

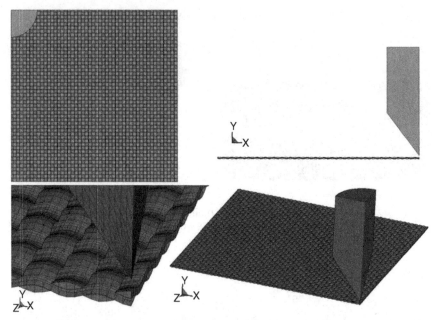

7.3 A finite element model of impactor–target interaction from different view angles and magnifications (Talebi *et al.*, 2009).

chapter starts with a fundamental overview of hydrocode modeling and its computational aspects, while Section 7.3 demonstrates the ability of computer-based models in solving ballistic-related problems. Suggestions for future trends are also presented and discussed in the final section.

7.2   Computational aspects

The governing equations of solids subjected to impact are commonly non-linear and cannot be described analytically; responses are therefore determined through numerical analysis of the given equations. Hydrocodes or wave propagation codes are computer-based codings or programs used to model dynamic events in solid mechanics, and particularly in shock mechanics. These enable researchers to analyze the time-dependent pattern of shock wave and acoustic propagation due to impact, penetration or explosion phenomena in fluids and solids. This is achieved by approximating a continuum point-wise (finite difference) or piece-wise (finite element), then solving the conservation equation coupled with material models (Goudreau and Hallquist, 1982; Anderson, 1987).

A typical hydrocode modeling process can be briefly described by the flow chart shown in Fig. 7.4. Hydrocode modeling is based on three main

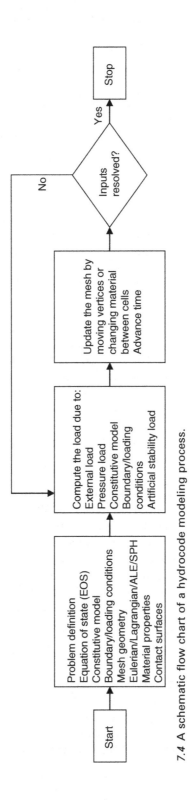

7.4 A schematic flow chart of a hydrocode modeling process.

fundamental concepts, which are the equation of state (EOS), Newtonian laws (motion) and the constitutive model (Anderson, 1987; Anderson *et al.*, 1994; Pierazzo and Collins, 2004). These concepts are used to resolve the forces acting on the mesh at each stage during the computational modeling process. The basic foundation of these codes is their basis on a spatial and time discretization, where the Newtonian laws of motion are solved through the governing conservation equations in terms of momentum, mass and energy (Anderson, 1987). The following expressions for the conservation equations are addressed as the Euler equations for dynamic, compressible fluid flow without thermal conduction, viscous forces and gravity.

Conservation of momentum $\quad \dfrac{dv_i}{dt} = f_i + \dfrac{1}{\rho}\dfrac{\partial \sigma_{ji}}{\partial x_j}$ 7.1

Conservation of mass $\quad \dfrac{d\rho}{dt} + \rho\dfrac{\partial v_i}{\partial x_i} = 0$ 7.2

Conservation of energy $\quad \dfrac{dI}{dt} = -\dfrac{p}{\rho}\dfrac{\partial v_i}{\partial x_i} + \dfrac{1}{\rho}\Pi_{ij}\dot{\varepsilon}'_{ij}$ 7.3

In the above equations v_i is the velocity; ρ is the material density; σ_{ij} is the Cauchy stress tensor; I is the specific internal energy, which is composed of a hydrostatic part, the pressure p, and a deviatoric part, Π_{ij}; f_i is the external body forces per unit mass; and ε'_{ij} is the deviatoric strain rate. With an equation of state (EOS) establishing the relationships between pressure, density and internal energy and a constitutive material strength model, a complete set of equations for the hydrodynamic behavior is formed (Pierazzo and Collins, 2004).

In a real-life scenario, where there are multiple variables and the problem is complex, the equations must be solved simultaneously. Numerical techniques with the advancement of high performance computing provide the only means of solving the number of mathematical operations required for the solution. All hydrocodes operate some form of the conservation equations mentioned above, but the effectiveness of the hydrocode relies on the sophistication of the constitutive model and of the equation of state employed by the user (Anderson, 1987; Pierazzo and Collins, 2004).

7.2.1   Spatial discretization

Discretization of a continuous physical constituent mainly requires a computer-based analysis. During the discretization procedure, the continuum, or an entity which has the property of being continuous, is replaced by a computational mesh. Three fundamental techniques exist for discretizing the differential equations: finite element schemes, finite-difference schemes and smoothed particle hydrodynamic (SPH) techniques. Basically, the three schemes present different algorithms for solving a similar problem; nevertheless, each has its own advantages and disadvantages (Hamouda and Hashmi, 1996).

Finite difference scheme

In the finite difference approach, the spatial derivatives in the differential equations are substituted by a different set of equations. For instance, function F, the partial derivative $\partial F/\partial x$, becomes $\Delta F/\Delta x$ where the differences are computed at grid points. The first derivative of F at x_n can be represented by a variety of difference formulae (Pierazzo and Collins, 2004):

$$\frac{\partial F}{\partial x}\bigg|_{x_n} = \frac{F_{n+1} - F_n}{\Delta x}$$

$$\frac{\partial F}{\partial x}\bigg|_{x_n} = \frac{F_n - F_{n-1}}{\Delta x} \qquad\qquad 7.4$$

$$\frac{\partial F}{\partial x}\bigg|_{x_n} = \frac{F_{n+1} - F_n - 1}{2(\Delta x)}$$

These formulae correspond to forward, backward and central difference equations, respectively. Pierazzo and Collins (2004) state that the finite-difference method is well founded and simple to implement. However, it does require the grid to be structured (cells arranged in rows and columns). Thus, an adaptive meshing algorithm involving certain coordinated mapping techniques must be applied in order to solve problems involving complicated geometries. Furthermore, there is no direct method for evaluating the accuracy of a solution, and the scheme is prone to certain types of numerical instability, which require either artificial correction or correction by trial and error. The accuracy of the solution usually increases with decreasing cell size; however, limits on the time-step point mean that small cell sizes imply shorter time-step points, leading to long computer run times as mentioned by other researchers in their works (Hamouda and Hashmi, 1996; Woodward, 1996).

Finite element scheme

The finite element scheme was initially formulated on a physical basis for the analysis of problems in structural mechanics. Nevertheless, it was soon accepted that the approach could be employed to solve a variety of problems (Reddy, 2006; Moaveni, 2003; Zienkiewicz and Taylor, 1989). Whereas the finite difference method is a point-wise discretization of the problem space, finite element methodology divides the problem space into elements.

The elements can be rectilinear or curved and, unlike in the finite difference method, need not be arranged in a structured grid. Hence, complicated problem geometries are handled better with a finite element approach.

Interpolation functions are used to represent the variation of a variable over the element. Each element is associated with a set of nodes, whose initial locations are known. The displacement of these nodes is the basic unknown of the problem. The equations governing the displacements of these nodes are calculated on an element-to-element basis and then combined. As a consequence, finite element codes may be parallelized as a way to reduce run time. Once combined, the system of equations relating the forces and displacements at each node is solved by inverting the 'stiffness matrix', which represents the constitutive relationship between stress and strain. One advantage of this method is that when the displacements have been derived, they can be substituted back into the original equations to check for consistency. Any inconsistency is a direct measure of the inaccuracy of the solution and can be corrected for during the simulation.

Smoothed particle hydrodynamics

Smoothed particle hydrodynamics (SPH) was formulated to study problems in astrophysics with respect to fluid masses traveling at random in three dimensions in the absence of boundaries (Monaghan, 1988). Monaghan (1988) gave a typical example of the numerical simulation of the fission of a rapidly rotating star. SPH involves the motion of a set of points, the velocity and thermal energy of which are known at any given time. A mass is also assigned to each point and, as a result, the points are referred to as particles. Therefore, to move the particles to the correct point during a time step, it is necessary to introduce certain forces which an element of fluid would experience. These forces are fundamentally constructed by applying sophisticated interpolation formulas to determine properties such as density at a given point. SPH codes offer an appealing option to the more tenable techniques of finite-difference and finite-element, due to the simplicity of the algorithm. In most cases, users tend to program the SPH according to their own requirements. The method is inherently Lagrangian and thus possesses most of the benefits of this formalism. Nevertheless, SPH codes do not collapse when large displacements are involved, since the particles are not connected to each other, in contrast to what occurs with a typical finite element scheme (Lagrangian and Eulerian) (Hamouda and Hashmi, 1996).

Although currently popular, and undergoing continual advances in development, particularly in ballistic impact-related research, SPH codes do suffer from several major shortcomings. Currently, there are no robust methods for describing complicated material rheologies such as strength, elasticity, etc. Moreover, by their very nature, SPH codes do not handle certain types of boundary conditions well, further limiting their potential use. Lastly, in problems such as impact calculations where the density varies dramatically (from very dense target rock to low density vapor), SPH suffers because the low density material is represented by too few particles to effectively simulate the problem. SPH codes are good for fluid flow problems involving relatively small density differences and primarily inflow or outflow boundary conditions (Aktay and Johnson, 2007; Hamouda and Hashmi, 1996).

7.2.2 Time-integrating methods

The time-stepping techniques are the most important procedure in most structural dynamics problems. Thus, there have been a considerable number of studies on the topic (Predebon *et al.*, 1991; Rosinsky, 1985; Herrmann, 1975; Liu *et al.*, 1988) and a concise description will be discussed in this chapter. There are primarily two time iteration methods which differ from the classical closed-form solutions available to analysts, namely implicit and explicit formulations of the systems of equations, which describe the mechanics.

The process for the discretized equation of motion is the explicit method, where the solution at some duration $t + \Delta t$ in the computational cycle is based on the knowledge of the equilibrium condition at time t. The advantage of applying the explicit method is that there is no requirement to calculate stiffness and mass matrices for the whole system; therefore, the solution can be carried through on the element level and relatively little storage is required. The disadvantage of the method is that it is conditionally stable in time, and the time step must be carefully selected: the size of the time step must be sufficiently small to precisely treat the high-frequency modes that dominate the response in wave propagation problems (Hamouda and Hashmi, 1992).

In an implicit scheme, the solution at any time $t + \Delta t$ is attained with the understanding of the accelerations at the same moment. Implicit methods are categorically stable; nevertheless, such stability is achieved at the expense of solving a set of equations at each time step. The often-mentioned implicit finite element codes are ABAQUS, ANSYS, NASTRAN, MARC and LUSAS. In general, it may be said that the implicit integration method is more effective for static or low frequency problems while the explicit integration method is the best for high speed impacts (Hamouda and Hashmi, 1996).

7.2.3 Problem description

The characterization of the deformed solid can be stated in either Lagrangian or Eulerian coordinates (Malvern, 1968). In Lagrangian coordinates, every point in the deformed solid is referred to some reference state, and any discretization, such as finite element mesh or finite difference zoning used in the analysis, deforms with the material. Vaziri *et al.* (1989) have compiled a list of some Lagrangian and Eulerian coordinates, which are shown in Tables 7.1 and 7.2, respectively.

In Eulerian coordinates, the points are fixed in space and the discretization process does not affect the material. The Eulerian formulation, therefore, has no mechanism for tracking material history (Hamouda and Hashmi, 1992). Predebon *et al.* (1991) reported that the Lagrangian formulation follows material particle paths which allow a precise historical description of the material. This facilitates the integration of the history-dependent material description and so far, the most sophisticated material

Table 7.1 Evaluation of Lagrangian hydrocodes

Code	Year	Developer	Organization
HEMP	1964	M.L. Wilkins	Lawrence Livermore Laboratories
HEMP-3D	1975	M.L. Wilkins	Lawrence Livermore Laboratories
HEMP-DS	1983	M.L. Wilkins	Lawrence Livermore Laboratories
C-HEMP	1987	L. Seanman *et al.*	Southwest Research Institute
TOODY	1967	W. Herrmann	Sandia National Laboratories
HONDO	1974	S.W. Key	Sandia National Laboratories
EPIC-2	1976	G.R. Johnson	Honeywell Inc.
EPIC-3	1977	G.R. Johnson	Honeywell Inc.
EPIC-2 (Erosion/ Plugging)	1987	B.E. Ringers	BRL
EPIC-3 (Erosion)	1985	Ted Belytschko	BRL
DYNA 2D/3D	1976	J.O. Hallquist	Lawrence Livermore Laboratories
DYNA 2D (Erosion)	1989	J.O. Hallquist	Livermore Software Tech. Corp.
DEFEL	1984	W. Flis	DYNA East Group
PEPSI	1984	R. Hunkler and G. Paulus	ISL France
PRONTO 2D	1987	L.M. Taylor and D.P. Flanagan	Sandia National Laboratories
ZEUS	1987	J.A. Zukas and S.B. Segletes	Computational Mech. Consult. Inc.

Source: Vaziri *et al.*, 1989.

Table 7.2 Evaluation of Eulerian hydrocodes

Code	Year	Developer	Organization
PIC	1954	M. Evans and F. Harlow	Los Alamos Laboratories
SHELL	1959	W. Johnson	General Atomic Corp.
SPEAR	1963	W. Johnson	General Atomic Corp.
OIL	1965	J. Walsh and W. Johnson	General Atomic Corp.
TOIL/TROIL	1967	W. Johnson	General Atomic Corp.
DROF	1971	W. Johnson	Systems, Science and Software (S-Cubed)
DROF-9	1971	W. Johnson	S-Cubed
TRIDROF	1976	W. Johnson	Computer Codes Consultants (CCC)
SOIL	1977	W. Johnson	CCC
LASOIL	1987	W. Johnson	Los Alamos Laboratories
RPM	1976	J. Daienes *et al.*	General Atomic Corp.
HELP	1971	L. Hageman and J. Walsh	S-Cubed, BRL
HELP-75	1975	L. Hageman *et al.*	S-Cubed
METRIC	1976	L. Hageman *et al.*	S-Cubed
CHART-D	1969	S.L Thompson	Sandia National Laboratories
CSQ	1975	S.L Thompson	Sandia National Laboratories
CSQ-II	1979	S.L Thompson	Sandia National Laboratories
CHT	1987	J. McGlaun *et al.*	Sandia National Laboratories
HULL	1971	R. Durrett and D. Matuska	Orland Technology (OTI)
HULL-78	1978	R. Durrent and J. Osborn	OTI
EPHULL	1988	R. Bell	S-Cubed
MESA	1989	D. Mandell *et al.*	Los Alamos Laboratories

descriptions have been performed with Lagrangian codes. Predebon *et al.* (1991) also conducted simulations of cylinder impact tests in Lagrangian code (HEMP) and in Eulerian code (CSQ). They found that the final dimensions of the simulated cylinder in the Lagrangian code are 2.8% higher than that simulated using the Eulerian code (Hamouda and Hashmi, 1992).

Generally, the Lagrangian formulation is the most suitable approach for impact of solid bodies, because the surfaces of the bodies will always correspond with the discretization and are therefore well defined. The disadvantage is that the numerical mesh can become badly compressed and altered in many aspects, which leads to an unfavorable outcome for the integration time-step and low accuracy. These problems can be minimized with the use of an eroding sliding interface and rezoning. The tunnel approach is another numerical technique that can be applied to counter similar problems (Rosinsky et al., 1988). A hydrocode can use either type of formulation to explain the focused condition; which mode of description should be selected is dependent on the nature of the problem under consideration.

Coupled Eulerian Lagrangian (CEL)

The CEL method was formulated with the aim of combining the advantages of the Lagrangian and Eulerian formulations. The advantage of this methodology is that either method can be applied, autonomously, to different regions of a problem according to the physics being modeled. The CEL approach is practical for problems involving two materials, one of which is more deformable than the other. In the more deformable region the problem can be modeled as Eulerian (in that it behaves like a fluid motion), while regions undergoing small deformation can be modelled with the Lagrangian method. The disadvantage of this method is the possible computational penalty associated with the Eulerian–Lagrangian interface. AUTODYN (now known as ANSYS-AUTODYN) is one of the commercial codes which use the CEL approach (Herrmann, 1975; Grujicic et al., 2009a).

Arbitrary Lagrangian Eulerian (ALE)

The ALE was initially formulated for fluids. The ALE description treats the computational mesh as a reference structure that may be moving with an arbitrary speed which differs from both the particle velocity (Lagrangian) and zero velocity (Eulerian). The complexity in developing the algorithms required for continuous rezoning has restricted the use of this technique. It has also been observed that the material interface, free surfaces and material history are too complex to be numerically compiled using the ALE technique (Steinberg et al., 1987).

7.2.4 Rezoning (re-meshing)

Rezoning is configuring a new mesh out of the previously developed mesh in the model. The new mesh can be manually characterized and defined, or an automatic mesh generator may be employed. Mesh rezoning not only alters the deformed mesh of a Lagrangian computation. Generally, the aim is to create a fine zoning for a better resolution in sections where large stress variations exist from zone to zone. Rezoning is not a direct task, because it involves the computation of the new mesh quantities by interpolating from those of the former mesh without substantial loss of accuracy in the response predictions (Hamouda and Hashmi, 1996).

7.2.5 Contact

Contact algorithms are applied in finite element simulations to prevent surfaces from interpenetrating in Lagrangian calculations. Eulerian codes do not require the contact algorithms, since the treatment of mixed zones (zones with multiple material properties) is carried out implicitly by the contact. The contacting surfaces are often passed on to a 'slid-line' even in three dimensions. From the model object, one surface is assigned as the master surface and the other as the slave. The related nodes on the surfaces will also follow the initially designated master and slave nodes. This termi-nology is based on the assumption that there are two distinct surfaces. Unique contact algorithms are necessary if a surface is to be allowed to come into contact with itself, because the two surface algorithms for deter-mining the contact points will work with a single surface (Hallquist *et al.*, 1985).

7.2.6 Commercially available finite element packages

Commercial finite element analysis (FEA) packages are most commonly used in engineering-related industries due to their cost-efficiency when compared to costly experimentation and because they enable the designer to virtually test any material which has not been physically developed. Many finite element codes employ the explicit integration scheme to solve highly transient, nonlinear problems. In the case of ballistic impact on textile-based material, finite element codes can also deal with complex interactions between the impactor and yarn/fabric, penetration, contact and friction between yarns/fabric plies, and the deformation and failure of the yarn. Hence, these packages are able to give a clearer understanding of the impact scenario of textile-based materials.

The most widely known commercially available explicit-based FEA soft-ware is LSDYNA 3D, which was developed from a three-dimensional solver code known as DYNA3D. The code was developed by Dr John O. Hallquist, a scientist at the Lawrence Livermore National Laboratory (LLNL), in 1976. It comes with a basic pre-post program called LS-PREPOST and is largely utilized in research publications.

ABAQUS packages, another piece of commercial software, also use the explicit time-integration approach in order to provide nonlinear, transient, dynamic response analysis of solids and structures. The code was developed by Hibbitt, Karlsson & Sorensen, Inc., and is configured specifically to process advanced, nonlinear continuum and structural analysis require-ments. ABAQUS/Explicit has the standard ABAQUS user interface on vector/parallel computers, as well as on UNIX workstations. P ABAQUS/ Explicit also provided extensive metal-based plasticity and elasticity models, including rate dependence and adiabatic heat generation in its packages. A user interface subroutine for user specification of material behavior is also available.

There are other types of codes such as LUSAS, MSC DYTRAN, PAM-CRASH and ANSYS/AUTODYN which have general capability for modeling contact problems, with or without friction. An extensive element library containing elements for 2D and 3D problems (plane stress and strain, axisymmetric, beams, membranes and shells) is embedded in most of these packages. The loading type in the codes can be categorized as distributed and concentrated forces, moments, displacements, velocities, etc.

7.3   Numerical modeling of single and multiple layer fabric

Numerical modeling of penetration into high strength fabric by finite element and finite difference methods can be divided into three categories (Talebi, 2006), namely:

- Modeling by assuming yarns as pin joint orthogonal elements
- Assuming the whole fabric as thin shell elements (unit cell approach)
- Modeling yarns as solid isotropic or orthotropic materials.

The most widely used method for modeling the ballistic impact of textile structures is the finite element model based on pin-jointed orthogonal cables (Roylance and Wang, 1980; Roylance *et al.*, 1995; Shim *et al.*, 1995, 2001; Lim *et al.*, 2003; Tan *et al.*, 2003; Shahkarami *et al.*, 2002; Johnson *et al.*, 1999; Billon and Robinson, 2001). This approach has been found to be practical in approximating the dynamic behavior of textile materials, but the discrete nature of such models has intrinsic oversimplifications that may decrease their predictive ability. Shahkarami and Vaziri (2005) have reported that these models do not really consider all the vital information related to the textile weave architecture, yarn-to-yarn contact conditions (or intra-layer interactions, e.g., surface finish and friction) and layer-to-layer contact conditions in multi-layer fabrics (or inter-layer interactions, e.g., friction, bonding, stitching and spacing effects).

Examples of studies using the pin-jointed orthogonal approach were reported by Tan *et al.* (2003), Talebi (2006), Talebi *et al.* (2009) and Lim *et al.* (2003). Figure 7.5 shows the pin joint finite element model built by Tan *et al.* (2003) in order to study the effect of yarn crimping, the sliding contact between yarns and yarn breakage using an assembly of visco-elastic bar elements at two target clamping settings. They concluded that the finite element model was able to reproduce a damage response similar to that observed in actual experiments, despite the relatively low number of degrees of freedom. Talebi (2006) and Talebi *et al.* (2009) adopted a similar approach

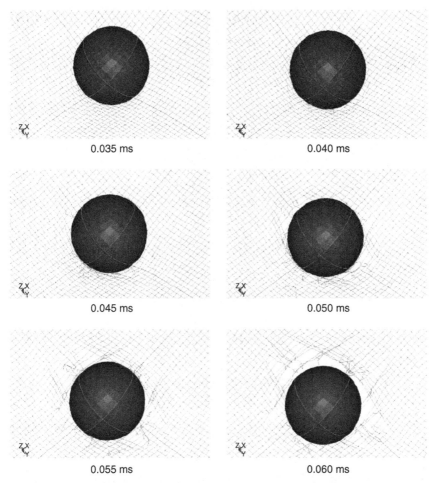

7.5 The 'wedging-through' action of a projectile into the fabric during impact (Tan *et al.*, 2003).

in studying the effect of the damage response of single and multiple fabric materials when subjected to projectile impact. As observed in Fig. 7.6, for higher velocities, Talebi (2006) found that the fabric deflection (transverse) is limited to the neighboring zone of the impacted point. The reason for this is that there is less time for stress waves to travel across the fabric area. According to Lim *et al.* (2003), when a projectile at a certain velocity hits the fabric, stress waves at high strain rates propagate from the impacted zone and shift along the corresponding yarns to the edges of the fabric, and will eventually be reflected back. In the fabric, each corresponding yarn at crossover points reacts in terms of partly transmitting and partly reflecting

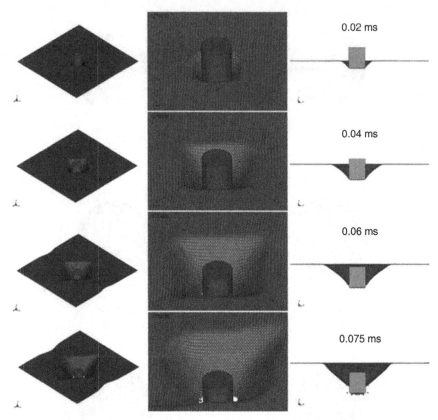

7.6 Sequence of fabric deformation and damage for impact velocity of 420 m/s (Talebi, 2006).

the stress wave along its boundary. Figure 7.6 also illustrates the pyramidal shape deformation similarly observed in high-speed photographs of the ballistic impact experiments reported by Tan *et al.* (2003). The models that do not incorporate many detailed properties of the woven fabric may not be able to predict a deformed shape of this type. For low velocity impacts the transverse deflection wave reached the fixed edges prior to complete penetration, resulting in the formation of an octagonal wave front (Talebi, 2006). Figure 7.7 presents the finite element models built by Lim *et al.* (2003) compared to the actual experiment.

The behavior of fabric under transverse impact involves a complex relationship involving many factors: this is discussed in great detail by Cunniff (1992), Roylance and Wang (1980) and Cheeseman and Bogetti (2003). When a local target material is subjected to a high velocity ballistic impact, it causes the material to act in a fluid-like manner (hydrodynamic effect), followed by wave propagation in the textile structure. If a single yarn is

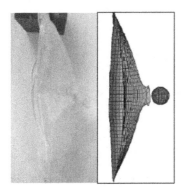
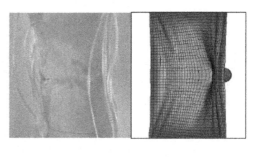

7.7 Finite element model simulation displacement profile compared to experimentation for impact velocity of 206 m/s (Lim *et al.*, 2003).

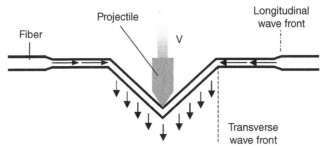

7.8 Schematic response of projectile impact into a single fiber/yarn materials (Cheeseman and Bogetti, 2003).

struck transversely, a longitudinal and transverse wave will be formed and will propagate from the point of impact as shown in Fig. 7.8. The longitudinal tensile wave moves through the material at the speed of sound along the fiber axis of the material. As the longitudinal wave propagates, the material behind the longitudinal wave front flows in the direction of the impact point, which has been deflected in the direction of motion of the impacting projectile. This transverse-like movement of the fiber is called the transverse wave, and it propagates at a velocity lower than that of the longitudinal wave (Cheeseman and Bogetti, 2003).

Furthermore, it is reported by Shim *et al.* (1995), Tan *et al.* (2003) that prior to the transverse wave reaching the fixed edges, the high-impacting projectile may have perforated the fabric system. The projectiles do not break all the yarns during the penetration process, but a wedging process occurs until a small gap is made (due to fiber breakage in several yarns). During the process, the projectile pushes aside and slides past the other yarns, forming a small hole in the fabric. Shahkarami and Vaziri (2005), however, reported that the adaptation of the pin-jointed technique for

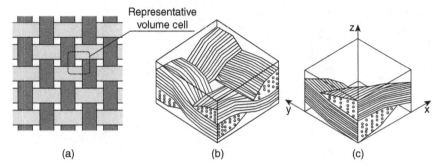

7.9 (a) Weave architecture; (b) RVE; (c) quarter of RVE according to Tabiei and Yi (2002).

application in impact problems has certain limitations in terms of the validity of the result due to oversimplification. More specifically, such models do not adequately take into account the crucial details associated with the yarn-to-yarn contact conditions (or intra-layer interactions, e.g. friction and surface finish), fabric weave architecture and layer-to-layer contact conditions in multi-layer fabrics (or inter-layer interactions, e.g. friction, stitching, bonding and spacing effects).

Another approach that is extensively used for the analysis of fiber and textile reinforced composites is the unit-cell method. A repetitive unit cell (RUC) or representative volume element (RVE) for a plain weave fabric as shown in Fig. 7.9 is discussed in detail by Tabiei and Yi (2002). Tabiei and Ivanov (2002) presented a unit-cell model where the single yarns interlacing at certain braid and crimp angles are embedded in a shell element. Kawabata (1989) introduced a simple analytical model to capture the biaxial behavior of a symmetrically loaded yarn crossover. Luo (2000) presented a constitutive model for the unit-cell of fabric-reinforced flexible composites under biaxial loads using a strain energy approach, accounting for the deformation of the yarn through both rigid body displacement and cross-sectional changes. Simple unit cell models make use of a thin shell-like element to represent the fabric system in impact simulations. Examples of simulations of this type have been reported by Lim *et al.* (2003), Sun *et al.* (2009) and Grujicic *et al.* (2009a).

Lim *et al.* (2003) used LS-DYNA 3D software to simulate the response of a Twaron fabric panel when subjected to ballistic impact. To account for the strain rate-dependent properties of fabric, a three-dimensional element spring-dashpot model was used, based on data obtained from experiments. The ballistic limit, energy absorption and transverse deflection profiles of the fabric, as well as the residual velocity of the projectile, were predicted and evaluated with the data from the experiment. However, the various limitations of the model adopted in this study, in terms of representing fabric behavior such as frictional effects between yarns and unraveling and fraying of yarns, restricted the model's use.

In other related works, Stubbs (1989) presented a model based on an assembly of truss elements which is able to simulate the response of coated fabrics to multiaxial loading, approximating interlacing yarns and the effect of coating material. Shahkarami and Vaziri (2005) proposed a new unit cell computational method based on the complete behavior of the smallest repeating unit in the fabric in order to predict the impact response of fabric panels. The unit-cell is assembled and calibrated using measured properties of the fabric which include both geometrical (crimp, fabric weave architecture, voids, etc.) and mechanical properties. The three-dimensional finite element analysis of the unit cell is used to impart a reference date for the response of the mechanical properties for calibrating the analytical model in the equivalent shell representation. This specific shell element takes advantage of a simple physics-based analytical constitutive relationship to predict the behavior of the fabric's weft and warp yarns under general applied displacements in these directions.

More detailed modeling techniques have been carried out by other researchers, such as the studies by Duan *et al.* (2005a, b, c, 2006) and Blankenhorn *et al.* (2004) among others. A finite element model to simulate the whole process of a multi-layered fabric panel being fully penetrated by a conically cylindrical steel projectile using LS-DYNA codes as shown in Fig. 7.10. This was carried out on the basic understanding that the crimps

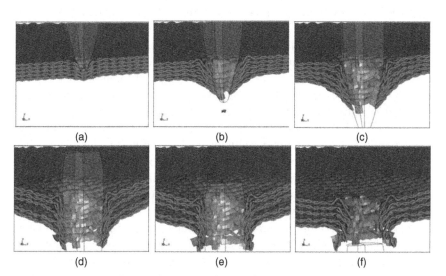

7.10 Simulation of the five-layered fabric target perforated by the projectile (local magnification) at different time intervals.

of warp and weft yarns were considered in the description of the actual structure. Weibull constitutive equations of filament yarns at high strain rate were used for material behavior to simulate the ballistic impact response. The authors reported that it was established that in the yarn-structural hierarchy, the ballistic impact between the steel projectile and the fabric target could be accurately simulated. Experimental and theoretical results were broadly in agreement on evaluation of the residual velocities of the projectile, and deformation and damages of the fabric target after ballistic perforation. However, although their model takes strain rate effects into consideration, they assumed that the yarns were isotropic materials, when in reality the yarn typically behaves in an orthotropic manner.

Duan *et al.* (2005a, b, c, 2006) used a numerical model to study the perforation of square plain-woven fabric targets in different clamping situations. Dunn found that the ballistic resistance of such systems is sensitive to boundary conditions and yarn orientation. Targets that are unclamped on two edges can absorb more impact energy than those with all four sides clamped. Orientating the yarns to the clamped edges can improve energy absorption significantly. Moreover, modeling indicates that friction dramatically affects the local fabric structure at the impact region by hindering the lateral mobility of principal yarns.

7.4   Conclusions

The main characteristics of the modeling of ballistic impact on textile structures have been reviewed. Hydrocodes can be very useful tools in high velocity impact-related research if the user understands their limitations and is familiar with the required operations. The developed codes can provide a detailed insight into the physical processes and can be used to perform analytical experiments virtually. Computational model experiments can be a cost-effective option when compared to laboratory experiments and their capability to virtually simulate the physical impact event has been demonstrated. Some examples of modeling of the problem of penetration of a textile structure have been presented in order to illustrate the capability of some commercially available computer codes.

Nevertheless, with respect to the design of soft body armor, where stitching of multiple plies of fabric is carried out, no numerically based studies or models for simulating the effect of stitching parameters (i.e. stitching density and area, etc.) have been conducted. The effects of stitching in terms of energy absorption of body armor are still uncertain. There may be advantages and disadvantages of stitching in soft body armors: the stitching process is often a time-consuming and costly step in manufacturing; furthermore, it is assumed by most armor designers that stitching can create weak points (at the stitched points), and also reduce armor flexibility, resulting in comfort issues. Research has been carried out on this subject (Ahmad *et al.*, 2008; Karahan, 2008; Karahan *et al.*, 2008) which has examined the behavior of stitched fabrics, but no numerical models have yet been developed to quantify and virtually simulate the effect of stitching on body armor performance.

As indicated in the review, the two most popular methods are the finite element and finite difference methods, although finite difference methods are not as popular today as they have been in the past. In recent years, finite element methods have become the common tool for modeling impact and penetration events. Today, boundary element methods and smooth particle hydrodynamics have emerged as promising new approaches, particularly for ballistic-related simulations.

However, according to Oden *et al.* (2003), one of the primary concerns of specialists in the field of computational mechanics is the issue of the (un) reliability of the results obtained from these computer simulations. With a lack of experimentally based validation and accuracy, the authenticity of the output is obviously diminished. Nowadays, in many areas of application, accurate simulations are routinely obtained using high performance computers while others are, at best, qualitative and only capable of describing trends in physical events. The particular interest in reliability has contributed to the creation of a challenging technological area marked out merely by validation and verification.

One of the major reasons behind the reliance on numerical analysis for the advancement of computer technology is the increasing industrial competitiveness in the reduction in cycle time for design and manufacturing and validation. This reduction has a critical effect on the potential for virtually assessing the design, the ability to evaluate the problem entirely on the computer, without carrying out time-consuming experiments. For textile-based ballistic products, intense environments, for example expensive live-fire tests, can be removed from the process and used only for validation purposes. While great progress has been made in the computational field in the past two decades, simulation model results should not be accepted without proper validation. In order to develop an empirical, analytical and numerical model, many tests must be performed, since many of the physical phenomena cannot now be modeled on the basis of first principles. The models that have been developed may have taken into account key factors but, in reality, there are numerous potential unknowns. Ballistic modeling of textile structures coupled with computer technology has become a key enabling discipline that has led to greater understanding and advances in defense science and technology. It has been the basis of more significant developments in recent years and will continue to be crucial to the development of the body armor industries, from safety to security issues, and to the understanding of the diverse physical systems occurring in nature and in society.

7.5   References and further reading

Ahmad, M.R., Wan Ahmad, W.Y., Salleh, J. and Samsuri, A. (2008), Effect of fabric stitching on ballistic impact resistance of natural rubber coated fabric systems, *Materials and Design*, 29, 1353–1358.

Aktay, L. and Johnson, A.F. (2007), FEM/JSPH coupling techniques for high velocity impact simulations, *Computational Methods in Applied Sciences*, 5, 147–167.

Anderson Jr, C.E. (1987), An overview of the theory of hydrocodes, *International Journal of Impact Engineering*, 5, 33–59.

Anderson, C., Cox, P., Johnson, G.R. and Maudlin, P. (1994), A constitutive formulation for anisotropic materials suitable for wave propagation computer program, *Computational Mechanics*, 15, 201–223.

Barauskas, R. and Abraitiene, A. (2007), Computational analysis of impact of a bullet against the multilayer fabrics in LS-DYNA, *International Journal of Impact Engineering*, 34, 1286–1305.

Benson, D.J. (1987). Adding an ALE capability to DYNA 2D: Experiences and conclusions, *Proc. 1st Int. Conf. on Effects of Fast Transient Loadings*, Lausanne, Switzerland, 219–231.

Billon, H.H. and Robinson, D.J. (2001), Models for the ballistic impact of fabric armor, *International Journal of Impact Engineering*, 25(4), 411–422.

Blankenhorn, G., Schweizerhof, K. and Finckh, H. (2004), Improved numerical investigation of a projectile impact on a textile structure, *4th European LS-DYNA Conference*, Ulm, Germany.

Cheeseman, B.A. and Bogetti, T.A. (2003), Ballistic impact into fabric and compliant composite laminates, *Composite Structures*, 61, 161–173.

Chocron-Benloulo, I.S., Rodríguez, J. and Sánchez-Gálvez, V. (1997a), A simple analytical model to simulate textile fabric ballistic behavior, *Textile Research Journal*, 67(7), 52.

Chocron-Benloulo, I.S., Rodríguez, J. and Sánchez-Gálvez, V. (1997b), A simple analytical model for ballistic impact in composites, *Journal de Physique IV*, 7, 821–826.

Cismasiu, C. Silva, M.A.G. and Chiorean, C.G. (2005), Numerical simulation of ballistic impact on composite laminates, *International Journal of Impact Engineering*, 31, 289–306.

Cunniff, P.M. (1992), An analysis of the system effect in woven fabrics under ballistic impact. *Textile Research Journal*, 62, 495–509.

Duan, Y., Keefe, M., Bogetti, T.A., Cheeseman, B.A. and Powers, B. (2004), A numerical investigation of the influence of friction on energy absorption by a high strength fabric subjected to ballistic impact, *International Journal of Impact Engineering*, 65, 1–15.

Duan, Y., Keefe, M., Bogetti, T.A. and Cheeseman, B.A. (2005a), Modeling the role of friction during ballistic impact of a high-strength plain-weave fabric, *Composite Structures*, 68, 331–337.

Duan, Y., Keefe, M., Bogetti, T.A. and Cheeseman, B.A. (2005b), Modeling friction effects on the ballistic impact behavior of a single-ply high-strength fabric, *International Journal of Impact Engineering*, 31, 996–1012.

Duan, Y., Keefe, M., Wetzel, E.D., Bogetti, T.A., Powers, B., Kirkwood, J. and Kirkwood, K.M. (2005c), Effects of friction on the ballistic performance of a high strength fabric structure, in: *WIT Transactions on Engineering Sciences, Impact*

Loading of Lightweight Structures, M. Alves and N. Jones (editors), WIT Press, Southampton, UK, 49, 1–10.

Duan, Y., Keefe, M., Bogetti, T.A., Cheeseman, B.A. and Powers, B. (2006), Finite element modeling of transverse impact on a ballistic fabric, *International Journal of Mechanical Sciences*, 48, 33–43.

Franzen, R.R. Orphal, D.L. and Anderson Jr, C.E. (1997), The influence of experimental design on depth-of-penetration (DOP) test results and derived ballistic efficiencies. *International Journal of Impact Engineering*, 19(8), 727–737.

Goudreau, G.L. and Hallquist, J.O. (1982), Recent developments in large-scale finite element Lagrangian hydrocode technology, *Computer Methods in Applied Mechanics and Engineering*, 33(1–3), 725–757.

Grujicic, M., Glomski, P.S., He, T., Arakere, G., Bell, W.C. and Cheeseman, B.A. (2009a), Material modeling and ballistic-resistance analysis of armor-grade composites reinforced with high-performance fibers, *Journal of Materials Engineering and Performance*,18(19), 1169–1182.

Grujicic, M., Bell, W.C., Arakere, G., Hea, T. and Cheeseman, B.A. (2009b), A mesoscale unit-cell based material model for the single-ply flexible-fabric armor, *Materials and Design*, 30, 3690–3704.

Gu, B. (2003), Analytical modeling for the ballistic perforation of planar plainwoven fabric target by projectile, *Composites Part B: Engineering*, 34, 361–371.

Hallquist, J.O., Goudreau, G.L. and Benson, D.J. (1985), Sliding interfaces with contact-impact in large-scale Lagrangian computations, *Computer Methods in Applied Mechanics and Engineering*, 51(1–3), 107–137.

Hamouda, A.M.S. and Hashmi, M.S.J. (1992), Simulation of the impact of a tool steel projectile into copper, mild-steel, stainless-steel (304) test specimens, in: *Structures Under Shock and Impact*, P.S. Bulson (editor), Computational Mechanics Publications, 51–61.

Hamouda, A.M.S. and Hashmi, M.S.J. (1996), Modelling the impact and penetration events of modern engineering materials: characteristics of computer codes and material models. *Journal of Materials Processing Technology*, 56, 847–862.

Hayhurst, C.J. and Clegg, R.A. (1997), Cylindrically symmetric SPH simulations of hypervelocity impacts on thin plates, *International Journal of Impact Engineering*, 20, 337–348.

Herrmann, W. (1975), Nonlinear transient response of solids, in: *Shock and Vibration Computer Programs, Reviews and Summaries* B. Pilkey, (editor), Shock and Vibration Information Center, Naval Research Laboratory, Washington, DC.

Ivanov, I. and Tabiei, A. (2004), Loosely woven fabric model with viscoelastic crimped fibres for ballistic impact simulations, *International Journal for Numerical Methods in Engineering*, 61, 1565–1583.

Johnson, G.R., Beissel, S.R. and Cunniff, P.M. (1999), A computational model for-fabrics subjected to ballistic impact, *18th International Symposium on Ballistics*, 962–969.

Karahan, M. (2008), Comparison of ballistic performance and energy absorption capabilities of woven and unidirectional aramid fabrics, *Textile Research Journal*, 78, 718–730.

Karahan, M., Abdil Kus, A. and Eren, R. (2008), An investigation into ballistic performance and energy absorption capabilities of woven aramid fabrics, *International Journal of Impact Engineering*, 35, 499–510.

Kawabata, S. (1989). Nonlinear mechanics of woven and knitted materials, in: *Textile Structural Composites*, W.-W. Chou and F. Ko (editors), Elsevier Science, Amsterdam, pp. 67–116.

Kirkwood, K.M., Kirkwood, J.E., Lee, Y.S., Egres, R.G., Wetzel, E.D. and Wagner, N.J. (2004), Yarn pull-out as a mechanism for dissipating ballistic impact energy in Kevlar® KM-2 fabric: Part II: Predicting ballistic performance, *Textile Research Journal*, 74, 939–948.

Lee. B.L., Walsh, T.F., Won, S.T., Patts, H.M., Song, J.W. and Mayer, A.H. (2001), Penetration failure mechanism of armor-grade fiber composites under impact, *Journal of Composite Materials*, 35(18), 1605–1629.

Lim, C.T., Shim, V.P.W. and Ng, Y.H. (2003), Finite element modeling of the ballistic impact of fabric armor, *International Journal of Impact Engineering*, 28, 13–31.

Liu, W.K., Chang, H., Chen, J.-S. and Belytschko, T. (1988), Arbitrary Lagrangian-Eulerian Petrov-Galerkin finite elements for nonlinear continua, *Computer Methods in Applied Mechanics Engineering*, 68, 259–310.

Luo, S.Y. (2000), Finite elastic deformation, in: *Comprehensive Composite Materials, Vol. 1*, A. Kelly and C. Zweben (editors), Elsevier Science, Oxford, 683–718.

Malvern, L.E. (1968), *Introduction to the Mechanics of a Continuous Medium*, Prentice-Hall, Englewood Cliffs, NJ.

Mamivand, M. and Liaghat, G.H. (2010), A model for ballistic impact on multi-layer fabric targets, *International Journal of Impact Engineering*, 37, 806–812.

Moaveni, S. (2003), *Finite Element Analysis: Theory and Application with ANSYS*, 2nd edition, Prentice-Hall, Englewood Chits, NJ.

Monaghan, J.J. (1988), An introduction to SPH, *Computer Physics Communications*, 48, 89–96.

Oden, J.T., Belytschko, T., Babuska, I. and Hughes, T.J.R. (2003), Research directions in computational mechanics, *Computational Methods Application in Mechanical Engineering*, 192, 913–922.

Parga-Landa, B. and Hernandez-Olivares, F. (1995), An analytical model to predict impact behavior of soft armours, *International Journal of Impact Engineering*, 16(3), 455–466.

Park, M., Yoo, J. and Chung, D.-T. (2005), An optimization of a multi-layered plate under ballistic impact, *International Journal of Solids and Structures*, 42(1), 123–137.

Phoenix, S.L. and Porwal, P.K. (2003), A new membrane model for the ballistic impact response and V_{50} performance of multi-ply fibrous systems, *International Journal of Solids and Structures*, 40, 6723–6765.

Pierazzo, E. and Collins, G. (2004), A brief introduction to hydrocode modeling of impact cratering, in: D. Dypvik, M. Burchell and P. Claeys (editors), *Cratering in Marine Environments and on Ice*, Springer, New York, 323–340.

Predebon, W.W., Anderson, C.E. and Walker, J.D. (1991), Inclusion of evolutionary damage measures in Eulerian wavecodes, *Computational Mechanics*, 7, 221.

Reddy, J.N. (2006), *An Introduction to the Finite Element Method*, 3rd edition, McGraw-Hill, New York.

Rosinsky, R. (1985), *Lagrangian Finite Element Analysis of Penetration of Earth Penetrating Weapons*, Lawrence Livermore National Laboratory Report UCID-20886.

Rosinsky, R. Schwer, L.E. and Day, J. (1988), Axisymmetric Lagrangian technique for predicting earth penetration including penetrator response, *International Journal for Numerical and Analytical Methods in Geomechanics*, 12(3), 235–262.

Roylance, D. and Wang, S.S. (1980), Penetration mechanics of textile structures, in: *Ballistic Materials and Penetration Mechanics*, R.C. Laible (editor), Elsevier Scientific, Amsterdam, 273–293.

Roylance, D., Chammas, P., Ting, J., Chi, H. and Scott, B. (1995), Numerical modeling of fabric impact, *Proceedings of the National Meeting of the American Society of Mechanical Engineers (ASME)*, San Francisco.

Shahkarami, A. and Vaziri, R. (2005), An efficient unit-cell based shell element for numerical modeling of fabrics under impact, in: *WIT Transactions on Engineering Sciences, Impact Loading of Lightweight Structures*, M. Alves and N. Jones (editors), WIT Press, Southampton, UK, 49, 135–152.

Shahkarami, A., Vaziri, R., Poursartip, A. and Williams, K. (2002). A numerical investigation of the effect of projectile mass on the energy absorption of fabric panels subjected to ballistic impact, *20th International Symposium on Ballistics*, 802–809.

Shim, V.P.W., Tan, V.B.C. and Tay, T.E., (1995), Modeling deformation and damage characteristics of woven fabric under small projectile impact, *International Journal of Impact Engineering*, 16(4), 585–605.

Shim, V.P.W., Lim, C.T. and Foo, K.J. (2001), Dynamic mechanical properties of fabric armor, *International Journal of Impact Engineering*, 25, 1–15,

Steinberg, D.J., Cochran, S.G. and Guinan, M.W. (1987), A constitutive model for metals applicable at high strain rates, *Journal of Applied Physics*, 51, 1498–1504.

Stubbs, N. (1989), Elastic and inelastic response of coated fabrics to arbitrary loading paths, in: *Textile Structural Composites*, T.-W. Chou and F. Ko (editors), Elsevier Science, Amsterdam, 331–354.

Sun, B., Liu, Y. and Gu, B. (2009), A unit cell approach of finite element calculation of ballistic impact damage of 3-D orthogonal woven composite, *Composites Part B: Engineering*, 40(6), 552–560.

Szosland, J. (2003), Modeling the structural barrier ability of woven fabrics, *AUTEX Research Journal*, 3(3), 1–13.

Tabiei, A. and Ivanov, I. (2002), Computational micro-mechanical model of flexible woven fabric for finite element impact simulation, *International Journal for Numerical Methods in Engineering*, 53, 1259–1276.

Tabiei, A. and Ivanov, I. (2004), Materially and geometrically non-linear woven composite micro-mechanical model with failure for finite element simulations, *International Journal of Non-Linear Mechanics*, 39, 175–188.

Tabiei, A. and Jiang, Y. (1999), Woven fabric composite material model with material nonlinearity for nonlinear finite element simulation, *International Journal of Solids and Structures*, 36(18), 2757–2771.

Tabiei, A. and Nilakantan, G. (2008), Ballistic impact of dry woven fabric composites: a review, *Applied Mechanics Reviews*, 61(1), 10801–10813

Tabiei, A. and Yi, W. (2002), Comparative study of predictive methods for woven fabric composite elastic properties, *Composite Structures*, 58, 149–164.

Tabiei, A., Jiang, Y. and Witao, Y. (2000), Novel micromechanics-based woven fabric composite constitutive model with material nonlinear behavior, *AIAA Journal*, 38(8), 1437–1443.

Talebi, H. (2006), *Finite Element Modeling of Ballistic Penetration into Fabric Armor*, Master's thesis, Universiti Putra Malaysia.

Talebi, H., Wong, S.V. and Hamouda, A.M.S. (2009), Finite element evaluation of projectile nose angle effects in ballistic perforation of high strength fabric, *Composite Structures*, 87, 314–320.

Tan, V.B.C. and Ching, T.W. (2006), Computational simulation of fabric armor subjected to ballistic impacts, *International Journal of Impact Engineering*, 32(11), 1737–1751.

Tan, V.B.C., Lim, C.T. and Cheong, C.H. (2003), Perforation of high-strength fabric of projectiles of different geometry, *International Journal of Impact Engineering*, 28, 207–222.

Van Gorp, E.H.M., Van der Loo, L.L.H. and Van Dingenan, J.L.J. (1993), A model for HPPE-based lightweight add-on armor, Symposium of Ballistic Research, Québec.

Vaziri, R., Pageau, G. and Poursartip, A. (1989), *Review of Computer Codes for Penetration Modelling of Metal Matrix Composite Materials*, Technical Report, University of British Columbia, Metals and Materials Engineering Group.

Walker, J.D. (1999), Constitutive model for fabrics with explicit static solution and ballistic limit, *Proceedings of the 18th International Symposium on Ballistics*, San Antonio, TX, 1231–1238.

Woodward, R.W. (1996), Modelling geometrical and dimensional aspects of ballistic penetration of thick metal targets, *International Journal of Impact Engineering*, 18(4), 369–381.

Zaera, R. and Sánchez-Gálvez, V. (1998), Analytical modelling of ballistic impact of normal and oblique ballistic impact on ceramic/metal lightweight armours, *International Journal of Impact Engineering*, 21(3), 133–148.

Zienkiewicz, O.C. and Taylor, R.L. (1989), *The Finite Element Method, 4th edition. Volume 1: Basic Formulations and Linear Problems*, McGraw-Hill, London.

Zohdi, T.I. (2002), Modeling and simulation of progressive penetration of multilayered ballistic fabric shielding, *Computational Mechanics*, 29, 61–67.

8

Modeling and simulation techniques for garments

S.-K. WONG, National Chiao Tung University, Taiwan

Abstract: The chapter begins by discussing the modeling and simulation techniques for garments. It reviews the three approaches for modeling garments: geometry based, physically based and hybrid. Computer graphics techniques and collision detection methods are then described for interactive garment simulation. To render garments realistically, they are discretized as polygons, with patterns repeated over the whole surface of the garments, and a bidirectional reflectance distribution function is then used for computing the light interaction with the surface of the garments. The future development of modeling and simulation of garment materials is discussed at the end of the chapter.

Key words: garment modeling, garment simulation, garment visualization.

8.1 Introduction

The simulation of garments is widely applied in the film and game industries, garment design, fashion design and virtual garment shopping malls. Garments cover a large percentage of a human body in general. In order to model humans to a satisfactory level, a crucial element is the realistic animation of garments. Cloth is characterized by its strong resistance to stretch but weak resistance to bending. Folds and wrinkles are the important characteristics of garments. They have an attractive visual appearance. The techniques of garment simulation should be stable and fast so that they can be applied in different environments, and interactive performance can be achieved.

To simulate garments, a proper modeling approach is adopted to model the structure of garments, either geometrical, physical or hybrid. If the final shape of garments is required without considering the dynamics process, geometrically based approaches may be employed due to their simplicity and the low cost of computation. On the other hand, physically based approaches may be employed for computing the mechanical behavior of garments. Collision detection and collision response should also be computed for handling the interaction between garments and other objects. After the shape of the garments is computed, the garments are rendered for visualization. In general, the garments are discretized into quads or triangles. Interactive or real-time applications are achieved for garment

173

models of moderate complexity on multicore platforms with graphics processing units.

8.2   Model development

Garment modeling techniques have been studied extensively over the last three decades to reproduce either the shape or the mechanical behavior of garments. In general, garment modeling techniques are classified into three categories (Ng, 1996): geometrical, physical and hybrid.

8.2.1   Geometrical techniques

In the geometrically based approach, mathematical functions are employed for describing the shape of garments either globally or locally that satisfies certain constraints (Decaudin *et al.*, 2005), for example fixed corners, hanging cloth and compressed cloth along the axial direction. Sampling points on the surfaces of cloth are then computed based on the constraints. These points are connected as either triangles or quadrilaterals. After that a relaxation process is carried out to refine the shape so as to meet the characteristics of garments.

Weil (1986) suggested a method for representing hanging cloth. The shape of cloth is computed by fitting catenary curves:

$$y = a\cosh\left(\frac{x}{a}\right) \qquad\qquad 8.1$$

where a is the scaling factor. A grid of points is computed over the surface formed by the curves and then they are relaxed according to the stiffness of the cloth. The stiffer the cloth, the less is the bending angle at a grid point. Finally, to obtain a finer mesh, splines are fitted to the grid points.

Agui *et al.* (1990) presented a method for modeling a sleeve on a bending arm. They represented the cloth as a hollow cylinder consisting of a series of circular rings. Folds are formed by displacing the rings along the axial direction. Hinds and McCartney (1990) obtained the shape of the human body by digitizing the upper torso of a mannequin. A garment is represented by 3D surfaces (panels). The panels can be edited by the garment designer. A fold is created by using a harmonic function superimposed on the panels. Ng and Grimsdale (1995) proposed a method for modeling folds within the gaps between the cloth and the skin layer. Sinusoidal functions are mapped along the fold lines near the bend region of the sleeve so as to generate folds. A set of rules was then developed to generate fold lines automatically.

Most of the geometrically based techniques compute the shape of garments based on some specific rules. In general such techniques are used for modeling the draped garments. The formulation does not take into account the dynamics interaction between garments and other objects. Hence, it is difficult to adopt geometrically based techniques for modeling interaction. The shape of garments would be affected by the other objects if the garments interact with them. Collision events should be detected and collision response should be integrated so as to model the reaction of garments for interaction. Recently, Stumpp *et al.* (2008) employed shape matching for modeling garments so that interaction can be computed. They introduced the inextensible fiber for connecting the adjacent triangles. Moreover, triangles are clustered for modeling the effect of bending. High stretching, shearing, and low bending resistance could be modeled. The deformed region can be restored back to its original shape. Thus, it can be used for modeling the interaction between cloth and other objects with arbitrary shapes.

Geometrically based techniques are efficient in computing the shape of garments. However, there are limitations in using geometrically based techniques. For example, it is difficult to incorporate dynamic constraints (Hong *et al.*, 2005) and friction and to represent local structures. Moreover, accuracy in modeling the mechanical properties of garments cannot be easily achieved.

8.2.2 Physically based methods

The mechanical properties of garments can be modeled by employing the physically based techniques. There are two approaches: particle-based and continuum. Particle-based methods are popular due to their low computational cost and easy implementation (Eberhardt *et al.*, 1996; House *et* al., 1996; Fan *et al.*, 1998). Many different numerical integration schemes can be used for solving the motion equations of particle systems (Press *et al.*, 1992). On the other hand, continuum mechanics are used for mapping authentic materials and performing accurate simulation of garments (Shanahan *et al.*, 1978; Choi and Ko, 1996; Hu and Chan, 1998; Teng *et al.*, 1999; Yu *et al.*, 2000; Chen and Hu, 2001). A set of partial differential equations are obtained and then solved by using finite difference methods (FDM) or finite element methods (FEM).

Feynman (1986) proposed an energy-based method for modeling draped cloth. The energy equation is obtained from the theory of elastic plates. The energy terms include elasticity, bending and gravitational energy. The energy of a point P(i, j) is computed by summing up these three kinds of energy:

$$E(\mathrm{P}(i, j)) = k_s E_{\mathrm{elast}}(i, j) + k_b E_{\mathrm{bend}}(i, j) + k_g E_{\mathrm{grav}}(i, j) \qquad 8.2$$

where k_s, k_b and k_g are elasticity, bending and density constants, respectively. The energy at point $P(i, j)$ is calculated based on the eight surrounding points. The steepest descent method is used in computing for the energy minima. A multigrid technique is adopted for efficiently computing the solution.

Terzopoulos and Fleischer (1988) introduced an approach for handling general deformable flexible objects, such as garments. Lagrange's equations of motion are used to obtain the motion equation of a point of an object. The curvature energy at a point is evaluated. To obtain the solutions, the motion equation is first discretized by a finite-difference or finite-element method. A large system of ordinary differential equations is required to be solved. Aono (1990) used elasticity theory and D'Alembert's principle to produce a wrinkle propagation model. The model included anisotropy and viscoelasticity.

Carignan et al. (1992) presented a method to compute garment animation. The model is based on the work of Terzopoulos and Fleischer (1988) and the curvature termed is replaced with a new energy term. An inelastic collision is considered for resolving the interaction between cloth and other objects.

Breen et al. (1992, 1994) used particles to model the draping behavior of woven cloth. This method treats the crossing points of the warp and weft threads as particles. There are two stages. Firstly, particles are allowed to fall freely. Any collision with the objects or the ground is determined during this step. Secondly, an energy minimization process is applied to the inter-particle energy function to generate the fine detail in the shape of the cloth. The total energy of the particle is defined as the sum of the repulsion energy among particles, the stretching energy, inter-thread bending out of plane, inter-thread bending in plane, and gravitational energy. The shape is then obtained after the energy minimum of cloth is computed.

Okabe et al. (1992) developed a system which processes 2D paper patterns for creating garments in 3D space. The 2D patterns are reassembled as a 3D garment and wrapped around the underlying body. An energy-based approach is then applied to compute the final shape of the cloth. To speed up the process, an adaptive coefficient gradient method is used to compute the energy minima.

Yang (1994) represented the cloth as a set of connected points and then used Newton's law of motion to describe the motion of points:

$$F(r, t) = mF(r)\frac{d^2 r}{dt^2}$$ 18.3

where F is the resultant force at the point r. The curvature viscosity force term is included for modeling the viscous nature of cloth. The internal force exerted on a point is computed based on 12 neighboring points.

Provot (1995) used a classic mass-spring model to simulate cloth under constraints. He classified the springs as flexion, structural and sheer springs, and used Newton's law of dynamics in the simulation. When simulating hanging cloth, unrealistic deformations would occur at the fixed points. To reduce this effect, the stiffness of springs should be increased, and a smaller time step is required for maintaining the stability of the system. It would take a longer time to compute the simulations as more steps are evaluated. He proposed an inverse kinetic approach to resolve the problem. The idea is that if any springs exceed the predetermined threshold of the stretch ratio, such as 15%, the particles of the springs are pulled towards each other along the axial direction.

Volino *et al.* (1995) used elasticity theory of an isotropic surface and Newtonian dynamics to simulate deformable objects, including cloth. They represented cloth as a set of particles arranged as vertices of triangles. Their model takes a two-pass approach. The first pass resolves interparticle constraints and external constraints. The second pass handles collisions by enforcing the law of conservation of momentum on the system.

Bridson *et al.* (2003) proposed a robust method for garments preserving wrinkles and folds while the garments are animated. It is important to preserve such features as designed by the animators so as to convey certain visual information. They developed a mixed explicit/implicit time integration scheme and a physically correct bending model with (potentially) nonzero rest angles for pre-shaping wrinkles. The fine details of features are gradually formed during animation.

8.2.3 Hybrid methods

Hybrid techniques combine geometrically based and physically based techniques. The rough shape of a garment is computed based on geometrically based techniques, and physically based methods are then employed to refine the final shape of the garment. Hybrid approaches allow animators to sketch the global shape of garments, and are computationally efficient.

Kunii and Gotoda (1990) proposed a system based on energy. The cloth is represented as a grid of points with springs connecting the points. There are two kinds of energy: metric and curvature, respectively. The shape of the cloth is obtained by using a gradient descent method to find the energy minima. After that, singularity theory is used to characterize the resulting wrinkles. A sinusoidal curve is used to construct the surface:

$$f(s) = x \sin\left(\frac{n\pi}{\lambda} s\right) \qquad\qquad 8.4$$

where x is a small constant, n is a natural number, λ is the length of the curve, s is the parameter for arc length ($0 \le s \le \lambda$), and $f(s)$ is the vertical displacement of the point s.

Taillefer (1991) extended Weil's method (Weil, 1986) and addressed hanging cloth. Assume P_1 and P_2 are two hanging points. The offset of each horizontal fold from the plane is computed by:

$$Z = Z_{max} \frac{\arctan\left(10\left(L - \frac{P_1 P_2}{L}\right)\right)}{\frac{\pi}{2}} \qquad 8.5$$

The vertical folds are modeled using a relaxation process with constraints including stretching, bending, gravitational and self-repulsion energies.

Tsopelas (1991) adopted a hybrid technique to model folds. Garments are treated as thin cylindrical tubes under axial loads. Garment folds are then refined by using thin-walled deformation theory. This process focuses on regions where folds are most likely to appear, such as at knees or seams. Dhande *et al.* (1993) used a swept surface to model draping effects of a fabric. They used a generatrix and a directrix curve to generate a biparametric surface. An elastic model includes fabric bending and shear properties are considered.

Decaudin *et al.* (2005) exploited geometric constraints to model the buckling features of garments. The wrinkles of garments are generated automatically after the input geometry of garments is processed. The new geometry of the garments can be flattened without distortion as they are piecewise developable. The animation of the garment can be obtained by changing the positions of the control points. Subsequently, the buckling mesh folds and unfolds. To incorporate dynamic effects, each control point is associated with a point mass. The control points are connected by three different springs based on the method of Provot (1995). The maximal elongation constraint is enforced by limiting the stretch ratio of the springs. A modified spring model is applied to the points so as to control the length of the springs.

Cordier *et al.* (2001) and Cordier and Magnenat-Thalmann (2002) described a method that makes use of predetermined conditions between the cloth and the model so as to simplify the method for handling complex collisions and computing dynamics of garments. Garments are partitioned into segments and animated according to their positions in relation to the model.

8.3   Computer graphics techniques for garment structure and appearance

In computer graphics, there are a variety of approaches for simulating garments. Some of the methods have been briefly reviewed in the previous section. Besides the modeling of garments, the interaction is considered

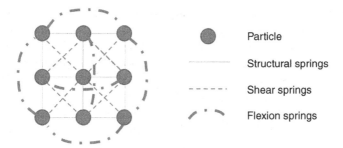

	Particle
	Structural springs
	Shear springs
	Flexion springs

8.1 The mesh model and three spring types in Provot's cloth model.

between garments and other objects. Collision detection (Lafleur *et al.*, 1991; Gottschalk and Lin, 1996; Klosowski *et al.*, 1998; Bergen, 1999; Vasiley *et* al., 2001; Zhang and Yuen, 2002; Wong and Baciu, 2005, 2007; Curtis *et al.*, 2008) and collision response (Volino *et al.*, 1995) are two other major issues when garments interact with themselves and other objects. After the motion of garments is resolved, garments are rendered.

Provot (1995) proposed a mass-spring network for modeling cloth. The cloth is modeled as a grid of points and the points are connected with three different springs: flexion, structural and bending. The model is illustrated in Fig. 8.1. The structural springs connect the neighboring particles in horizontal and vertical directions, the shear springs connect the particles in the diagonal direction, and the flexion springs connect the particles in an interleaving pattern in horizontal and vertical directions. In every simulation time-step, the total force exerted on each particle is computed. The total force is the sum of the spring force, the gravity force, the air resistance force, the viscous damping force and the force exerted by fluid. An explicit Euler integration method is then employed for solving the motion of the points. However, the stretch ratio of springs may exceed 100% if their stiffness is too low. On the other hand, the motion of cloth is unstable for higher stiffness of springs. To tackle the problem a smaller time-step size should be used but the computation cost would be high. In order to handle this problem, an inverse kinetic approach was proposed for pulling back points along springs if the stretch ratio exceeds 15%. In order to handle interaction between cloth and other objects, continuous collision detection (Provot, 1997) is employed so as to detect geometry collisions. For each triangle pair, there are 15 feature pairs including nine edge–edge pairs and six point–triangle pairs. A cubic equation is solved for each feature pair for the time when the feature pair becomes coplanar. The shortest distance between the feature pair is then computed for checking proximity. If the distance of the feature pair is less than or equal to a predefined threshold, the feature pair collides. Subsequently, the velocities of the colliding particle pairs are modified. To model friction, the velocities of particles are modified based on Coulomb friction laws. After modifying the velocities of particles, collision detection is performed again. There may be secondary collisions where cloth may penetrate itself. To tackle this problem, Provot introduced the concept of rigid impact zones by grouping unresolved colliding pairs. The rigid impact zone is treated as a rigid body. Since then, many methods have adopted rigid impact zones for cloth simulation.

Baraff and Witkin (1998) proposed a particle-based system for simulating cloth using a second-order semi-implicit integration method. The method performs stably with a large time-step compared to those using the explicit Euler integration method. The computation is simple as it requires only the Jacobian of the particle forces. The method results in a linear system with a large sparse matrix. It is then solved by using a modified conjugate gradient method. Collision constraints are integrated into the integration system so that constrained particles move stably towards their desired positions. However, there is an unwanted damped effect due to the intrinsic properties of the implicit integration methods.

Kang et al. (2000) proposed a method to approximately compute the inverse of the matrix of the linear system for fast and stable simulation of cloth. However, the cloth hardly moves due to unwanted damping. Wrinkles and folds are not formed naturally. Volino and Magnenat-Thalmann (2000a) proposed a simple and fast cloth simulation based on the work of Baraff and Witkin (1998). The method does not have an explicit storage of the sparse matrix. It simplifies the conjugate gradient process. Although the implicit Euler step brings stability, it introduces unwanted damping into the system. They therefore proposed a method to recover the unexpected movements caused by the unwanted damping forces.

Bridson et al. (2002) proposed a robust method for handling collisions in cloth simulation. There are three stages: (1) repulsion force, (2) continuous collision response, and (3) rigid impact zones. Inelastic repulsion impulse is applied if primitives are approaching each other. A spring-based repulsion force is applied if the interacting primitive pairs are separating. Interacting pairs are pushed away from each other in the first two stages. The process is performed iteratively a certain number of times so as to reduce the number of colliding pairs. To resolve the remaining colliding pairs, the authors reformulate the approach for handling rigid impact zones (Provot, 1997) by grouping the unresolved colliding pairs. Each rigid impact zone is treated as a rigid body. The motion of the colliding pairs in a rigid impact zone is computed based on the motion of the rigid impact zone.

Choi and Ko (2001) developed a particle-based system in which there are three different types of springs connecting the particles. The method improves the realism of wrinkles of cloth when the implicit integration methods are adopted. It resolves the post-buckling instability as well as the numerical instability without introducing any fictitious damping. The motion of the cloth is computed based on a semi-implicit integration method with a second-order backward difference formula (BDF). Let $\Delta^{-1}x = x^{n+1} - x^n$. Then the kth order BDF is given by

$$\frac{d}{dt} = \frac{1}{\Delta t} \sum_{q=1}^{k} \frac{1}{q} (\Delta^{-1})^a \qquad 8.6$$

Let $k = 2$. Then $\dot{x}$ is given by

$$\dot{x} = \frac{1}{\Delta t} \left(\frac{3}{2} x^{n+1} - 2x^n + \frac{1}{2} x^{n-1} \right) \qquad 8.7$$

The state equation of motion can be described as

$$\frac{1}{\Delta t} \begin{pmatrix} \frac{3}{2} x^{n+1} - 2x^n + \frac{1}{2} x^{n-1} \\ \frac{3}{2} v^{n+1} - 2v^n + \frac{1}{2} v^{n-1} \end{pmatrix} = \begin{pmatrix} v^{n+1} \\ M^{-1} f^{n+1} \end{pmatrix} \qquad 8.8$$

The term f^{n+1} is computed as

$$f^{n+1} = f^n + \frac{\partial f}{\partial x}(x^{n+1} - x^n) + \frac{\partial f}{\partial v}(v^{n+1} - v^n) \qquad 8.9$$

Finally, a linear system equation for $\Delta^{-1}x$ is obtained:

$$\begin{aligned} &\left(1 - \Delta t \frac{2}{3} M^{-1} \frac{\partial f}{\partial v} - \Delta t^2 \frac{4}{9} M^{-1} \frac{\partial f}{\partial x} \right)(x^{n+1} - x^n) \\ &= \frac{1}{3}(x^n - x^{n-1}) + \frac{\Delta t}{9}(8v^n - 2v^{n+1}) \\ &+ \frac{4\Delta t^2}{9} M^{-1} \left(f^n + \frac{\partial f}{\partial v} v^n \right) - \frac{2\Delta t}{9} M^{-1} \frac{\partial f}{\partial v}(x^n - x^{n-1}) \end{aligned} \qquad 8.10$$

The linear system is sparse and unbanded. It is solved by using a preconditioned conjugate gradient method. A voxel-based collision detection algorithm which is similar to the method proposed by Zhang and Yuen (2000) is then employed for resolving collision events. The idea is that the space enclosing the cloth is voxelized and the cloth particles are registered to voxels and their neighboring voxels. To detect cloth–object collision events, the cloth particles are determined whether or not they are lying beneath the surface of the solid objects. If there is collision, a constraint is added for the involved particles. To detect self-collision events, the particle–particle pairs are checked for proximity. If there is collision, a repulsive force between them is applied.

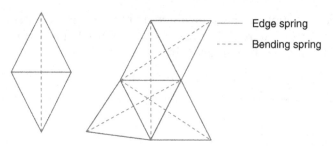

8.2 The constitutive model of cloth used in the work of Selle *et al.*
(2009).

Selle *et al.* (2009) proposed a history-based framework that can simulate cloth consisting of up to two million triangles. They adopted a mass-spring constitutive model. The arrangement of particles of the cloth model is irregular (see Fig. 8.2). There are two types of springs: edge springs and bending springs. Edge springs connect particles connected by edges, and bending springs connect the unshared particles of adjacent triangles. Hence, the model of Selle *et al.* can handle arbitrary cloth meshes. The force for a spring is given by

$$F = \left[E\left(\frac{\|u(t)\|}{\|u(0)\|} - 1 \right) + d(V_1(t) - V_2(t))^{\mathrm{T}} \frac{u(t)}{\|u(t)\|} \right] \frac{u(t)}{\|u(t)\|} \qquad 8.11$$

where $u(t) = X_2(t) - X_1(t)$, E is the Young's modulus, d is the damping parameter and $V(t)$ is the velocity of a particle.

In the method of Selle *et al.*, they employ impulse and repulsion for eliminating the number of colliding pairs. The numerical computation and collisions are handled based on distributed memory parallelism using message passing. Assume that Δt is the time step, v is the velocity of particles, $a(t, x, v)$ is the acceleration and $x'^{n+\frac{1}{2}} = (x'^n + x'^{n+1})/2$. The algorithm is outlined as follows:

A. Compute repulsion pairs and their orientation history.
B. Perform k_i time integrations (inner loop):
    1. $v'^{n+\frac{1}{2}} = v^n + \Delta t/2a\left(t^{n+\frac{1}{2}}, x^n, x^{n+\frac{1}{2}}\right)$
    2. Modify $v'^{n+\frac{1}{2}}$ with elastic and inelastic self-repulsion
    3. $v'^{n+1} = x^n + \Delta t v'^{n+\frac{1}{2}}$
    4. Collide with body objects to obtain x^{n+1} and v^n*
    5. $v^{n+\frac{1}{2}} = v^n* + \Delta t/2a\left(t^{n+\frac{1}{2}}, x^{n+\frac{1}{2}}, v^{n+\frac{1}{2}}\right)$
    6. Extrapolate $v'^{n+1} = 2v^{n+\frac{1}{2}} - v^n*$
    7. Modify v'^{n+1} to v^{n+1} for friction with objects
    8. Modify v^{n+1} with friction and inelastic self-repulsion.
C. Detect and resolve moving collisions.

In step A, the repulsion pairs are collected. In step B, an iteration process is performed k_i times and consists of eight steps. Step B.1 is a backward Euler method for computing the velocity $v'^{n+\frac{1}{2}}$ at the half-step and then $v'^{n+\frac{1}{2}}$ is modified with elastic and inelastic self-repulsion in step B.2. The cloth positions are advanced in step B.3. The cloth–object collision algorithm is applied for handling interaction between cloth and body objects in step B.4. The velocity $v^{n+\frac{1}{2}}$ at the half-step is computed by employing a backward Euler method in step B.5 and then an extrapolation method is applied to compute v'^{n+1} in step B.6. Friction and inelastic self-repulsion are handled in the last two steps B.7 and B.8. Finally, collisions are completely resolved in step C.

The inelastic repulsion used in step B.2 is computed as $l_c = mv_N/2$, where v_N is the normal velocity, and m is the total mass of the involved particles. The elastic repulsion is computed as $I_r = -\min\left(\Delta t\, kd,\, m\left(\dfrac{0.1d}{\Delta t} - v_N\right)\right)$, where k is the spring stiffness, d is function of the repulsion thickness, and Δt is the simulation time-step.

Volino *et al.* (2009) proposed a method to employ the non-linearized Green–Lagrange tensor for garment simulation in which adapted strain–stress laws are used for describing the garment behavior. The method is simple and accurate for modeling the nonlinear anisotropic characteristic of garments under large deformations. It avoids intermediate computations, such as coordinate transformations.

8.4   Rendering of garment appearance and model demonstration for garments

This section presents the approaches for modeling garment appearance and then presents four examples for garment simulation. To render garments realistically, they are discretized as polygons, patterns are repeated over the whole surface of garments, and a bidirectional reflectance distribution function is then used for computing the interaction of light with the surface of garments.

8.4.1 Rendering of garment appearance

Weil (1986) modeled garments as a grid of points. Splines are fitted with the points. After that the garments are treated as a collection of line segments (threads). Ray tracing techniques are developed for rendering each thread of garments individually. The method is expensive but it preserves the fine detail of garments. Each line is treated as a shape with thickness and depth. The threads are cylindrical in shape. If the distance between a ray and the line segment is smaller than a given threshold, the ray intersects the thread. Then shading information is computed for the corresponding pixels. The exact distance between the ray and the line segment is not computed. An approximated value for the horizontal or vertical distance from the ray to the line is computed for speeding up the process. Consider the case in which the scene is normalized to a rectangular view volume. Rays are cast perpendicular to the viewing plane into the scene. There are two cases for computing the distance between a ray at location (x, y) of the image plane and a line segment from (x_1, y_1, z_1) to (x_2, y_2, z_2). Let

$$dx = (x_2 - x_1)$$
$$dy = (y_2 - y_1)$$
$$dz = (z_2 - z_1)$$

Case 1: $|dy| > |dx|$

$$t = \frac{y - y_1}{dy}$$

Then the distance d(ray, lineseg) is computed as $x - (x_1 + tdx)$ and the z-intersection is $z_1 + tdz$.

Case 2: $|dy| \leq |dx|$

$$t = \frac{x - x_1}{dx}$$

Then the distance d(ray, lineseg) is computed as $y - (y_1 + tdy)$ and the z-intersection is $z_1 + tdz$. If the distance d(ray, lineseg) is smaller than or equal to a given threshold, the distance is normalized to the interval $[-1, 1]$. There are three normal vectors that are used for computing the final shade. The first two normal vectors are determined by projecting the line segment onto the $z = 0$ plane and are the two opposing vectors perpendicular to the projected line segment, i.e. $(-dy, dx, 0)$ and $(dy, -dx, 0)$. The third normal vector is computed as the cross-product of the first two normal vectors (see Fig. 8.3) and is perpendicular to the thread. A shade value is computed for each normal vector. The final shade is then computed by interpolating these three shade values.

Daubert et al. (2001) presented an efficient method for garment visualization. The presentation of the detail of garments is based on a bidirectional reflectance distribution function (BRDF). The garment is point sampled

8.3 The three normal vectors used for computing three shade values for a thread.

into a 2D array. The fabric pattern is repeated across the whole garment geometry.

The spatially varying BRDF representation is adopted for capturing the material properties. A 2D array is used for storing the samples in two spatial dimensions and the Lafortune reflection model is employed for modeling the reflection and lighting effect. The parameters for the Lafortune reflection model and the normal and tangent vectors are stored for each entry. The BRDF $f_r(\vec{l}, \vec{v})$ of an entry is computed as

$$f_r(\vec{l}, \vec{v}) = T(\vec{v}) \bullet f_l(\vec{l}, \vec{v})$$
8.12

where $\vec{l}$ is the light direction, $\vec{v}$ is the viewing direction, $f_l(\vec{l}, \vec{v})$ is the Lafortune model and $T(\vec{v})$ is the lookup table. The lookup table $T(\vec{v})$ stores color and alpha values for a set of specific viewing directions. For an arbitrary viewing direction $\vec{v}$ not stored in the lookup table, the color and alpha values are assembled by using interpolation. The Lafortune model $f_l(\vec{l}, \vec{v})$ is computed as

$$f_l(\vec{l}, \vec{v}) = l_z\left(\rho + \sum_i \left[(l'_x, l'_y, l'_z)\begin{pmatrix} C_{x_i} & 0 & 0 \\ 0 & C_{y_i} & 0 \\ 0 & 0 & C_{z_i} \end{pmatrix}\begin{pmatrix} v'_x \\ v'_y \\ v'_z \end{pmatrix}\right]^{N_i}\right)$$
8.13

where ρ is the diffuse part, and C_x, C_y, C_z and N are parameters of each lobe. Let S be the local coordinate system defined by the average of the normal vectors and tangent vectors of the sample points. Then the position (l'_x, l'_y, l'_z) and the direction $\begin{pmatrix} v'_x \\ v'_y \\ v'_z \end{pmatrix}$ are the light position and viewing direction in the coordinate system S, respectively. Transparency is modeled by using the alpha value. The data acquisition process of the method of

Daubert *et al.* is summarized in Algorithm One. Mainly, the normal vectors, alpha values and radiance are computed in the data acquisition process. After that an optimization process is performed for each entry of the 2D array of BRDFs so as to compute the parameters for the Lafortune model.

Algorithm One: Data acquisition
```
for each v⃗ do
  Compute sampling points
  RepeatScene(vertex color = normal vectors)
  Store normal vectors
  Store alpha
  for each l⃗ do
    Compute lighting
    RepeatScene(vertex color = lighting)
    Store radiance
  end for
end for
Average normal vectors
```

The method developed by Daubert *et al.* (2001) improves the performance further by using the capability of graphics hardware. For example, normal vectors and lighting are coded as color values. Their rendering algorithm has four stages:

1. Compute the normal vectors of pixels.
2. Compute a BRDF f_r for evaluating the color for each pixel.
3. Compute f_r with light and viewing direction in the coordinate system.
4. Store the result to the image buffer.

In step 1, the normal of each vertex is color coded for estimating the normal of a pixel. This is done by performing Gouraud shading. In this way, the normal is coded as the color of the pixel in framebuffer. A certain BRDF is computed for each pixel in step 2. Step 3 maps the light and viewing direction into the geometry's local coordinate system. There are two mappings: the first maps the world view and light to the cloth geometry's local coordinate system, and the second transforms these values to the pattern's local coordinate system when evaluating the BRDF. After that in step 4, red, green, blue and alpha values are written to the image buffer, such as framebuffer. The OpenGL mip-mapping technique is used for creating different levels of mip-maps so as to tackle the aliasing artifacts and improve the rendering process.

Durupınar and Güdükbay (2007) proposed a procedural method for the visualization of knitted and woven fabrics. The method is view-dependent for changing the level of detail as well as improving the performance. If the distance is far, texture-mapped and antialiased line primitives are drawn. The line size is proportional to the proximity of the viewer to the garments. For intermediate distance ranges, the knitwear is generated with line primitives and with cylinders separately. These knitwear images are blended to produce the final image of the cloth. The blending weights of the two images are determined according to the distance of the viewer to the garments.

8.4.2 Model demonstration

Some snapshots of garment simulation are demonstrated. A piece of cloth interacts with an object with a sharp feature in Fig. 8.4. A cloth interacts with a solid sphere in Fig. 8.5. Figure 8.6 shows a square cloth interacting with a deformable object under the effect of a wind drag force. Three layers of garments and a garment with long sleeves are shown in Figs. 8.7 and 8.8, respectively. Finally, Fig. 8.9 shows a cloth interacting with a spinning ball and many self-collisions occur.

8.4 A square cloth and an object with sharp features.

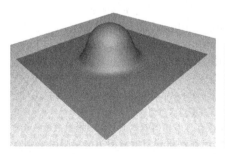 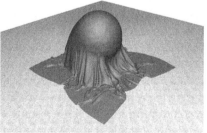

8.5 A square cloth and a ball.

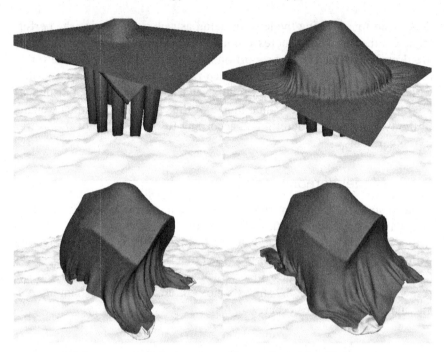

8.6 A square cloth interacting with a deformable object.

8.5   Considerations for real-time applications

A real-time application requires high-performance computation and seeks a balance between realism and speed. A simulation system of garments consists of several components: garment modeling, motion integration computation, collision detection, collision response and visualization. The computation cost of the components should be balanced in order to maximize the performance. The following items should be considered: number of particles and number of elements, modeling methods, integration methods, collision detection, collision response and visualization methods.

The higher the number of particles or the number of elements, the higher is the cost in performing the numerical integration, collision detection and visualization. According to the Courant condition (Provot, 1995), a smaller time-step is required for a smaller size of elements if physically based techniques are used. One way to tackle this problem is to use meshes of low resolution and then adaptively increase the resolution when necessary (Zhang and Yuen, 2001). If geometrically based techniques are adopted, the cost in the relaxation process would be more expensive for meshes of higher resolution. To render cloth, primitives (lines, triangles or quadrilaterals) are sent onto graphics processing units for processing. It takes longer to render a higher number of primitives.

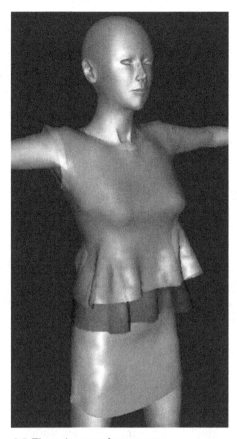

8.7 Three layers of garments.

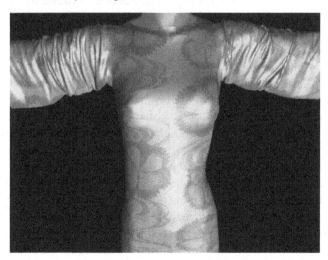

8.8 A garment with long sleeves.

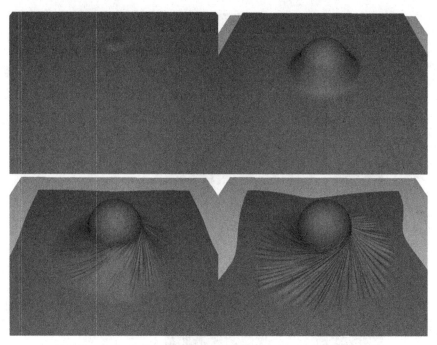

8.9 A square cloth and a spinning ball.

The modeling methods affect the choice of integration methods. For geometrically based techniques, the cost in computing the solution is fast. If physically based methods are employed, the integration methods should be stable and fast. Furthermore, if there is interaction between the garments and the other objects, the integration methods should be integrated with the collision constraints.

Collision detection should be employed for modeling the interaction between garments and other objects. Accurate collision detection is expensive for meshes with high resolutions, such as when using continuous collision detection. Collision detection is reported to contribute over 50% of the total cost. If exact interaction is not required, approximate collision detection can be adopted. Moreover, adaptive methods may be applied for adaptively increasing the number of particles or the number of elements for modeling garments. Not only is the computation cost of motion integration reduced but also the cost of collision detection. Alternatively, if at the preprocessing phase, collision regions are known beforehand, then during simulation, collision detection is carried out only for the predefined set of regions. Collision constraints are enforced for colliding geometry. Such collision constraints may be integrated into the motion equation of the system so that the stable collision response can be computed.

After the motion of garments is computed, the garments are rendered for visualization. We may trade efficiency against realism. Texturing mapping or predefined layers of maps, such as a normal map, a displacement map and a BRDF (bidirectional reflectance distribution function) map can be used on the graphics processing units. If ray tracing is required, parallel processing can be applied on multicore platforms with the acceleration of graphics processing units (GPUs). Parallel processing can be carried out for numerical integration and collision detection such that real-time performance can be achieved for simulating complex garments.

8.6   Advanced modeling techniques

This section presents some advanced methods in modeling garments and then discusses the future development of simulating garment materials.

Thomaszewski *et al.* (2006) proposed a method to model folds and wrinkles of cloth with different types and resolutions. It is based on a corotational formulation of subdivision finite elements. This formation allows large deformation of cloth while retaining a linear system. The integration method results in a linear system of equations. Their method performs continuum-based strain limiting so as to prevent elements from too much elongation (see Algorithm Two). It is a post-processing step for computing corrected positions of vertices based on strain limiting and collision handling. For example, consider an element whose displacement vector u^s with a strain tensor $\varepsilon(u^s)$ violates a deformation limit. Then a corrected velocity v^s and strain tensor $\varepsilon(u^s + \Delta t\, v^s)$ should be computed.

Algorithm Two: Continuum-based strain limiting
while simulation notdone do
  1. Compute state $(x_0,\ v_0)$ at time t
  2. Compute candidate state $(x_n^{cand},\ x_n^{cand})$
  3. Compute v_n^{corr} based on strain limiting
  4. Compute new position $x_{\downarrow}n^{\uparrow}corr = x_{\downarrow}0 + (tv_{\downarrow}n^{\downarrow}corr)$
  5. Modify $(x_n^{corr},\ v_n^{corr})$ for handling collisions
  6. Compute the final velocity v_n by limiting strain
     rate
end while

Cloth does not stretch much under its own weight in a normal situation. Goldenthal *et al.* (2007) employed constrained Lagrangian mechanics and a novel fast projection method for modeling inextensible cloth. It simulates cloth with low strain along the warp and weft directions. The method acts as a velocity filter that is easily integrated into an existing simulation system.

Ye (2008) developed a method to apply an impulse for constraining the maximum length of springs for particle-based simulation of garments.

Simulating cloth realistically requires robust handling of collision events, particularly in a complex environment. Harmon *et al.* (2008) proposed a method for handling complex collisions in cloth simulation so as to reduce artificial dissipation. They developed a fail-safe method with an approximation of Coulomb friction that allowed the control of sliding dissipation. The proposed method effectively handles collisions occurring at the same time interval. An inelastic projection step is proposed for reducing the number of rigid impact zones. The method consists of three stages: repulsion forces, continuous collision response (CCR), and rigid impact zones (RIZ). Algorithm Three shows the major steps as follows.

> **Algorithm Three: Collision Response Algorithm**
> 1. Detect potentially colliding pairs
> 2. Apply repulsion forces to colliding pairs
> 3. **while** there are new collisions **do**
> 4. Insert constraint into its own impact zone (IZ)
> 5. Merge all new and existing IZs that share vertices
> 6. **for each** IZ **do**
> 7.   Reset IZ to pre-response velocities
> 8.   Apply inelastic projection
> 9.   Apply friction
> 10. **end for**
> 11. **end while**

8.7   Future developments in simulating garment materials

Garments are made of a nonlinear materials. With the development of new fabrics, new methods or parameters are required for accurate simulation of garments. One of the challenging problems is the simulation of multilayered garments. An accurate method for modeling the large contact regions of garments is difficult and they have a lot of colliding pairs. There are several future directions for improving techniques: modeling methods, visualization, continuous collision detection (Redon *et al.*, 2004; Tang *et al.*, 2008) and collision response (Volino *et al.*, 1995). Advanced methods should be developed for handling self-collision detection, as this incurs a significant amount of computation cost in collision detection (Volino and Magnenat-Thalmann, 1994). Although physically based techniques can be greatly accelerated by modern processors, they are costly. There is still a long way to go to build a system that can perform in real time on conventional computers, so that ordinary users can use the system for performing complex simulation and visualization in real time.

Elasticity theory considers macroscopic behavior but is still not good enough to model the mechanical behavior of garments. The mechanical properties of threads or fibers should be well studied for authentically reproducing the mechanical behavior of garments. However, a model large enough to represent every thread would require excessive memory and expensive computation. A unified approach should be developed to blend the two aspects of macroscopic and microscopic characteristics of garments.

Visualization of garments (Xu *et al.*, 2001) is also important for end-users, such as garment designers and customers. Fashion designers can design garments and visualize their appearance before they are manufactured. Customers can view virtual garments on Internet browsers or dedicated software without actually visiting the shops. Thus customers can try garments and purchase them more easily. Moreover, customers can sketch their garments (Turquin *et al.*, 2004) and the system can then compute a realistic simulation of such garments. The demand for the development of sketching techniques is increasing. The animation system lets designers and end-users try out virtual garments easily. Customers can see the wrinkles and folds for their made-to-measure virtual characters.

High-performance parallel computation techniques (Thomaszewski and Blochinger, 2007; Thomaszewski *et al.*, 2008) should be developed for modeling, numerical integration, collision detection (Kim *et al.*, 2009), collision response and visualization. Parallel computation can be performed not only on distributed PC clusters but also on multicore platforms and on platforms with multiple graphics processing units. Simultaneous multi-threading techniques can be implemented on multicore platforms so that each thread is executed on a core. Moreover, hundreds of threads can be executed at the same time on an advanced graphics processing unit at a low price. A lot of effort has been invested in parallel computation techniques on multicore platforms with graphics processing units. In the future, the realistic simulation of complex multilayered garments and complex interaction between garments and other models can be achieved. Ordinary users can afford to buy the required equipment.

8.8   Conclusions and sources of further information and advice

We have reviewed several different aspects of garment simulation. In general there are three categories of techniques: geometrically based, physically based and hybrid. To model garments interacting with objects, collision detection and collision response should be handled. In some complicated simulation environments, there are many collision events occurring at the same time. Robust methods for handling them are essential for animating the garments. Moreover, parallel computation can be carried out for improving the simulation performance on multicore platforms with graphics processing units (GPUs).

We list further information based on the work of Jiang *et al.* (2008) on cloth simulation for applications and software.

8.8.1 Garment CAD/CAM

Fashion designers can design garments with different patterns and fabrics easily on an interactive garment design system before the real garments are produced. Garments can be dressed on an animated character so as to visualize the garments. Not only are the real materials saved, but the production time is reduced significantly. On the other hand, customers can try garments on a virtual try-on system before they purchase them. Customers can easily select their favorite garments and combination.

8.8.2 Game engines

Virtual actors play an important role in games. Tools are developed so that different styles of garments can be designed intuitively. The realistic animation of garments greatly improves the gaming experience for the players. The methods should be fast so as to maintain the real-time performance in a game.

8.8.3 E-commerce

The purchasing patterns of people are changing due to the popularity of web-based browsers. People are likely to order garments via the Internet as it provides great convenience. Visitors can choose their garments or even patterns to make their own garments (made to measure) on Internet browsers. After they select the garments, real-time simulation is performed for evaluating them. At the same time, the visitors can change parameters to suit their body measurements.

8.8.4 Computer animations and fashion shows

Visual realism is important and a balance between accuracy and performance should be considered. Particularly, garments should be controllable so that certain kinds of shapes or animations can be produced.

8.8.5 The production of cloth simulation

MIRACloth Software (Volino and Magnenat-Thalmann, 2000b) is a garment simulation software package developed at MlRALab, University of Geneva. There are two major components: 2D design and 3D simulation system. Panels can be edited in the 2D design software and then assembled in the simulation system.

My Virtual Model Inc. has two core technologies – My Virtual Model™ Dressing Room and My Fit – that enable consumers to 'try on' clothes on the Internet. Users can manage their virtual wardrobe and post the information on their personal homepage. It is available at http://www.mvm.com/cs/ (accessed 31 March 2010).

NVIDIA PhysX can compute the motion of a variety of interacting objects, such as rigid bodies, deformable objects, cloth, and breakable objects. It adopts the most advanced physics simulation techniques and delivers realistic physics simulation. PhysX SDK is free to download. It is available at http://developer.nvidia.com/object/physx_downloads.html (accessed 31 March 2010).

Havok Cloth is a user-friendly development tool with which game artists can animate character garments and clothing environments. It is performance-optimized so that high performance can be achieved on the latest hardware. It maximizes the productivity of artists, animators and programmers. It is available at http://www.havok.com/index.php?page=havok-cloth (accessed 31 March 2010).

8.9   References

Agui T, Nagao Y and Nakajma M (1990), An expression method of cylindrical cloth objects – an expression of folds of a sleeve using computer graphics, *Transactions of the Society of Electronics, Information and Communications* (in Japanese), 23, 1095–1097.

Aono M (1990), A Wrinkle Propagation Model for Cloth, *CG International'90: Computer Graphics Around the World*, 95–115.

Baraff D and Witkin A (1998), Large steps in cloth simulation, *SIGGRAPH*, 43–54.

Bergen G (1999), Efficient collision detection of complex deformable models using AABB trees, *Journal of Graphics Tools*, 2, 1–14.

Breen D, House D and Getto P (1992), A physically-based particle model of woven cloth, *The Visual Computer*, 8, 264–277.

Breen D, House D and Wozny M (1994), A particle-based model for simulating the draping behavior of woven cloth, *Textile Research Journal*, 64(11), 663–685.

Bridson R, Fedkiw R, and Anderson J (2002), Robust treatment of collisions, contact and friction for cloth animation, *ACM Transactions on Graphics*, 21(3), 594–603.

Bridson R, Marino S, and Fedkiw R (2003), Simulation of clothing with folds and wrinkles, *ACM SIGGRAPH/Eurographics Symposium on Computer Animation (SCA 2003)*, 28–36.

Carignan M, Yang Y, Magnenat-Thalmann N and Thalmann D (1992), Dressing animated synthetic actors with complex deformable clothes, *SIGGRAPH*, 26(2), 99–104.

Chen S F and Hu J (2001), A finite-volume method for contact drape simulation of woven fabrics and garments, *Finite Elements in Analysis and Design*, 37, 513–531.

Choi K and Ko H (1996), Finite-element modeling and control of flexible fabric parts, *IEEE Computer Graphics and Applications*, 16(5), 71–80.

Choi K J and Ko H S (2001), Stable but responsive cloth, *ACM Transactions on Graphics*, 21(3), 604–611.

Cordier F and Magnenat-Thalmann N (2002), Real-time animation of dressed virtual humans, *Computer Graphics Forum*, 21, 327–335.

Cordier F, Volino P and Magnenat-Thalmann N (2001), Integrating deformations between bodies and clothes, *Journal of Visualization and Computer Animation*, 12(1), 45–53.

Curtis S, Tamstorf R and Manocha D (2008), Fast collision detection for deformable models using representative-triangles, in *Proceedings of the Symposium on Interactive 3D Graphics*, 61–69.

Daubert K, Lensch H, Heidrich W and Seidel H (2001), Efficient cloth modeling and rendering, in *Proceedings of the 12th Eurographics Workshop on Rendering Techniques*, 63–70.

Decaudin P, Thomaszewski B and Cani M P (2005), Virtual garments based on geometric features of fabric buckling, INRIA.

Dhande S G, Rao P M, Tavakkoli S and Moore C L (1993), Geometric modeling of draped fabric surfaces, in *Proceedings ot the IFIP CSI International Conference on Computer Graphics*, 349–356.

Durupinar F and Güdükbay U (2007), Procedural visualization of knitwear and woven cloth, *Computers & Graphics*, 31(5), 778–783.

Eberhardt B, Weber A and Strasser W (1996), A fast, flexible, particle-system model for cloth draping, *IEEE Computer Graphics and Applications*, 16(5), 52–59.

Fan J, Wang Q, Chen S, Yuen M and Chan C (1998), A spring-mass model-based approach for warping cloth patterns on 3D objects, *Journal of Visualization and Computer Animation*, 9(4), 215–227.

Feynman C (1986), *Modeling the Appearance of Cloth*, Master Thesis, Deptartment of EECS, Massachusetts Institute of Technology, Cambridge, MA.

Goldenthal R, Harmon D, Fattal R, Bercovier M and Grinspun E (2007), Efficient simulation of inextensible cloth, *ACM Transactions on Graphics*, 26(3).

Gottschalk S and Lin M (1996), OBBTree: A hierarchical structure for rapid interference detection, *SIGGRAPH*, 171–180.

Harmon D, Vouga E, Tamstorf R and Grinspun E (2008), Robust treatment of simultaneous collisions, *ACM Transactions on Graphics (SIGGRAPH)*, 27(3).

Hinds B K and McCartney J (1990), Interactive garment design, *The Visual Computer*, 6(2), 53–61.

Hong M, Choi M, Jung S and Welch S (2005), Effective constrained dynamic simulation using implicit constraint enforcement, *IEEE International Conference on Robotics and Automation*, 4520–4525.

House D, DeVaul R and Breen D (1996), Towards simulating cloth dynamics using interacting particles, *International Journal of Clothing Science and Technology*, 8(3), 75–94.

Hu J and Chan Y-F (1998), Effect of fabric mechanical properties on drape, *Textile Research Journal*, 68, 57–64.

Jiang Y, Wang R and Liu Z (2008), A survey of cloth simulation and applications, *Ninth International Conference on Computer-Aided Industrial Design and Conceptual Design*, 765–769.

Kang Y, Choi J, Cho H and Park C (2000), Fast and stable animation of cloth with an approximated implicit method, *Proceedings of the Conference on Computer Graphics International*, 247–256.

Kim D, Heo J P, Huh J, Kim J and Yoon S E (2009), HPCCD: Hybrid parallel continuous collision detection, *Computer Graphics Forum (Pacific Graphics)*, 28(7), 1791–1800.

Klosowski J, Held M, Mitchell J, Sowizral H and Zikan K (1998), Efficient collision detection using bounding volume hierarchies of k-DOPs, *IEEE Transactions on Visualization and Computer Graphics*, 4(1), 21–36.

Kunii T and Gotoda H (1990), Singularity theoretical modeling and animation of garment wrinkle formation processes, *The Visual Computer*, 6(6), 326–336.

Lafleur B, Magnenat-Thalmann N and Thalmann D (1991), Cloth animation with self-collision detection, *IFIP Conference on Modeling in Computer Graphics Proceedings*, 179–197.

Ng H (1996), *Techniques for Modeling and Visualization of Cloth*. PhD thesis. University of Sussex, UK.

Ng H and Grimsdale R (1995), GEOFF – A geometrical editor for fold formation, *Lecture Notes in Computer Science*, 1024, 124–131.

Okabe H, Imaoka H, Tomiha T and Niwaya H (1992), Three dimensional apparel CAD system, *Computer Graphics*, 26(2), 105–110.

Press W, Teukolsky S, Vetterling W and Flannery B (1992), *Numerical Recipes in C: The Art of Scientific Computing*, second edition, Cambridge University Press, Cambridge, UK.

Provot X (1995), Deformation constraints in a mass-spring model to describe rigid cloth behaviour, in *Proceedings of Graphics Interface '95*, 147–154.

Provot X (1997), Collision and self-collision handling in cloth model dedicated to garment design, in *Proceedings of Graphics Interface '97, Computer Animation and Simulation*, 177–189.

Redon S, Lin M, Manocha D and Kim Y (2004), Fast continuous collision detection for articulated models, *ACM Symposium on Solid and Physical Modeling*, 145–156.

Selle A, Su J, Irving G and Fedkiw R (2009), Robust high-resolution cloth using parallelism, history-based collisions, and accurate friction, *IEEE Transactions on Visualization and Computer Graphics*, 15(2), 339–350.

Shanahan W, Lloyd D and Hearle J (1978), Characterizing the elastic behavior of textile fabrics in complex deformations, *Textile Research Journal*, 48(9), 495–505.

Stumpp T, Spillmann J, Becker M and Teschner M (2008), A geometric deformation model for stable cloth simulation, in *Proceedings of the Conference on Virtual Reality Interactions and Physical Simulations*, 39–46.

Taillefer F (1991), Mixed modeling, in *Proceedings of Compugraphics '91*, 467–478.

Tang M, Curtis S, Yoon S and Manocha D (2008), Interactive continuous collision detection between deformable models using connectivity-based culling, in *Proceedings of the 2008 ACM Symposium on Solid and Physical Modeling*, 25–36.

Teng J, Chen S and Hu J (1999), A finite-volume method for deformation analysis of woven fabrics, *International Journal of Numerical Methods in Engineering*, 46, 2061–2098.

Terzopoulos D and Fleischer K (1988), Modeling inelastic deformation: Viscoelasticity, plasticity, fracture, *Computer Graphics*, 22(4), 269–278.

Thomaszewski B and Blochinger W (2007), Parallel simulation of cloth on distributed memory architectures, *Parallel Computing*, 33(6), 377–390.

Thomaszewski B, Wacker M and Straßer W (2006), A consistent bending model for cloth simulation with corotational subdivision finite elements, *ACMSIGGRAPH/Eurographics Symposiom on Computer animation*, 107–116.

Thomaszewski B, Pabsta S and Blochinger W (2008), Parallel techniques for physically based simulation on multi-core processor architectures, *Computers and Graphics*, 32(1), 25–40.

Tsopelas N (1991), Animating the crumpling behavior of garments, *Eurographics Workshop on Animation and Simulation*, 11–24.

Turquin E, Cani M and Hughes J (2004), Sketching garments for virtual characters, *Eurographics Workshop on Sketch-Based Interfaces and Modeling*.

Vassilev T, Spanlang B and Chrysanthou Y (2001), Fast cloth animation on walking avatars, *Computer Graphics Forum*, 20(3), 260–267.

Volino P and Magnenat-Thalmann N (1994), Efficient self-collision detection on smoothly discretized surface animations using geometrical shape regularity, *Computer Graphics Forum*, 13, 155–166.

Volino P and Magnenat-Thalmann N (2000a), Implementing fast cloth simulation with collision response, *Computer Graphics International 2000 (CGI'00)*.

Volino P and Magnenat-Thalmann N (2000b), *Virtual Clothing, Theory and Practice*, Springer-Verlag, Berlin.

Volino P, Courchesne M and Magnenat-Thalmann N (1995), Versatile and efficient techniques for simulating cloth and other deformable objects, *SIGGRAPH*, 137–144.

Volino P, Magnenat-Thalmann N and Faure F (2009), A simple approach to nonlinear tensile stiffness for accurate cloth simulation, *ACM Transactions on Graphics*, 28(4).

Weil J (1986), The synthesis of cloth objects, *Computer Graphics*, 20(4), 49–54.

Wong S K and Baciu G (2005), Dynamic interaction between deformable surfaces and nonsmooth objects, *IEEE Transactions on Visualization and Computer Graphics*, 11(3), 329–340.

Wong S K and Baciu G (2007), Robust continuous collision detection for interactive deformable surfaces, *Computer Animation and Virtual Worlds*, 18(3), 179–192.

Xu Y, Chen Y, Lin S, Zhong H, Wu E, Guo B and Shum H (2001), Photorealistic rendering of knitwear using lumislice, *SIGGRAPH*, 391–398.

Yang Y (1994), *Cloth Modeling and Animation using Viscoelastic Surfaces*, doctoral dissertation, MIRALab, University of Geneva.

Ye J (2008), Simulating inextensible cloth using impulses, *Computer Graphics Forum*, 27(7), 1901–1907.

Yu W, Kang T and Chung K (2000), Drape simulation of woven fabrics by using explicit dynamic analysis, *Journal of the Textile Institute*, 91(2), 285–301.

Zhang D and Yuen M (2000), Collision detection for clothed human animation, *Pacific Conference on Computer Graphics and Applications (PG 2000)*, 328–337.

Zhang D and Yuen M (2001), Cloth simulation using multilevel meshes, *Computers and Graphics*, 25(3), 383–390.

Zhang D and Yuen M (2002), A coherence-based collision detection method for dressed human simulation, *Computer Graphics Forum*, 21(1), 33–42.

Part III
Computer-based technology for apparel

9

Human interaction with computers and its use in the textile apparel industry

G. BACIU and S. LIANG,
The Hong Kong Polytechnic University, Hong Kong

Abstract: This chapter reviews the application of human computer interaction (HCI) techniques to the use of computer design of textile apparel. The chapter discusses ways of improving HCI in both 2D and 3D garment design.

Key words: human computer interaction, 2D and 3D garment design.

9.1   Introduction

The development of fashion and esthetics in the history of human civilization has surpassed the level of mere functional attributes in garment design. The garment industry has influenced our lives and cultures, through shape and color, with multiple notions of style and expressions of health, beauty and candor. Today, fashion is a metaphor of elegance and beauty surrounding our senses – a mechanism for expressing ourselves in the social context of our daily activities.

Successful fashion hinges on creative ideas and much imagination. *Designers* first conceptualize a theme for a new garment. Shape, material, color, movement and flow give a piece of clothing its unique character as a personal extension of our presence. Designers generally use sketches and drawings to express their initial visualization of a new garment. Sketches of various views of the garment provide the clues needed for the next person involved in the garment design process: *the pattern technologist*, or the *tailor*.

Tailors create two-dimensional patterns from the designer's drawings. These patterns, once cut out and sewn together according to the tailor's specifications, provide the final step towards the realization of the designer's intended garment.

With the quick development of digital technology, the design, modeling and animation of digital objects have recently been identified as one of the grand challenges in the field of computer-aided design technology. Many applications, from the entertainment industry to simulators, require the design of complex 3D worlds including many elements of different nature,

203

from plausible landscape and buildings to plants, animals and complex characters, including specific details such as clothes, pose, expression and hair.

Apparel design with intelligent computer supports is one of the most popular and appealing domains that attract attention from professional stylists, manufacturers and researchers. As one of the main criticisms of computer animated synthetic actors is their lack of 'personality', it soon becomes obvious why tailoring synthetic garments fashionably is of great importance.

9.2   Principles of human computer interaction (HCI)

Before laying out the basic components of intelligent garment design technology, we start with some preliminary principles of human computer interaction. Our aim is not to emphasize technology, but rather to provide principles for its effective use in a design environment. Many of these principles have not been applicable until now due to the lack of interactive devices that adapt to the user rather than requiring a steep learning curve and substantial technical support in order to evolve with design cycles.

9.2.1 Definition of HCI

Human computer interaction, or HCI, is the study, planning, and design of what happens when you and a computer work together. As its name implies, HCI consists of three parts: the user, the computer itself, and the ways they work together.

The user

When we talk about HCI, we don't necessarily indicate a single user with a desktop computer. By 'user', we may mean an individual user, a group of users working together, or maybe even a series of users in an organization, each involved with some part of the job or development. The user is whoever is trying to get the job done using the technology. An appreciation of the way people's sensory systems (sight, hearing, touch) relay information is vital to designing a first-class product. For example, display layouts should accommodate the fact that people can be sidetracked by the smallest movement in the outer (peripheral) part of their visual fields, so only important areas should be specified by moving or blinking visuals. Specifically, in the field of the apparel industry, the user indicates the garment designers or the tailors.

The computer

When we talk about the computer, we are referring to any technology ranging from small portable computing devices, desktop computers, to large-scale computer systems – even a process control system or an embedded system could be classed as the computer. In the context of the apparel industry, the working devices for the design, modeling and animation of a garment would all be referred to as 'the computer'.

The interaction

There are obvious differences between humans and machines. This is usually referred to as the 'semantic gap', which means that humans usually perceive high-level concepts and feelings while machines could only calculate the low-level computable features or execute definite instructions. This difference in the level of understanding results in the semantic gap between human and machine in perception and cognition.

In spite of these, HCI attempts to ensure that they both get on with each other and interact successfully. In order to achieve a usable garment design system, you need to apply what you know about humans and computers, and consult with likely users throughout the design process. You need to find a reasonable balance between what can be done by the designers and the computer, and what would be ideal for designing a working space.

9.2.2 Goals of HCI

The goals of HCI are to produce usable and safe systems, as well as functional systems. In order to produce computer systems with good usability, developers must attempt to:

- *Understand* the factors that determine how people use technology
- *Develop* tools and techniques to enable building suitable systems
- *Achieve* efficient, effective, and safe interaction.

Underlying the whole theme of HCI is the belief that people using a computer system should come first. Their needs, capabilities and preferences for conducting various tasks should direct the way in which the interaction systems are being built. People should *not* have to change the way they use a system in order to fit in with it. Instead, the system should be designed and built to match the user's requirements.

9.1 Development of a HCI framework.

9.2.3 Basic principles of HCI

The development of a HCI system includes four steps: (1) analysis, (2) design, (3) prototype, and (4) feedback. The flow shown in Fig. 9.1 shows the corresponding processing framework.

1. Requirements analysis
   - Establish the goals for the HCI from the standpoint of the users.
   - Agree on the users' needs and aim for usability requirements.
   - Appraise existing versions of the systems (if any).
   - Carry out an analysis of the competition.
   - Complete discussions with potential users and questionnaires.
2. Conceptual proposal
   - Outline system design and architecture at an abstract level.
   - Perform a task analysis to identify essential features.
3. Prototyping
   - Create visual representations (mock-ups) or interactive representations (prototypes) of the system.
   - Evaluate usability using a proven method.
   - Using the results, create more mock-ups or improve the prototypes.
   - Repeat this process until the design and usability goals are met.
4. Feedback
   - Evaluate functionality through testing, quality assurance, usability testing, and field testing.
   - Use the evaluation results to improve the product.
   - Repeat this process until the business goals are met.

- Maintain and tweak with user feedback.
- Use the feedback to create new requirements, and begin major design improvements (system iteration).

9.3   Methods for improving human interaction with computers for textile purposes

With the rapid development of the clothing industry, garment design with computer supports becomes more and more appealing and promising, attracting attention from professional designers and researchers as well as manufacturers [1].

There have been great efforts in introducing HCI into the textile domain. Dedicated computer-aided design (CAD) software for manipulating a variety of geometric models has been developed. More specifically, the process of conventional CAD-supported garment design [2] is shown in Fig. 9.2.

9.3.1 Importance of HCI in textile manufacturing

The importance of HCI in the future of textile development is not to be taken lightly. It has been shown that a large percentage of the design

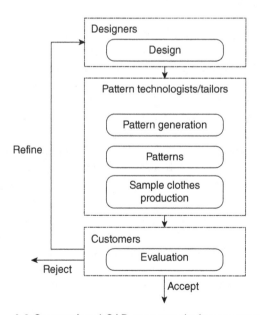

9.2 Conventional CAD garment design process.

and programming effort of projects goes into the actual design of apparel generation systems. The interface is a fundamental part of making the site more successful, safe, useful, functional and, in the long run, more pleasurable for the user.

The tools and techniques that have been developed in this field have contributed immensely towards decreasing costs and increasing productivity. Savings have been created through decreased task time, fewer user errors, greatly reduced user disruption, reduced burden on support staff, the elimination of training, and avoidance of changes in maintenance and redesign costs. Studies have shown that, by estimating all the costs associated with usability engineering, the benefits can amount to 5000 times the project's cost. HCI is a textile imperative now, and it will continue to be so in future.

9.3.2 Typical garment design methods

Up to now, there have been three typical garment design methods, namely the A, B and C approaches [1], shown in Fig. 9.3.

A: 2D garment design

The traditional 2D CAD systems provide such 2D grading tools to generate patterns of different sizes from the basic pattern set, and then use them to make garments, such as in path A in Fig. 9.3.

In traditional manufacturing and textile garment production, pattern technologists (tailors) design the 2D garment pieces, or so-called panels, according to the stylists' design concepts before they are sewn together to form a complete dress [3].

Several commercial CAD tools [4–6] are available for the design of 2D pattern pieces, supporting the garment panel design process. These are from the traditional manufacturers and producers of textile garments: for instance, using TexWinCAD™ by the Italian F.K. Group company [7], a module for pattern design is created in which piece contours are drawn as closed polygonal curves connecting vertex points, provided with reference hole points and notches (Fig. 9.4).

However, the traditional 2D design process is inflexible and time-consuming, and not easy to use. Along with the development of computer graphics and simulation techniques, more advanced software has been developed which combines 3D techniques to ease the design process of garment modeling.

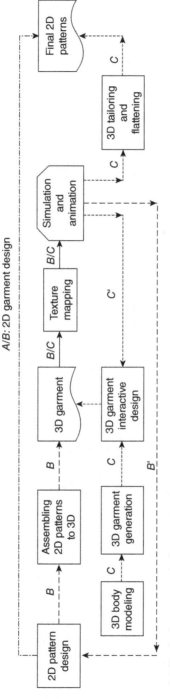

A/B: 2D garment design

9.3 Typical garment design methods.

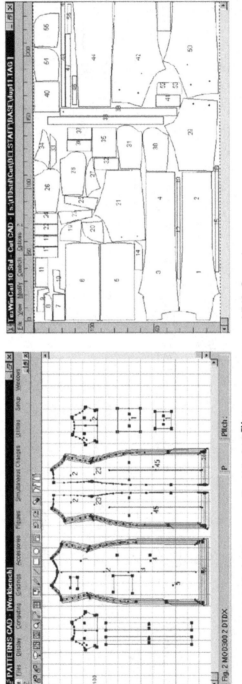

9.4 A commercial 2D CAD system: TexWinCAD™ (Image courtesy of F.K. Group).

B: 2D/3D garment design

Along with the development of garment simulation techniques [8–10], some commercial CAD systems test the garment design results by assembling 2D patterns and draping them on a virtual human body [11–17], such as in path B in Fig. 9.3. Such solutions are not intuitive enough and need the designers to have rich experience and accomplished skills. In order to make the design process more intuitive, Volino *et al.* [18] provided an interactive design environment to edit the patterns in either 2D or 3D with immediate preview of the garment draping result. However, the 2D patterns still need to be modified many times for generating a 3D garment that will fit.

Some researches incorporate sketches to facilitate the garment creation process [19–21]. Typically, these systems use 2D sketch drawings to reconstruct 3D garment models. Wang *et al.* [19] presented a feature-based approach for 3D garment modeling with sketch input. Turquin *et al.* [20] proposed a sketch-based interface for a virtual tailor. Decaudin *et al.* [21] proposed a fully geometric method for clothing design with 2D sketch drawings. All of these systems require users to compose a complete sketch as their input, and therefore cannot recognize partial sketch shapes according to domain knowledge or existing design results.

C: 3D garment design

With the development of the 3D laser scan and surface reconstruction technology, more and more studies are focusing on designing a garment directly on a 3D scanned body, and 3D CAD technology is gradually diffusing into garment design and manufacturing applications. The 3D approach consists of several elements, which include 3D body modeling, 3D garment generation, interactive 3D garment surface modeling and pattern design, and 3D/2D pattern conversion, such as in path C in Fig. 9.3. Compared to other elements, few studies have focused on providing effective tools for 3D garment interactive design.

9.4   Future trends

These processes characterize the garment panels, and the relations or connections among panel parts. However, operating a CAD system requires extensive prior training, and it is neither flexible to operate nor efficient to carry out validity assessment based on prior domain knowledge. This is because there is a gap in current CAD technology for garment design, as it is mainly conceived for 2D/3D geometric modeling for cloth shapes but generally does not provide a high-level interactive and easy-to-use design environment with esthetic and functional features.

9.4.1 Challenges in the apparel industry

For intelligent garment design in the apparel industry, there are some problems that need to be solved.

How to design garments more effectively?

Designing a complex garment surface needs effective tools to edit the outline and cross-sectional shape directly and easily. For a complex pattern design, it is necessary to draw not only seam lines and dart lines but also other curves according to the craft requirements.

How to design garments validly and reasonably?

For the apparel industry, constraints have to be considered for convenient editing and satisfying some craft requirements. Moreover, the users need to be guided to tailor the pattern reasonably, and thus enhance the pattern quality.

How to make the interactive tools fit the process of garment design?

This includes outline modeling and inner region tailoring. It is meaningful in the apparel industry to make the design process separate and suitable for cooperation between fashion designers and pattern makers.

9.4.2 New features

In order to solve or alleviate these challenges, both a new type of intelligent design interface and a garment panel knowledge base are necessary. The designing of clothes and garments for synthetic characters presents some interesting interface design issues. Users of the software require ways of duplicating already developed or imagined styles in a natural way. Provision should exist for controlling not only the way a garment appears, but also the way it behaves, for example, seams, physical properties, etc.

We would like to make the interactive design tools fit the process of garment design, while suitable for cooperation between designers and pattern makers, which leads to a realization in computer terms of what tailors and dress designers have done for centuries: cut flat pieces of cloth according to patterns, and pin them around mannequins to see how they fit.

The interface should provide the designers and pattern technologists with both efficiency and flexibility, as well as effectiveness for designing new garment panels or editing existing mother patterns to generate a costumed garment. These 2D pattern shapes are then combined together with physical parameters such as material properties and mannequin metrics to simulate the visual effect of the designed garment.

There are several new features that are promising for existing CAD garment design systems, which are described as follows.

Sketch interaction

The existing CAD and simulation software is not easy to use and usually takes some years to learn. As a consequence, computer artists still rely on real media such as paper and pen to quickly express their ideas of a shape at the early stages of creative design.

As Rob Cook, the vice-president of Pixar Studios, recently emphasized in his keynote talk at SIGGRAPH Asia 2008, the big challenge in our field is now to make the (mathematical) tools as invisible for artists as the technical methods were for the public. This talk reviewed some recent advances in this direction: it presented the use of 2D sketching interfaces, combined with adapted geometric models, for quickly designing complex shapes. What makes real sketching so popular is that most people who design objects and regularly use modeling programs already use sketches and artwork extensively in early phases of the creative process. They are expected to provide flexible, informal interaction between computers and users that do not hinder creative thinking. Therefore, people are dedicated to developing algorithms for interactive garment design, modeling and animation based on sketchy editing, which aim to preserve the original creative concept during the early design phase while at the same time achieving real-time 3D graphics generation and simulation. However, the convenience of sketch-based interactive techniques is suited only to conceptual design, not the apparel industry. Figure 9.5 shows the working space of the garment design system with a sketching interface.

Gesture-based interaction

We interact in the real, physical world in many different ways, such as by seeing (visual), hearing (audio), taste, smell (olfaction) and touch (tactile). However, our interaction with the virtual world is comparatively limited mostly to visual and audio, and in some cases tactile as in some video game controllers.

Gesture-based interaction enlarges the degree of freedom of interaction by enabling more flexible means of freehand gesture communication. Although it is very natural to gesture in the real world in order to interact with and model real objects or simply to transmit information, performing the same activities in the virtual environment is not self-revealing.

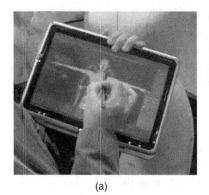
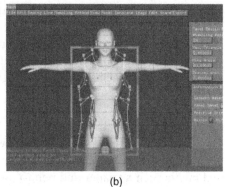

(a)   (b)

9.5 Sketching interface for garment design: (a) sketching interface; (b) designed garment shape.

This contrasts with standard graphic user interfaces such as buttons, menus or selection lists. Gesturing comes naturally in real-world environments, but in virtual environments guidance must be available in the form of visual reminders of the interaction techniques.

Luckily, with the new emergence of electronic multi-touch devices, we are now able to explore more of this type of haptic input communication. The multi-touch devices allow a user to interact with the system using multiple fingers simultaneously, in which more information and more degrees of freedom are collected and encoded. This in turn provides users with a more flexible and natural way of interaction.

Three most important tasks with virtual objects and inside virtual environments can be distinguished as selection, manipulation and navigation.

- *Selection* is the action of specifying one or more objects from a set. The usual goals of selection are to indicate action on an object, make an object active, travel to an object's location, and set up manipulation. Selection performance is affected by several factors: the object's distance from the user, the object's size, and the density of objects in an area. Commonly used selection techniques include the virtual hand metaphor [22], ray casting [23], occlusion/framing [24], naming, and indirect selection. Selection tasks may be classified according to feedback (graphical, tactile and audio), object indication (object touching, pointing, indirect selection) and the indication to select (button, gesture, voice).

- *Manipulation* is the task of modifying object properties (for example color, shape, orientation, position, behavior, etc.). The goals of manipulation may be object placement, tool usage, etc. Figure 9.6 shows gesture-based garment shape editing with two fingers.

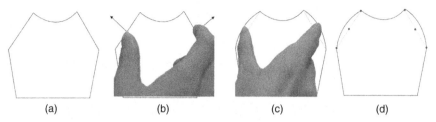

9.6 Gesture-based garment shape editing: (a) the garment shape; (b) the initial position of the fingers; (c) the final position of the fingers; (d) the final garment shape.

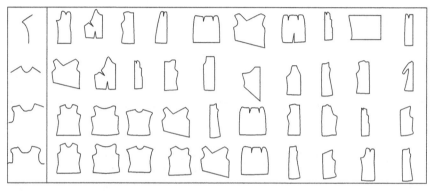

9.7 Shape retrieval and dynamic association of garment shapes.

- *Navigation* and travel are the tasks of controlling viewpoint and camera location via movement and wayfinding.

Shape retrieval and dynamic association

In current CAD garment systems, there are requirements to complete a shape input in order to operate, which may result in time and efficiency lost throughout the designing process. In the inputting process, we could use partial shape retrieval/matching with dynamic feedback to predict the user's intention for the rest of the drawing. This could save user effort in the further inputting process. This may be solved with shape retrieval and matching techniques, which achieve a non-rigid garment design with dynamic association, prediction or suggestion in which professional designers, researchers and manufacturers will certainly be interested [3]. Figure 9.7 gives an illustration of shape retrieval and dynamic association along with increase of the degree of completeness of an input garment shape.

The feature of shape retrieval and dynamic association requires partial recognition and matching of incomplete garment shapes. Partial shape matching is the key technique for developing a partial matching system for

9.8 Examples from the garment shape database.

the garment design process, which is the task of matching sub-parts or regions. These parts are not predefined to be matched and can be any sub-shape of a larger part in any orientation or scale. Partial matching is a much harder problem than global matching, since it needs to search for and define the sub-parts prior to measuring similarities.

Garment database management

It is desirable to build a garment knowledge base with geometric panel shapes, metrics and physical fabric materials. These shapes, metrics and material parameters could specify, facilitate or verify a concrete design during the design process (online) or after the design is accomplished (offline).

In order to make sure the designed garment is valid or usable, it has to be marked with specific metrics. Usually, the basic two-dimensional 'model' is created from certain measurements, also called a sloper or block. It is designed to fit perfectly and is used as a template or prototype to create patterns for a complete wardrobe. A correctly engineered sloper *always* works because it provides consistency of fit for everything wearable, from undergarments to outerwear.

There are seven common basic sloper pieces available: front and back bodices, sleeve, front and back skirt, front and back trousers. Geometric parameters of slopers could be settled according to certain measurements or fitted to the standard sizes, usually ranging from size 8 to size 20. Figure 9.8 shows some examples of panel shapes in a pre-collected garment database.

9.5   Conclusions

The apparel industry plays an important role in everyday life. In recent years, garment computer-aided design (CAD) systems have been rapidly developed and have become the basis of the clothing design process. Typical commercial CAD platforms are built towards rigid shape generation and editing. Ways of making CAD systems efficient and effective to save the cost of both time and materials are attracting attention from both computer science professionals and textile manufacturers.

In this chapter, we have discussed the principles of human computer interaction, and its usage in the textile domain, i.e. computer-aided design (CAD) systems. We have also highlighted new features for the future development of garment design systems, such as sketch interface, gesture-based interaction, shape retrieval and dynamic association, as well as garment database management. By providing new tools that truly enhance the creative process and reduce the design and production cycles, the cost and material savings for the garment industry will be enormous.

9.6   Acknowledgment

We acknowledge the partial support of Hong Kong RGC GRF grants PolyU 5335/06E, PolyU 5101/07E and PolyU 5101/08E.

9.7   References

[1] J. Wang, G. Lu, W. Li, L. Chen and Y. Sakaguti, 'Interactive 3D garment design with constrained contour curves and style curves', *Computer-Aided Design*, vol. 41, pp. 614–625, 2009.

[2] M. Fontana, A. Carubelli, C. Rizzi and U. Cugini, 'ClothAssembler: a CAD module for feature-based garment pattern assembly', in *Proceedings CAD*, 2005.

[3] W. Aldrich, *Fabric, Form and Flat Pattern Cutting*, Blackwell Science, 1996.

[4] Gerber: http://www.gerber-technology.com

[5] Lectra: http://www.lectra.com

[6] Pad: http://www.padsystem.com

[7] E. Saber, Y. Xu and A. M. Tekalp, 'Partial shape recognition by sub-matrix matching for partial matching guided image labeling', *Pattern Recognition*, vol. 38, pp. 1560–1573, 2005.

[8] M. Kazhdan, T. Funkhouser and S. Rusinkiewicz, 'Shape matching and anisotropy', in *ACM Transactions on Graphics*, August 2004.

[9] R. Gal and D. Cohen-Or, 'Salient geometric features for partial shape matching and similarity', *ACM Transactions on Graphics*, 2006.

[10] L. J. Lateck, V. Megalooikonomou, Qiang Wang and D. Yu, 'An elastic partial shape matching technique', *Pattern Recognition*, vol. 40, pp. 3069–3080, 2007.

[11] F. Durupinar and U. Güdükbay, 'A virtual garment design and simulation system', in *11th International Conference on Information Visualization*, 2007.

[12] N. Metaaphanon and P. Kanongchaiyos, 'Real-time cloth simulation for garment CAD', in *SIGGRAPH*, 2005.

[13] R. Bridson, S. Marino and R. Fedkiw, 'Simulation of clothing with folds and wrinkles', in *Eurographics/SIGGRAPH Symposium on Computer Animation*, 2003.

[14] K.-J. Choi and H.-S. Ko, 'Stable but responsive cloth', in *SIGGRAPH*, 2002.

[15] D. L. Zhang and M. M. F. Yuen, 'Cloth simulation using multi-level meshes', *Computer Graphics*, vol. 25, pp. 383–389, 2001.

[16] Z. G. Luo and M. M. F. Yuen, 'Reactive 2D/3D garment pattern design modi- fication', *Computer Aided Design*, vol. 37, pp. 623–630, 2005.

[17] P. Volino, M. Courchesne and N. Magnenat-Thalmann, 'Versatile and efficient technique for simulating cloth and other deformation objects', *SIGGRAPH*, 1995.

[18] P. Volino, F. Cordier and N. Magnenat-Thalmann, 'From early virtual garment simulation to interactive fashion design', *Computer Aided Design*, vol. 37, pp. 593–608, 2004.

[19] C. C. L. Wang, Y. Wang and M. M. F. Yuen, 'Feature based 3D garment design through 2D sketches', *Computer-Aided Design*, vol. 35, pp. 659–672, 2002.

[20] E. Turquin, J. Wither, L. Boissieux, M.-P. Cani and J. F. Hughes, 'A sketch-based interface for clothing virtual characters', *IEEE Computer Graphics and Applications*, vol. 27, pp. 72–81, 2007.

[21] P. Decaudin, D. Julius, J. Wither, L. Boissieux, A. Sheffer and M.-P. Cani, 'Virtual garments: a fully geometric approach for clothing design', in *Eurographics Computer Graphics Forum*, vol. 25, 2006.

[22] I. Poupyrev, M. Billingshurst, S. Weghorst and T. Ichikawa, 'The go-go interac- tion technique: Non-linear mapping for direct manipulation in VR', in *ACM Symposium on User Interface Software and Technology*, 1996.

[23] R. A. Bolt, 'Put-that-there: Voice and gesture at the graphics interface', in *SIGGRAPH*, 1980.

[24] J. S. Pierce, A. Forsberg, M. J. Cowway, S. Hong, R. Zeleznik and M. R. Mine, 'Image plane interaction techniques in 3D immersive environments', in *ACM Symposium on Interactive 3D Graphics*, 1997.

10

3D body scanning: Generation Y body perception and virtual visualization

M.-E. FAUST, Philadelphia University, USA and
S. CARRIER, University of Quebec at Montreal, Canada

Abstract: In this research we investigated Hong Kong Generation Y females' body perception and virtual visualization. On one hand research demonstrated that women are dissatisfied with their body and encouraged to be thin while comparing themselves with an 'ideal shape'. On the other hand, research showed that retailers cannot retain their consumers simply by offering products; consumers are looking for a retail experience. Our results showed that Hong Kong young female adults are not satisfied with their body (breast and lower part); as for Western women, their ideal shape is a slim '**X**' shape; they are dissatisfied with garment fit, and they (HK Generation Y females) believe, after being 3D body scanned, that a smart card with their body measurements would improve their garment shopping experience.

Key words: 3D body scanner, smart shopping card, body cathexis, fit garment, body shape.

10.1 Introduction

Most industries invest in the latest technology to maintain their competitive advantage. The apparel industry acknowledges that quick response and marketing standardized products is not sufficient, especially, to satisfy young consumers. One must develop product differentiation while consistently exceeding consumers' expectation. Beyond style, price, and garment construction the consumer's retail experience and consumer's perception of a good fit are now perceived as strong elements of differentiation for the apparel industry. Moreover, shopping guru Underhill (2009) stated that shopping is the kind of activity where people like to experience before choosing or rejecting an item. If the experience is good, if the garment fits well, the consumer will feel like they look good and feel great, and then will purchase the garment!

Indeed, according to Smathers and Horridge (1978–79) garment fit is one of the major factors providing confidence and comfort to the wearer. Good fit contributes to individual psychological and social well-being.

219

Survey results by Kurt Salmon (in Ashdown, 2003, p. 1) show that 55% of consumers are unable to find appropriate fitting garments. Anderson (2004) reported that the primary factor causing consumers' returns from catalog or website purchases is the 'fit'. Consumers often purchase multiple garments of the same item in different sizes before finding the best fit (Faust *et al.*, 2006). Moreover some women mentioned that they have to try on as many as 20 pairs of jeans to find one that fits (*Consumer Reports*, 1996). Dissatisfaction with fit is one of the most frequent complaints with ready-to-wear purchases (Faust *et al.*, 2006).

Lack of fit relates to sizing systems (Ashdown, 2003). Women find it difficult to purchase off-the-rack clothing primarily because manufacturers/retailers label their garments with one single number and do not provide appropriate fit information (Workman, 1991; Chun-Yoon and Jasper, 1996). This lack of information costs the apparel industry millions of dollars annually in lost sales and generates consumer dissatisfaction (Tamburrino, 1992). Furthermore, nowadays manufacturers are targeting one specific body shape (**X, A, H**) yet still label their garments with the same numeric code system (Faust and Carrier, 2009).

As mentioned, numerous studies (Kurt Salmon, 2000; Goldsberry *et al.*, 1996; LaBat and DeLong, 1990) showed that garment fit is a problem for both the individual and for the industry. It is an issue for individuals trying to find a specific type of garment such as jeans. But it is also a challenge for specific target markets such as plus size women or those older than 55 years old. Moreover, women tend to blame themselves for not having the right size and shape (the ideal size and shape) if they don't fit into a garment (LaBat and DeLong, 1990). Understanding fit satisfaction from the consumer's point of view is therefore crucial as each consumer determines what constitutes a good fit.

For that reason the starting point of our research was to investigate whether Hong Kong young female adults were satisfied with their body (body cathexis) and/or if they compare themselves with an 'ideal body shape'. Then we investigated how satisfied they were with the fit of ready-to-wear garments. Lastly, we investigated whether Hong Kong young female adults would use technologies to enhance their shopping experience: *a priori* using the 3D body scanner, and *a posteriori* with a smart shopping card. This last part is based on our literature review which mentioned that people are expecting a retail experience.

Our research showed that the use of technological tools such as the 3D body scanner and a smart shopping card would improve the consumer's shopping experience. It would allow rapid access to one's size and shape data, improving at the same time the garment fit, enhancing the consumer's self-esteem.

10.2   Literature review

In this section, the literature review related to the body shape, body image, body cathexis and the ideal perceived body shape is presented. Then it is followed by aspects of a good garment fit and a brief explanation of the sizing systems. Last but not least, the functionality of the 3D body scanner and the RFID smart shopping card concept with the idea of enhancing the shopping experience is introduced.

10.2.1 Body shape

Numerous journal papers, magazines and books are being published each year on the subject of finding the best garment for different and specific type of shapes (Hamel and Salvas, 1992; Arbetter and Lind, 2005; Alfano and Yoham, 2009). In order to define 'your' body shape, some authors have set body proportions (Rasband and Liechty, 2006; [TC]², 2004). Rasband and Liechty (2006) stated that figure types are recognized according to the specific areas on the body where weight tends to accumulate, regardless of height (Table 10.1).

For others such as Arbetter and Lind (2005) and Alfano and Yoham (2009), both from *InStyle*, to define what is 'your figure' one needs look at a set of scales between length of the legs in relation to the torso with the width of the shoulders in relation with the hips, etc. (Table 10.2). These are given without any specific measurements, and figures are summarized in the table. For Goldsberry *et al.*, (1996) or D'Souza (2009), proportions and body shapes can also be related to the women's age group.

10.2.2 Body image, self-concept and self-esteem

Thompson *et al.* (1999) define body image as 'the internal representation of one's own outer appearance'. Similarly, Kaiser (1997, p. 98) defines body image as referring to 'the mental picture one has of his or her body at any given moment in time'. Schilder (1950, p. 11) also has a similar interpretation, defining it as 'the picture of our own body which we form in our mind, that is to say, the way in which the body appears to ourselves'.

For Grogan (1999) body image is a person's perceptions, thoughts, and feelings about his or her body. It relates to self-concept, which can be described as the global perception of who one is (Kaiser, 1997). Kalish (1975) defines self-concept as the total image one has about oneself (based on one's actual experiences and interpretation of those experiences). Along with self-concept, self-esteem is the way someone feels toward and perceives oneself (Laurer and Handel, 1977).

Table 10.1 Figure types

Figure type	Description	2D view
Triangular	Smaller at waist and larger below waist. Weight is concentrated below the waist in the buttocks, low hips and thighs	
Inverted triangular	Larger above waist and smaller below waist. Weight is concentrated in the shoulders, upper back, bust or all above the waist	
Rectangular	Nearly same width at shoulders, waist and hips. Weight tends to be fairly evenly distributed	
Hourglass	Larger at bust (larger than average) area and at hip area (rounded) with smaller waist. Weight is fairly evenly distributed over or underneath the waist line	
Diamond	Narrow shoulders and hips in combination with a wide midriff and waist. Weight is concentrated in the midriff, waist and abdomen	
Rounded	Full-rounded all over. Upper back and upper arms, bust, midriff, waist, abdomen, buttocks, hips, and upper legs are larger and rounding. Weight is noticeably above or ideal range	

Source: Rasband and Liechty, 2006.

Table 10.2 Body types

Short waist/ long legs	Narrow shoulders	Small bust	Tummy	Heavy arms	Flat bottom
Long waist/ short legs	Broad shoulders	Full bust	Boyish	Curvy	Large bottom

Sources: Arbetter and Lind, 2005; Alfano and Yoham, 2009.

As a result, women who do not meet the ideal standards (discussed later) demonstrate a greater dissatisfaction with their bodies than women closer to the ideal. More dieting and body dissatisfaction were reported by girls furthest from the ideal figure, larger girls especially (Huon, 1994).

10.2.3 Body cathexis

Body cathexis is used by researchers to evaluate body perception (LaBat and DeLong, 1990; Secord and Jourard, 1953). It is an attribute that refers to the evaluative dimension of body image (Jourard, 1958); the degree of feeling of satisfaction (dissatisfaction) with various parts of one's body (Secord and Jourard, 1953). Many studies report that body cathexis is also closely related to a person's self-image, self-concept or self-esteem (Secord and Jourard, 1953; Wendel and Lester, 1988).

Harter (1985) stated that dissatisfaction with a particular aspect of one's body may lead to lower self-esteem. Consequently, poor body image may be expected to impact negatively on someone's self-esteem, especially when physical attractiveness is a significant attribute to an individual. Salem (1990) studied several variables (satisfaction with body, importance of meeting ideal body, and perceived discrepancy from ideal body) in relation to depression and self-esteem, and found a positive relationship between body dissatisfaction and depression.

10.2.4 Ideal body image

An ideal body image is developed in each society; it varies from time to time and is readily recognized by its members (Roach and Eicher, 1973).

Several researchers have focused on consumers' perception related to ideal body image in a social and cultural setting. Schilder (1935) stated that the amount of time and effort people take to change their body's appearance indicates the importance of the body image in a culture. Females often use clothes to modify their appearance to meet idealized body image in their culture (Jourard and Secord, 1955). Cultural standards impact on one's perception as they are as important as any other standards in a society. They affect one's body image positively or negatively depending on how one perceives or compares his or her body to these standards (Fallon, 1990).

Ancient Greeks

Ancient Greeks believed that the world was beautiful because there is a certain measure, proportion, order and harmony between elements (Gaut and Lopes, 2001). The golden ratio or golden proportion, a ratio of 1:1.618,

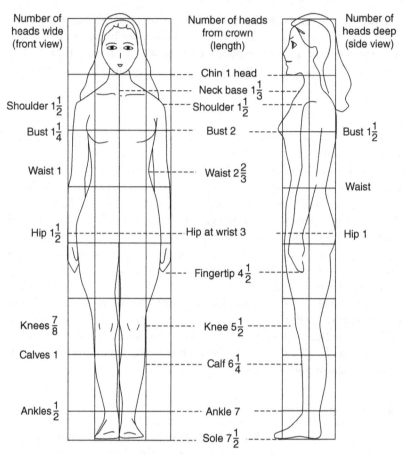

Number of heads wide (front view)

Shoulder $1\frac{1}{2}$

Bust $1\frac{1}{4}$

Waist 1

Hip $1\frac{1}{2}$

Knees $\frac{7}{8}$

Calves 1

Ankles $\frac{1}{2}$

Number of heads from crown (length)

Chin 1 head

Neck base $1\frac{1}{3}$

Shoulder $1\frac{1}{2}$

Bust 2

Waist $2\frac{2}{3}$

Hip at wrist 3

Fingertip $4\frac{1}{2}$

Knee $5\frac{1}{2}$

Calf $6\frac{1}{4}$

Ankle 7

Sole $7\frac{1}{2}$

Number of heads deep (side view)

Bust $1\frac{1}{2}$

Waist

Hip 1

10.1 Ideal Greek proportions of female figure (Horn and Gurel, 1981).

has been considered as the perfect ratio of beauty. This classic average of body proportions has been widely considered as 'the ideal' for centuries (Horn and Gurel, 1981). For the Greek the ideal female body dimensions (Fig. 10.1) are measured in the unit of head length.

The height should be seven and half head lengths, with the fullest part at the hipline and wrist level dividing the total length exactly in half. The neck should be about one-third of the length of the head, and the shoulder lines slopes at a distance of a half head length from the level of the chin. The fullest part of the bust or chest should be located two head lengths from the crown. The waistline, which coincides with the bend of the elbow, should be two and two-thirds of a head length from the crown. The knees should be at five and a half head lengths from the crown and the ankles at seven head lengths from the crown, etc.

Western cultures

In Western cultures, a fat body shape was considered appealing and fashionable from the fifteenth to the seventeenth centuries. The ideal woman was portrayed as plump, big-breasted and maternal (Boucher, 1996). By the nineteenth century, the ideal of feminine beauty went for a voluptuous, corseted figure, to idealizing an hourglass shape such as that of Marilyn Monroe (Stoppard and Younger-Levis, 1994). Western culture then shifted to thinness coupled with somewhat inconsistent large breasts and a more toned, muscular physique (Thompson *et al.*, 1999). Since Twiggy (to today), Western models on the podium, such as Lara Stone with her bust measurement of 83.8 cm (33 in), waist of 60.9 cm (24 in) and hip of 88.9 cm (35 in) are still imposing a very thin trend, although since the death of models such as Ana Reston and Luisel Ramos (who died of anorexia) and the arrival of Crystal Renn and Velvet D'Amour selected by Jean-Paul Gaultier for his *haute couture* defile, where he stated 'diversity is what fashion needs', the small-minded definition of beauty is somehow changing and is bringing controversy in the fashion and modern world. Crystal Renn, considered as one of the most successful models in the industry, with her 175.5 cm (5′9″) height and her bra of 38 C, her waist of 76.2 cm (30 in) and her hip of 106.7 (42 in), is becoming very successful even if she is still labeled as a 'plus size model' (Renn and Ingall, 2009).

Proportion and ratio (WHR)

The ideal is also described by proportion, or relationship of one segment of the body to another, adding that balance is important, to get an ideal symmetrical figure (Kefgen and Touchie-Specht, 1986). Armstrong (1987) defined the best female figure as one where the shoulder width equals the hip width, with the waist measurement girth of 25.4 cm to 31.8 cm (10 to 12.5 inches) smaller than that of the bust and hip girth.

Evolutionary psychology suggested that female attractiveness was based on cues of health and reproductive potential. One cue for physical attractiveness is the waist-to-hip ratio (Singh, 1993) where women with a WHR of 0.7 were perceived as most attractive. Fan *et al.* (2004) also ascertain that WHR affects body attractiveness.

Asian women's perception of body image

Traditionally in Asian cities inner beauty was considered a more important attribute than external appearance, although by 1995 increasing rates of body dissatisfaction were reported in Hong Kong where women's body dissatisfaction rate reached 40% (Lee and Lee, 1996). Asian women are using strict dieting and exercising and are taking on the same behavioral patterns that have been seen by American women for the last few decades (Lai, 2000). Miss Hong Kong contestants have reported a decrease in body mass index (BMI) over the years similar to the statistics on Miss America contestants. In 1978, the average body mass index was 17.99, in 1985 it was 18.77, in 1990 it was 17.98 and in 2000 it was the lowest to date, 17.13, stated Leung *et al.* (2001). Analyses of Hong Kong Pageant data over a 25-year period indicated that the winners are taller and thinner and that a curvaceous body shape with a narrow waist set against full hips is important (Leung *et al.*, 2001).

Evidence from Southeast Asian research showed that young women have relatively high rates of encouragement to thinness, particularly from their mothers (Mukai *et al.*, 1994). Body dissatisfaction was higher in college women than after graduation. Women seem more likely to lose weight in college (82%) compared with after graduating (68%) (Heatherton *et al.*, 1997). College-age women perceive their bodies to be larger than what they feel men prefer (larger than they would like it) (Forbes *et al.*, 2001). Monteath and McCabe (1997) reported that 96% of college females perceived themselves larger than the societal ideal; 94% of women showed a desire to be smaller than their perceived actual size and 56% overestimated their size.

The increased body dissatisfaction of Asian women is a result of the increased pressure to fit the Western ideal (Haudek *et al.*, 1999). Mukai *et al.* (1998) stated that Asian culture places a strong value on conforming to the social norms of the European beauty standard.

Individuals of Chinese descent may be susceptible to culture clash. Increased exposure to Western cultures results in attitudinal shifts regarding both body size and facial features (Lee and Lee, 1996).

Compared to European women, Asian women have a greater concern about body shape and are more dissatisfied with their body (Haudek *et al.*, 1999). Other studies have shown that Asian women report some form of greater dissatisfaction with their bodies than any other ethnic group of women around the globe (Sanders and Heiss, 1998).

According to a Chinese article in Xinhuanet (2002), young Asian females envy European women's body shape. Asia's women hope to get taller and wish their body shape to be slender as well as to show curves further up and down, as the models. This seems to be the case in Tokyo where the average waist circumference of young women is only 60 cm (23.6 in). In Beijing, the situation is similar but the average height of young women is taller, 161.5 cm (63.6 in), the highest among Asian cities. In Hong Kong, women are looking for a slender body shape with their brassiere size increasing by a few inches. The exception may be in Taipei, where it appeared that young women have a strong preference for a 'V' shape. Most Asian

(a)   (b)

10.2 Chinese women presenting medals at the 2008 Olympic Games: (a) from *Xinmin Evening News* (2008) and (b) from Beijing Olympic Games (2008).

females' ideal build is stated to be a height of 160.5 cm (63.2 in), weight of 49 kg (107.8 lb), bust girth of 84 cm (33.1 in), waist girth of 62 cm (24.4 in) and hip girth of 86 cm (33.9 in).

For the 2008 Beijing Olympic Games, the *Xinmin Evening News* (2008) reports an unusual glimpse of what the government of China defines as the ideal beauty (see Fig. 10.2(a)). The perfect women to be presenting the medals had to be between 1.68 and 1.78 meters tall (5′6″ to 5′9″), have bones in every part of the body well proportioned and symmetrical, muscles elastic enough to display a healthy, beautiful body, full-figured, not fat and cumbersome, with long and slender limbs, thighs soft and smooth, calves high and slightly protruding and shoulders symmetrical, full and even, not drooping (translated from Chinese). In addition to Fig. 10.2(a) from the *Xinmin Evening News*, the authors found a photo taken from the 2008 Beijing Olympics Games (Fig. 10.2(b)).

Asian women are most concerned with their attractiveness; they are least satisfied with their specific body parts (e.g., face or breasts), and they are least satisfied with their physical condition (Kennedy *et al.*, 2004). When looking at specific body parts, Kennedy *et al.* (2004) found that Asian women were the most dissatisfied with the size of their bust. Lee and Lee's (1996) research revealed that Chinese women desired, to an even greater degree than their Western counterparts, an increase in the size of their breasts concurrent with a contradictory 'petiteness' for all other body parts. Chinese women consequently showed low self-esteem, because of the bio-logical improbability of full breasts and a slim body. Other studies con-ducted in North America have found that Chinese and Japanese women were more dissatisfied with their breasts than their Caucasian counterparts (Marsella *et al.*, 1981; Mintz and Kashubeck, 1999). Lastly, Asian women reported less satisfaction than European women with their height, eyes, bust, face and arms (Mintz and Kashubeck, 1999).

10.2.5 Fit

The definition of fit varies from time to time and is affected by the fashion culture, industrial norm and individual perception. Fit is directly related to the anatomy of the human body and it seems that the prominences in the human body create most of the fitting problems (Cain, 1950). Garments that fit well should be conforming to the human body and should have adequate ease of movement. They should show no wrinkles and have been cut and manipulated in such a way that they appear to be part of the wearer (Hackler, 1984). A combination of five factors including as ease, line and grain help for a good fit (Erwin and Kinchen, 1969). In addition, clothing fit is influenced by fashion, style and other factors such as design (Efrat, 1982). Lastly, a good fit should provide a tidy and smooth appearance and again should allow maximum comfort and movement for the wearer (Shen and Huck, 1993).

LaBat and DeLong (1990) stated that external factors such as the sizing systems also affect consumers' fit satisfaction, body image and body cathexis. Sizing and fit become important elements to customers.

10.2.6 Sizing

Before talking about sizing, one needs to know that ready-to-wear was originally a cheap interpretation of *couture* and women knew that altera-tions would be necessary to obtain a good fit. Nowadays, ready-to-wear clothes continue to be made for consumers with specific and proportional standard bodies whereas again most of the women's body proportions may deviate from these specific standards (Faust and Carrier, 2009).

The voluntary sizing system with numbers such as 6, 8, 10, etc., was intro-duced and reinforced in the 1940s. It was based on a national survey and previous practical and commercial use. It served as a guide to define dimen-sions for each desired size, although since it was voluntary, sizes on the garments continued to be marked according to what manufacturers/retail-ers believed was best. Moreover, vanity sizing (reducing the size number without changing the garment dimension) is widely used by the apparel industry. Some state that vanity sizing is a necessity since consumers in Western culture have a psychological need to feel slim. Vanity sizing entices women to buy and increases brand loyalty (Marina, 2005). Measurements for a size 10 a decade ago might be the measurements used to produce a size 8 today, stated Workman and Lentz in 2001. This practice has been used by some design firms for such a long time that the consumer may not notice 'she' is getting larger since 'she' fits into the same size.

Our previous research showed that measurements in women's pants vary from brand to brand for one specific same size (Faust and Carrier, 2009). More importantly, research showed that brands with higher prices are often labeled as smaller sizes (Kinley, 2003). As a result, the numbering system is meaningless (Ashdown, 1998, 2003) and as mentioned earlier many apparel companies and consumers feel confused, since designations of women's apparel sizes with numbers have no direct relationship to any part of the female body.

Moreover, good fit goes beyond a set of body measurements. Women with the same bust, waist, and hip measurements may have a different shape due to back curvature, posture, hip position, leg length, etc. (Rasband and Liechty, 2006). Technology such as the 3D body scanner may help to identify one's real figure and need.

10.2.7 3D body scanner

Contrary to manual measurements such as were used for the Women Measurements Garment Pattern Construction (WMGPC) survey in 1939–41 and other manual measurement methods, the 3D body scanner captures measurements of the whole body without physical contact. Either with white light projection or with laser technology, body measurements are captured within a few seconds (Fig. 10.3).

In the USA, [TC]² is one of the organizations that has developed a 3D body scanner. Their 3D body scanner, like many others such as *Human*

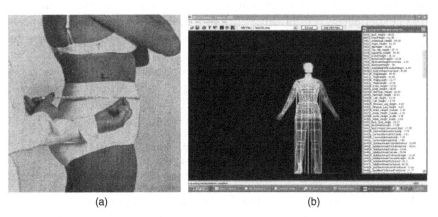

(a)   (b)

10.3 (a) WMGPC manual body measurements; (b) measurements extracted from [TC]²'s 3D body scanner

Solutions, Intellifit or *CAD Modelling Ergonomics,* scans the whole body in seconds and rapidly produces a true-to-scale 3D body model. Their software can extract hundreds of unique measurements, many of which can be customized. High-fidelity, accurate avatars can be created from body scan data, or through the use of [TC]²'s avatar engine with the input of a few basic measurements.

The uses of the 3D body models are unlimited, including made-to-measure clothing, clothing sizing standards development, 3D product development again including clothing, body shape analysis, mass customization national surveys, etc. (Shape analysis, 2010). As a matter of fact, at the end of the 1990s, many Asian, European and North American countries used the 3D body scanner to measure thousands of voluntary subjects of all ages (1992 to 1994 in Japan; 1999 to 2002 in the UK; 2002 and 2003 in the United States). Other applications are being implemented such as for airport security, health and fitness, etc. Last but not least, when linked to 3D garment CAD packages it enables virtual visualization.

This recent application of the 3D body scanner linked to software that enables virtual shopping promises to enhance the shopping experience.

10.2.8 A smart card based on a radio frequency identification device (RFID) to enhance the retail experience

According to Underhill (2009) shopping is the kind of activity that involves a human being experiencing and using his or her senses – sight, touch, smell, taste, and hearing. People buy things today more than ever based on trial and touch, i.e. consumers are choosing or rejecting items based on their retail experience. Not only are consumers looking for a good retail experience but moreover stores are no longer built to serve new consumers but to steal them from someone else. For Underhill (2009) the science of shopping is directly related to common sense; if consumers can examine comfortably and easily, they usually stay longer, and the longer a shopper remains in a store, the more they buy. Underhill (2009) highlights another important factor: accommodation. The author stated that:

> '...in New York, Los Angeles and other big cities ... no accommodation is made for Asian shoppers, despite their numbers and tendency to spend a lot of money ... there are no sizing conversion charts, ...' (Underhill, 2009, p. 35).

RFID is a general term to describe technologies that use radio waves to obtain and identify data. Kou *et al.* (2006) described it as a wireless technology that has been used to carry data between microchip-embedded transponders and readers. RFID tags can be attached to or incorporated into a product, animal, or person. The transponders or tags consist of a microchip that can be read. The reader, using one or more antennae, reads the data stored on the microchip by emitting radio waves to the tag and receiving signals back.

RFID is widely used in many areas such as purchasing and distribution logistics, automation and manufacturing. It was discussed in a previous study (Faust, 2003) but recently the Schmidt Company has introduced RFID technology in the retail clothing industry. The clothes with RFID tags can contribute to tools such as a smart fitting room and smart dress mirror. When a garment with an RFID tag is presented close to a mirror with an antenna, all mixed items without spatial limitation are displayed on the mirror. Salespersons don't need to physically bring garments to the consumers nor show them catalogs. RFID improves customer services by offering additional or alternative clothes and accessories on request without leaving the fitting room, i.e. try-on with ease. It captures customers' data before point of sales, i.e. buying behavior (mi-tu, 2007).

According to Wong (2007) who worked on the implementation of a system using smart cards with RFID, this helps customers to get a quick and satisfying choice of garments through the use of simple technology. It should not only create a new trend for retailers but also upgrade the quality of the service.

RFID is one of the most common contact-less smart cards on the market. The smart card can be used in an automated electronic transaction, storing data securely and hosting/running a series of security algorithms and functions. Lastly, it is not easily copied (Mayes, 2008).

This study focuses on the perception and the possibility of using a smart shopping card based on RFID technology, i.e. incorporating the consumer's size and shape to enhance the shopping experience and to improve the satisfaction with fit.

10.3   Methodology

The above literature review has provided a deeper understanding of the importance of body shape, body image, body cathexis and the effort women make to 'fit the ideal'. Since our objectives were to get Hong Kong young females' opinion of their actual body shape and their ideal body shape, and to investigate whether or not technologies could enhance their shopping experience, increasing at the same time their self-esteem, hypotheses were developed which then served to build a questionnaire.

The questionnaire was used to portray the sample chosen (demographic data). The questionnaire and 3D body scanning were used both to collect the data and to test (validate or reject) our hypotheses linked to (1) body perception, (2) the ideal body shape, and (3) garment fit and sizing satisfaction. Lastly, they were used to determine whether technologies such as the 3D body scanner and RFID with the use of a smart card for apparel shopping would enhance the shopping experience of these young Hong Kong women.

10.3.1 Sampling

Similar to Jourard and Secord's (1955) study, our sample was composed of young female adults. Most of them were single and between the age of 18 and 25. Unlike in Jourard's study they were not Westerners but Asian. Lastly as for Jourard and Secord (1955), it was a convenience sample.

Hypotheses

As mentioned in the literature, in the process of rating one's body, a person becomes involved in a subjective comparison between one's actual measurements (or perceived size) and ideal size. Results of previous studies showed that Westerners have a positive cathexis (wishing to be smaller in size) towards their whole body and different body parts whereas they have a negative cathexis (wishing to be bigger in size) for their bust. Therefore our first hypothesis was that young Hong Kong women, like Westerners, would prefer to be smaller in size with the exception of their bust.

According to the literature review, Asian women are no longer valorizing only inner beauty. The second hypothesis therefore was that young Hong Kong women have the same perception about ideal body shape as do Westerners.

Since Hollywood stars and beauty queens are often portrayed with three measurements (bust, waist and hip) and since Westerners tend to blame themselves because their shape is different from the ideal one, our third hypothesis was that like their western counterparts, young Hong Kong women perceived their bust, waist and hips as different from the ideal model.

According to the literature (LaBat and Delong, 1990; Schofield and LaBat, 2005), women satisfied with their whole body or part of their body seem to be more satisfied with garment fit. Consequently our fourth hypothesis was that as for Westerners, young Hong Kong women who have a positive appreciation of their body or part of it tend to be more satisfied with garment fit.

Lastly, according to Underhill (2009) retailers need to offer an experience to consumers. For many, as for Schmidt (2007), one way of doing so is by using technologies (3D body scanner/RFID). For that reason, our fifth and last hypothesis was that a 3D body scanner or RFID would enhance the retail experience for young Hong Kong women. To ensure that they not only understood the concept but realized what it involves, each subject was scanned before answering.

Questionnaire

The questionnaire consisted of different parts: (1) appraisal of body shape, body image and body cathexis, (2) identification of ideal body shape, (3) satisfaction with garment fit, and (4) positive reception utilizing technologies such as 3D body scanner and RFID smart card to enhance the shopping experience. Lastly, demographic data was collected.

The questionnaire for body shape, body image and body cathexis was built using scales similar to those presented in the literature. Subjects were asked to consider their appraisal (satisfaction/dissatisfaction) towards their personal body and parts of their body. Participants rated their satisfaction on a six-point Likert scale from (1) very dissatisfied to (6) very satisfied. Apart from the whole body, upper part and lower part, 19 body areas were identified. Then, to identify the ideal body shape, participants were asked to select from illustrations and from descriptions which one(s) represented their ideal body shape.

As for the third part, a modified version of fit (LaBat and Delong, 1990) was used to record responses about garment fit satisfaction. Evaluation was done on the same six-point Likert scale. Besides the whole body, upper part and lower part, the same 19 body areas were mentioned for garment fit evaluation.

After being scanned, women were asked to identify the type of garment they wear; whether they tried them on before buying and, if so, why; where they like to purchase their garments; and whether they would prefer to shop with their own avatar.

Lastly, as mentioned, the demographic data was collected, which is summarized in the first part of our results.

A pilot test was conducted to establish content validity of the questionnaire. Suggestions for improvement were noted, and minor changes led to our final questionnaire.

10.4   Results and findings

10.4.1 Sampling and demographics

As in Jourard and Secord's (1955) study, this research was done using a convenience sample. Half a century later and on the other side of the globe, our sample comprised 62 female students enrolled in classes of The Hong Kong Polytechnic University during 2008. These included females from the Institute of Textiles and Clothing. Most participants were aged 18 to 25

(98.3%). Most were single (again 98.3%) and 93.5% were undergraduate students. Monthly income ranges were HK$ 1000 or below (35.4%); between HK$ 1001 and HK$ 5000 (53.2%), and between HK$ 5001 and HK$ 15,000 (9.67%).

10.4.2 Body cathexis

Participants were asked to rate their current satisfaction/dissatisfaction with the following body parts: (1) lower part, (2) upper part, (3) whole body, (4) legs, (5) waist, (6) hips, (7) buttocks, (8) thighs, (9) knees, (10) feet, (11) abdomen, (12) arms, (13) shoulders, (14) neck, (15) back, (16) bust, and (17) height, as well as their (18) body build, (19) body profile, (20) torso, (21) posture and (22) weight. The response choices were arranged according to the Likert scale from 1 (very dissatisfied) to 6 (very satisfied). For the purpose of discussing the body cathexis, participants responding 1, 2 and 3 were combined as dissatisfied and responses 4, 5 and 6 were combined as satisfied.

Results showed that at least 50% of participants chose points 1, 2 and 3, indicating their dissatisfaction, for 16 of the 22 body parts: lower part (74.2%), whole body (62.9%), upper part (58.1%), thighs (82.2%), abdomen (74.2%), hips (67.7%), buttocks (64.6%), legs (64.6%), waist (64.5%), bust (61.3%), arms (59.7%), weight (59.7%), height (56.5%), posture (54.8%), back (50%) and torso (50%).

Results also showed that at least 50% of participants chose points 4, 5 and 6; they were satisfied with seven of the 22 body parts: neck (74.2%), knees (66.1%), shoulders (66.1%), feet (58.1%), body profile (54.9%), back (50%) and torso (50%).

These results show that women are more concerned about the parts of their body that show their shape (which are seen as most important in our actual world and which are the key measurement points in garment fit).

Results also indicated that if they could, participants would like to improve or modify their thighs (25.5%), their waist (18.6%), their hips (15.12%) because they are too fat, their bust (11.63%) because it is too small, and their height (10.47%) because they are too short. The result of the one-sample t-test indicated $t < t_{.05,61} = -1.67$; like Westerners, young Hong Kong women are dissatisfied with their (a) whole body, (b) upper part, (c) lower part, (d) bust, (e) waist and (f) hips. The frequencies were: lower part (74.2%), whole body (62.9%), upper part (58.1%), hips (67.7%), waist (64.5%) and bust (61.3%).

These results validate the hypothesis that Hong Kong young female adults of our sampling are as concerned about their body shape as were Westerners 50 years ago and probably are still today.

10.4.3 Body shape

The participants were asked to give their impression about the ideal body shape and their own body shape. Participants were asked to select the shape from illustrations (identified in the literature review (Table 10.1)): triangular, inverted triangular, rectangular, hourglass, diamond and rounded shapes. Results are presented in Table 10.3. This shows that the hourglass shape was, as for Westerners, the ideal body shape for the majority of participants (90.3%), validating our second hypothesis.

Results also revealed that most participants perceived their body shapes as triangular (40.3%). One quarter of our respondents perceived their bodies as the ideal figure, whereas the rectangular shape and the rounded shape accounted for 19.4% and 9.7% respectively.

Bust, waist and hip measurement of participants

The bust, waist and hip measurements were compared with the ideal bust, waist and hip measurements. Since this research was done in Hong Kong, the ideal measurements were based according to the report in the *Xinmin Evening News* (2008), which gave the ideal bust, waist and hip measurements as 84 cm (33.07 in), 62 cm (24.4 in) and 86 cm (33.85 in) respectively.

Bust
As seen from Table 10.4 the minimum bust measurement is smaller than ideal by 10.6 cm (4.19 in) whereas the maximum bust measurement is larger than ideal by 17.5 cm (6.9 in). One can see that the mean of bust measurement is very close to the ideal: 82 cm vs 84 cm. The frequency distribution showed that close to 65% of participants had a smaller bust than the ideal

Table 10.3 Ideal body shape and participants' body shape

Figure type	Ideal body shape		Perception of participants body shape	
	Frequency	Percent	Frequency	Percent
Triangular	0	0%	25	40.3%
Inverted triangular	5	8.1%	0	0%
Rectangular	1	1.6%	12	19.4%
Hourglass	56	90.3%	16	25.8%
Diamond	0	0%	3	4.8%
Rounded	0	0%	6	9.7%

Table 10.4 Mean body measurement (cm) of bust, waist and hips of participants

	Bust	Waist	Hips
Number of participants	61	61	61
Ideal (cm)	84	62	86
Minimum (cm)	73.4	61.5	81.6
Maximum (cm)	101.5	84.6	104.3
Mean (cm)	82	73.2	89.4
Standard deviation	5.4	5.5	6
Mean difference	2	−11.2	−3.5
t-Test	−2.923	15.819	6.840
P-value	.01*	.00*	.00*

*Significant at 0.05 level.

one. As for Westerners (Secord and Jourard, 1953) results showed a negative *t*-test. In other words young Hong Kong women are dissatisfied with their bust, wanting to be bigger.

Waist

Table 10.4 shows that the minimum waist measurement is close to the ideal, 61.5 cm vs 62 cm (24.2 vs 24.4 in), and the maximum waist measurement is larger than ideal by 22.6 cm (8.9 in). One can see that contrary to the bust, the waist mean of 73.2 cm (28.8 in) is significantly larger than ideal by 11.2 cm (4.4 in). Fewer than 5% of the participants had a waist measurement within 1 inch of the ideal waist. As for Westerners (Secord and Jourard, 1953) results showed a positive *t*-test, showing that again young Hong Kong women are dissatisfied with their waist, in this case wanting it to be smaller.

Hip

From Table 10.4 it can be seen that the minimum hip measurement is smaller than ideal by 4.4 cm (1.74 in), while the maximum hip measurement is larger than ideal by 18.3 cm (7.2 in). The mean of hip measurement is only slightly higher than the ideal and almost 35% of the sample fell within 1 inch of the ideal. As for the waist, the *t*-test showed a positive value, meaning that the participants in our survey would prefer to have smaller hips.

These results clearly demonstrate that, as for Westerners, our group of young Hong Kong women would prefer to be smaller in size when it comes to their waist and their hips but not their bust. These findings are important since perception of these young women is the link to their self-esteem.

10.4.4 Fit satisfaction

Participants were asked to rate their current fit satisfaction or dissatisfaction with the garments: pant length (in-seam), crotch, thighs, buttocks, hips, sleeve length, bust, waist, shoulders, shoulder blades, abdomen, skirt length, armscye, upper arm, midriff, lower arm, calf, elbow, neckline. The response choices were arranged in order from 1 (very dissatisfied) to 6 (very satisfied). Fit satisfaction of participants rating 1, 2 and 3 were combined again as dissatisfied and 4, 5 and 6 were combined as satisfied. At least 50% of participants indicated fit dissatisfaction with the following: buttocks (74.2%), thighs (72.6%), hips (62.9%), crotch (55%), calf (51.6%), and skirt length (50%), all referencing the lower body. On the other hand, at least 50% were satisfied with fit of the following: sleeve length (75.8%), shoulders (72.6%), upper arm (71%), shoulder blades (69.3%), armscye (67.7%), waist (62.9%), pant in-seams (56.5%), bust (56.5%), midriff (54.8%), abdomen (51.6%), and skirt length (50%). These findings were supported by previous literature showing that subjects were less satisfied with the fit of garments to the lower body (LaBat and DeLong, 1990).

10.4.5 Buying behavior of participants

Participants in our survey were asked to identify the type of clothes they wear (they could select as much as they want). They were also asked to identify the places they shopped for garments and moreover whether they tried before buying and, if so, why. Lastly, they were asked (after being scanned) if they would like to shop with their own avatar.

Results are summarized in Table 10.5 (percentages may not sum to 100% in some cases since participants were allowed to select more than one, or due to rounding). Table 10.5 shows that participants prefer to wear five types of clothing: casual pants (65%), shirts and blouses (52%), skirts (47%), dresses (45%), and polo or T-shirts (37%). Results also revealed that outdoor markets and walk-by kiosks were chosen by 76% of participants, and international apparel stores by 55%, whereas catalog or online shopping and tailor-made accounted for not even 5%.

Almost all participants (93.5%) want to try garments on before purchasing. Out of these the three main reasons stated were: (1) because they are concerned with the fit of the garment (82%), (2) because they want to buy the right size (63%), and (3) because they want to see how they look (44%). Only 18% mentioned that the major reason was to show their friends and relatives. No participant chose 'to get the salesperson's advice'. These results show that young women are concerned about fit and about their look. They imply that garment fit is crucial for body cathexis and underline that salesperson advice has limited value.

Table 10.5 Buying behavior of participants

Category	Frequency	Percent
Types of clothing which the participants were used to wearing		
Casual pants	40	65%
Shirts and blouses	32	52%
Skirts	29	47%
Dresses	28	45%
Polo or T-shirts	23	37%
Shorts	17	27%
Athletic wear	7	11%
Work wear	1	2%
Suits	0	0%
Place where the participants were used to purchasing clothing		
Outdoor markets and walk-by kiosks	47	76%
International apparel store	34	55%
Local shop	20	32%
Department store	16	26%
Retail designer shop	15	24%
Catalog or online shopping	2	3%
Tailor-made or made-to-measure	0	0%
Percentage of participants who tried on garments before buying		
Yes	58	93.5%
No	4	6.45%
Reasons why the participants tried on garments before buying		
Because they are concerned about		
the garment fit	51	82%
To buy the right size	39	63%
To see how respondents look	27	44%
To show it to friends and relatives	11	18%
To get salesperson's advice	0	0%
Smart card with respondents' size and shape would help to select the right size in store and should be better than the actual size label		
Yes	57	91.9%
No	5	8.06%
Smart card with respondents' size and shape could help selecting the right size online and should be better than the actual size chart online that retailers proposed		
Yes	58	93.5%
No	4	6.45%

More than 90% of participants stated they would like to use a smart card with their size and shape to help them select the right size garment *in situ* or online. They believe it would be better than relying on the actual size label or online size chart. These last answers were collected after females had been scanned. This implied that a virtual world may soon be a reality.

10.5   Conclusions and recommendations

Previous studies indicated that subjects were less satisfied with their lower body and this dissatisfaction may affect their body cathexis as a whole. Results from the present study validated that young Hong Kong female subjects were also least satisfied with their lower body, especially their thighs and hips, but in addition they would like to have a larger bust.

According to our literature review, Westerners' ideal female body shape is hourglass as described by Rasband and Liechty (2006). This research showed that young Hong Kong women are no different from their Westerner counterparts. They also perceived the hourglass as being the ideal female body shape.

Jourard (1958) stated that clothing can be used as a tool to reduce the discrepancy between an individual's ideal and actual body image, whereas Hwang (1996) found a relationship between subjects' satisfaction with fit and feelings towards their personal body.

More recently, Tyrangiel (2001) stated that ready-to-wear clothes are made for consumers with proportional bodies, and LaBat and DeLong (1990) mentioned that dissatisfaction is a result of trying to fit real bodies into garments sized according to an ideal body shape.

The findings of this study showed that the lower body dimensions of Hong Kong female participants were larger in comparison to the ideal, and the correlation of body proportion and fit satisfaction indicated that less well-proportioned participants were more dissatisfied with fit than well-proportioned participants. The results validated that the conventional sizing system based on ideal body proportions is too limited. As LaBat and DeLong (1990), Ashdown (2003) and Faust and Carrier (2009) suggested, a more diverse sizing system could improve fit satisfaction, hopefully increasing body cathexis and self-esteem.

Lastly, this study validated the potential use of an avatar integrated through RFID in a smart card by young female adults. Not only would it provide a novel experience but it would enhance their shopping, since this smart card system would link individual measurements with the best-fit garment in conformity to its shape. It should consequently improve the consumers' perception of their bodies (no need to compare with the ideal).

10.6   Limitations

The sample size was limited to 62 participants, i.e., a convenient sample. The reliability of the research may be affected by the small sample size. The females were mainly students from The Hong Kong Polytechnic University and the sample may not be representative of the Hong Kong female population. Finally, the ideal body standard used in this research was for Asian women. No specific ideal standard for young Hong Kong females was found. The ideal was based on the newspaper report concerning sizes for hostesses for the 2008 Beijing Olympics. This may affect the results of the analysis, since young Hong Kong females may have different body measurements and shapes than other Asians.

10.7 Future studies

This research studied and analyzed only Generation Y Hong Kong females; therefore it would be interesting to study the population from other cities in Asia. Also, consumers' perceptions may vary according to their age groups.

10.8 References

Alfano J and Yoham B (2009), *The New Secrets of Style, Your Complete Guide to Dressing Your Best Every Day*, InStyle editors, New York.
Anderson G (2004), If the clothes fit, buy 'em, available at www.retailwire.com/Print/PrintDocument.cfm?DOC_ID=10058 (2005, December).
Arbetter L and Lind M (2005), *In Style, Secrets of Style, the Complete Guide to Dressing Your Best Every Day*, InStyle editors, New York.
Armstrong H (1987), *Patternmaking for Fashion Design*, Harper and Row, New York.
Ashdown S P (1998), An investigation of the structure of the sizing systems: a comparison of three multidimensional optimized sizing systems generated from anthropometric data with the ASTM standard D5585-94, *International Journal of Clothing Science and Technology*, 10(5), 324–341.
Ashdown S P (2003), Sizing up the apparel industry, *Cornell's Newsletter: Topstitch*, Spring 2003.
Beijing Olympic Games (2008), Olympics organizers seek the perfect woman, *South China Morning Post*, 1, 16 February 2008, in *Olympic Beijing 2008* viewed April 2010, http://www.google.com/imgres?imgurl=http://english.people.com.cn/mediafile/200808/18/P200808182121131810315195.jpg&imgrefurl=http://english.peopledaily.com.cn/90001/90779/94837/6479500.html&usg=__MxEndQEAiTSLHZFqub1TXi3X3l4=&h=352&w=450&sz=44&hl=en&start=1&itbs=1&tbnid=p4t1Ki5dS262kM:&tbnh=99&tbnw=127&prev=/images%3Fq%3Dolympics%2Bbeijing%2Bhostest%2B2008%26hl%3Den%26gbv%3D2%26tbs%3Disch:1
Boucher F (1996), *Histoire du Costume en Occident*, Flammarion, Paris.
Cain G (1950), *The American Way of Designing*, Fairchild Publications, New York.
Chun-Yoon J and Jasper C R (1996), Consumer preferences for size description systems of men's and women's apparel, *Journal of Consumer Affairs*, 29(2), 429–441.
Consumer Reports (1996), Why don't these pants fit?, May, 38–39.
D'Souza N (2009), *Fabulous at Every Age; Your Quick and Easy Guide to Fashion*, Harpers Bazaar, Hearst Communications Inc., New York.

Efrat S (1982), *The development of a method of generating patterns for clothing that conform to the shape of the human body*, Ph.D. thesis, School of Textile and Knitwear Technology, Leicester Polytechnic, UK, pp. 234–235.

Erwin M D and Kinchen L A (1969), *Clothing for Moderns*, 4th edition, MacMillan, New York.

Fallon A (1990), Culture in the mirror: sociocultural determinants of body image, in Cash T F and Pruzinsky T (eds), *Body Images: Development, Deviance, and Change*, Guilford Press, New York.

Fan J, Liu F, Wu J and Dai W (2004), Visual perception of female physical attractiveness, *Proc Roy Soc Lond: Bio Sci*, 271, 347–352.

Faust M-E (2003), *L'utilisation des technologies de l'information et de la communication (TIC) lors de la fonction essayage vestimentaire*, M.Sc.A. thesis, Ecole Polytechnique de Montréal, Québec, Canada.

Faust M-E, Carrier S and Baptiste P (2006), Variations in Canadian women's ready-to-wear standard sizes, *Journal of Fashion Marketing and Management*, 10(1), 71–83.

Faust M-E and Carrier S (2008), A global size label: a step in mass customizing for the apparel industry, International Mass Customization Meeting (IMCM) + International Conference on Ergonomic, Technical and Operational Aspects of Product Configuration System (PETO), Copenhagen, 19–20 June 2008.

Faust M-E and Carrier S (2009), A proposal for a new size label to assist consumers in finding well-fitting women's clothing, especially pants: An analysis of Size USA female data and women's ready-to-wear pants for North American companies, *Textile Research Journal*, 79(16), 1446–1458.

Forbes G B, Adams-Curtis L E, Rade B and Jaberg P (2001), Body dissatisfaction in women and men: the role of gender typing and self esteem, *Sex Roles*, 44(7–8), 461–484.

Gaut B and Lopes D M (2001), chapter 20 in *The Routledge Companion to Aesthetics*, Routledge, London, 341–352.

Goldsberry E, Shim S and Reich N (1996), Women 55 years and older: Part II, Overall satisfaction and dissatisfaction with the fit of ready-to-wear, *Clothing and Textiles Research Journal*, 14(2), 121–132.

Grogan S (1999), *Body Image: Understanding Body Dissatisfaction in Men, Women, and Children*, Routledge, London.

Hackler N (1984), What is good fit? *Consumer Affairs Update*, published by Consumer Affairs Committee of Apparel Manufacturers Association, May, 2(1).

Hamel C and Salvas G (1992), *C'est moi! Ma personalité . . . mon style*. Communiplex Marketing Inc., Québec, Canada.

Harter S (1985), *The Self-perception Profile for Children: Revision of the Perceived Competence Scale for Children*, University of Denver, Co.

Haudek C, Rorty M and Henker B (1999), The role of ethnicity and parental bonding in the eating and weight concerns of Asian Americans and Caucasian college women, *International Journal of Eating Disorders*, 25, 425–433.

Heatherton T F, Mahamedi F, Striepe, M, Field A and Keel P (1997), A ten year longitudinal study of body weight, dieting, and eating disorder symptoms, *Journal of Abnormal Psychology*, 106, 117–125.

Horn M J and Gurel L M (1981), *The Second Skin*, 3rd edition, Houghton Mifflin, Boston, MA.

Huon G F (1994), Dieting, binge eating and some of their correlates among secondary school girls, *International Journal of Eating Disorders*, 15, 159–164.

Hwang J (1996), *Relationships between body-cathexis, clothing benefits sought, and clothing behavior; and effects of importance of meeting the ideal body image and clothing attitude*, unpublished doctoral dissertation, Virginia Polytechnic Institute and State University, Blacksburg, VA.

Jourard S M (1958), *Personal Adjustment*, MacMillan, New York, 150–173.

Jourard S and Secord P (1955), Body cathexis and the ideal female figure, *Journal of Abnormal Social Psychology*, 50, 243–246.

Kaiser S B (1997), *The Social Psychology of Clothing: Symbolic Appearances in Context*, 2nd revised edition, Fairchild Publications, New York.

Kalish R (1975), *Late Adulthood: Perspectives on Human Development*, Brooks/Cole Publishing Company, Monterey CA.

Kefgen M and Touchie-Specht P (1986), *Individuality in Clothing Selection and Personal Appearance*, 4th edition, MacMillan, New York.

Kennedy M A, Templeton L, Gandhi A and Gorzalka B B (2004), Asian body image satisfaction: ethnic and gender differences across Chinese, Indo-Asian and European descent students, *Eating Disorders: The Journal of Treatment and Prevention*, 12(4), 321–336.

Kinley T R (2003), Size variations in women's pants, *Clothing and Textiles Research Journal*, 21(1), 19–31.

Kou D, Zhao K, Tao Y and Kou W (2006), *Enabling Technologies for Wireless e-Business*, Springer, Berlin and Heidelberg.

Kurt Salmon Associates (2000), Annual consumer outlook survey, paper presented at a meeting of the American Apparel and Footwear Association Apparel Research Committee, Orlando, FL, November.

LaBat K L and DeLong M R (1990), Body cathexis and satisfaction with fit of apparel, *Clothing and Textiles Research Journal*, 8(2), 43–48.

Lai K Y C (2000), Anorexia nervosa in Chinese adolescents: does culture make a difference?, *Journal of Adolescence*, 23, 561–568.

Laurer R and Handel W (1977), *The Theory and Application of Symbolic Interactionism*, Houghton Mifflin, Boston, MA.

Lee A M and Lee S (1996), Disordered eating and its psychosocial correlates among Chinese adolescents in Hong Kong, *International Journal of Eating Disorders*, 20, 177–183.

Leung F, Lam S and Sze S (2001), Cultural expectations of thinness in Chinese women, *Eating Disorders*, 9(4), 339–350.

Marina A (2005), Communications clothing fit preferences of young female adult consumers, *International Journal of Clothing Science and Technology*, 17(1), 52–64.

Marsella A, Shizuru L, Brennan J and Kameoka V (1981), Depression and body image satisfaction, *Journal of Cross-Cultural Psychology*, 12, 360–371.

Mayes K E and Markantonakis, K (2008), *Smart Cards, Tokens, Security and Applications*, Springer, New York.

Mintz L B and Kashubeck S (1999), Body image and disordered eating among Asian American and Caucasian college students: an examination of race and gender differences, *Psychology of Women Quarterly*, 23, 781–796.

mi-tu (2007), *Case study – mi-tu, smart retail system – fashion*, Schmidt Electronics Group, Schmidt RFID.

Monteath S A and McCabe M P (1997), The influence of societal factors on female body image, *Journal of Social Psychology*, 137(6), 708–728.

Mukai T, Crago M and Shisslak C M (1994), Eating attitudes and weight preoccupation among female high school students in Japan, *Journal of Child Psychology and Psychiatry*, 35, 677–688.

Mukai T, Kambara A and Sasaki Y (1998), Body dissatisfaction, need for social approval and eating disturbances among Japanese and American college women, *Sex Roles*, 39, 751–763.

Rasband J A and Liechty L G (2006), *Fabulous Fit: Speed Fitting and Alteration*, 2nd edition, Fairchild Publications, New York.

Renn C and Ingall M (2009), *Hungry, a Young Model's Story of Appetite, Ambition, and the Ultimate Embrace of Curves*, Simon & Schuster, New York.

Roach M and Eicher J (1973), *The Visible Self: Perspectives on Dress*, Prentice-Hall, Englewood Cliffs, NJ.

Salem D (1990), *Black Women in Organized Reform, 1890–1920*, Carlson, New York.

Sanders N M and Heiss C J (1998), Eating attitudes and body image of Asian and Caucasian college women, *Eating Disorders*, 6, 15–27.

Schilder P (1935), *The Image and Appearance of the Human Body*, International Universities Press, New York, in Landry Sault N (1994), *Many Mirrors: Body Image and Social Relation*, Rutgers University Press, New Brunswick, NJ.

Schilder P (1950), *The Image and Appearance of the Human Body*, International Universities Press, New York, in Pylvänäinen P (2003), Body image: a tripartite model for use in dance/movement therapy, *American Journal of Dance Therapy*, 25(1), 39–55.

Schmidt RFID (2007), *Smart retail system, SRS, features and benefits*, Smart Dressing system brochure.

Schofield N A and LaBat K L (2005), Defining and testing the assumptions used in current apparel grading practice, *Clothing and Textiles Research Journal*, 23(3), 135–150.

Secord P F and Jourard S M (1953), The appraisal of body-cathexis: body-cathexis and the self, *Journal of Consulting Psychology*, 17(5), 343–347.

Shape analysis (2010), http://www.shapeanalysis.com/prod01.htm

Shen L and Huck J (1993), Bodice pattern development using somatographic and physical data, *International Journal of Cloth Science Technology*, 5(1), 6–16.

Singh D (1993), Adaptive significance of female physical attractiveness: role of waist to hip ratio, *Journal of Personal and Social Psychology*, 65(2), 293–307.

Smathers D and Horridge P E (1978–79), The effects of physical changes on clothing preferences of elderly women, *International Journal of Aging and Human Development*, 9(3), 273–278.

Stoppard M and Younger-Lewis C (1994), *Etre Femme: un Guide de Vie*, Association Médicale Canadienne, Selection du Reader's Digest, Westmount, Québec, Canada.

Tamburrino N (1992), Apparel sizing issues, part 1, *Bobbin*, 33(8), 44–59.

[TC]² (2004), *Size USA let's size up America ... the national sizing survey: body measurement and data analysis reports on the U.S. population*, report, [TC]² Cary, NC.

Thompson J, Heinberg L J, Altabe M and Tantleff-Dunn S (1999), *Exacting Beauty: Theory, Assessment, and Treatment of Body Image Disturbance*, American Psychological Assocation, Washington, DC.

Tyrangiel J (2001), Spree school, *Fortune*, 144, 17 September, 250–252.

Underhill P (2009), *Why We Buy: The Science of Shopping*, Simon and Schuster Paperbacks, New York.

Wendel G and Lester D (1988), Body-cathexis and self esteem, *Perceptual and Motor Skills*, 67, 538.

Women's measurements for garment and pattern construction (WMGPC), retrieved 28 November, 2004 from USDA Miscellaneous Publication (1941), http://museum.nist.gov/exhibits/apparel/role.h

Wong C (2007), Case Study – mi-tu, Smart Retail System – Fashion, Schmidt RFID.

Workman J E (1991), Body measurement specifications for fit models as a factor in clothing size variation, *Clothing and Textiles Research Journal*, 10(1), 31–36.

Workman J E and Lentz E S (2001), Measurement specifications for manufacturers' prototype bodies, *Clothing and Textiles Research Journal*, 18(4), 251–259.

Xinhuanet (2002), The Asian ideal body, 11 April 2002, retrieved 11 February 2008 from http://news.sina.com.cn/s/2002-04-11/0945542366.html

Xinmin Evening News (2008), China searches for Olympic medal ceremony hostesses with ideal looks (Xinhua), retrieved 15 February 2008 from http://www.chinadaily.com.cn/olympics/2008-02/15/content_6459345.htm

11

Computer technology from a textile designer's perspective

H. UJIIE, Philadelphia University, USA

Abstract: This chapter discusses computer technologies used in the domain of textile design from the textile designer's perspective. The chapter starts with a brief explanation of the history and role of CAD, CAM and CIM for textile weave, knit and print design. Then, it describes the main computer technologies, which encompass computer technologies as tools for (1) information gathering, (2) designing, (3) design editing and (4) presentation in textile design workflows. In addition, the chapter explains benefits and limitations in the textile design process from the initial design concept to final production and presentation.

Key words: computer aided design (CAD), computer aided manufacturing (CAM), computer integrated manufacturing (CIM), textile design, computer technology.

11.1 Introduction

Computer technology in the domain of textile design is categorized into manufacturing and designing systems that consist of hardware and software. CAM (computer aided manufacturing) and CIM (computer integrated manufacturing) are tools used to assist in textile production machineries. CAD (computer aided design) is a medium used to assist in the creative design processes. Historically, these technologies were initially developed and utilized as production systems, and it wasn't until the late 1980s that textile designers, for the first time, had access to computer systems in their design software (Ujiie, 2000).

The first computer system for textile design was introduced in 1967 by IMB and the International Business Machines Corporation, in New York City. This digital system, launched as a 'textile graphics' system, functioned to input technical production data of jacquard weaving and engraved roller printing directly within the computer's memory. The technical designers were able to draw simple shapes directly onto a CRT monitor with a corded graphic tablet, and assign technical production weave and engraving information (*Textile World*, 1967; Lourie and Lorenzo, 1967). Although this revolutionary system shortened the timeline of product development from the technical designing stage to the sampling stage, it did not initially contribute to any creative design outcomes. Similarly, throughout the 1970s to the mid-1980s, evolution in textile computer systems was geared toward production, especially in the woven textile sector. By the end of the 1980s, computers had become powerful technical tools to enable input of binary weave notations, and send information directly to punch machines for power weaving. At the same time, computing in the knit sector followed the same route, and more advanced computer technologies were implemented for operating knitting machinery (Noonan, n.d.).

The late 1980s became the benchmark for innovation in the field of textile design computer technology. For the first time, computer technologies were developed for creative textile design applications. More advanced developments in general computing had a dramatic impact on the maturation of designing software. These included faster microprocessors, high performance operating systems, CCD (charge-coupled devices), inkjet printing technology and peripheral devices. Thus, textile designing software systems were able to (1) input original artwork on paper, (2) color separate and edit, (3) simulate for the final production, and (4) assign technical information for production (Howard, 2010). However, these systems were only available as costly proprietary software systems that were operated on custom-built supercomputers, and only a limited number of textile industries could afford them. Nonetheless, by the late 1990s in the United States and Western European countries, most of the textile industries and educational institutions had begun to facilitate these textile design-computing systems (Ujiie, 2000). Moreover, because of the rapid progress in computer processing power, textile software companies began to reengineer their systems for Windows and Mac OS operating systems. Innovation in complex proprietary computer hardware and supercomputers operated by Unix platforms eventually led to developments that encouraged the popularity of the computer as a designing tool. Significantly, as the demand for computer designing software became popular, the price of software and hardware started to drop by as much as 20% of the original costs per station (Simmons, 2010).

Contrary to high-end industry proprietary textile design software, commercial off-the-shelf software has been adapted to assist independent textile designers. Starting in the early 1990s, textile designers, educators and practitioners, who were pioneers in their field, began to integrate early versions of the commercial off-the-shelf Adobe Photoshop and Illustrator software in their textile design activities (Ujiie, 2000). Unlike specific proprietary textile software, which is engineered specifically for textile design usage, commercial off-the-shelf software has certain functional limitations, and requires extra steps to get to certain procedures. However, it has gained a large market popularity for textile design applications, due its affordable price points and improvements in computer hardware processing power.

Throughout history, innovation in computer technology has created new methodologies and attitudes for textile designing processes. In contrast to analog processes, digital technology has led us to investigate its roles, advantages and limitations for further development.

11.2   Role of computer technology in textile design

The fundamental role of computer technology for textile designers is determined by the ability of the computer to assist the designers' creative process. Successful textile designers of today should facilitate specialized design skills to generate original products, and respond rapidly to market trends, while performing multi-tasking in diversified design requirements.

Generally speaking, computer technology has been misconceived as a tool for engineers and business people. This perception was directly reflected in early developments of computer technology in the CAM and CIM systems for textile design. In the past, computer technology was considered as a programmable machine to receive, store and manipulate data, whereas in contemporary industry and culture, the computer can be perceived as an important vehicle for creative textile design applications. It is important to note that theories on the perception of the computer as a design tool have led to the inception of the term 'metamedium'. This term implies that computer technology can be viewed as a dynamic and creative medium with many outcomes and processes (Kay, 1984). Notably, it is critical that textile designers generate original creative designs, and that textile CAD technologies support their creative process. Since the early 1990s, the role of computer technology in textile design has been investigated among textile designers, technology developers and educators worldwide. CITDA (the Computer Integrated Textile Design Association) was established in 1992 as one of the organizations in the US to bridge the gap of diversified perceptions of textile computer technologies. CITDA was formed among textile designers, educators, software developers, textile manufacturers, and hardware developers, and their conference was held annually. In the mid-2000s, CITDA was dissolved after it had accomplished one of its implied missions, which was to use computer-aided textile design technology as a metamedium (CITDA, 1993).

One of the roles of a textile designer is to design and produce a given number of textile productions in a give time (Wilson, 2001). Developments in computer technology enable textile designers to accomplish more design work in shorter time frames. Under the accelerated trend cycles for fashion and home furnishing industries, textile designers can respond rapidly to market needs and design alterations. At the same time, computer technologies can eliminate manpower to achieve the same tasks, and thus create cost-effective design product lines. In addition, new technologies have created new business paradigms and workflows. Mass customization and personalization of textile design products for the consumer are now achieved by textile computer technologies. In addition to advanced information and production technologies, textile design software and hardware enable designers to meet the needs of customers in a short period of time.

11.3   Main computer technologies in textile design

Today, the textile design process consists of several divided procedures:

1. generation of design concept
2. original design creation
3. design editing and alteration, and
4. presentation of 2D and 3D simulation.

Once the design procedures are completed, design data is transferred to a more technical division in the textile company to prepare for production. In the textile industry today, computer technology has been implemented throughout the entire production process: from the initial design concept to the final printing production stage. Industry and design studios can combine these procedures differently, depending on the end-use of the product and the type of customer. Significantly, now that low-cost off-the-shelf software from Adobe Photoshop and Illustrator (as well as their plug-in software) has become accessible, industry professionals have now begun to replace some of their own proprietary textile design software with the more economical popular versions. These changes have indirectly led to many textile companies adapting and changing their computer systems into more sequential textile production. Instead of providing singular complete design systems, the majority of proprietary textile design software companies now offer their systems in modules, which correspond to each design development stage. Such modules can include online trend forecast, color separation, coloring, design in repeat, 3D texture mapping, 3D weave simulations, presentation, and so on.

Developments in textile CAD systems not only result from software technologies, but are supported by inventions in computer hardware technology. Hardware in textile design studios implement (1) input devices: digital camera and scanner, (2) VDU (visual display unit) and CPU (central processing unit) units, (3) peripheral devices: keyboard, mouse and pressure-sensitive stylus, and (4) output devices of printers.

According to Mary Howard, a former principal of Athena Design System (textile CAD system), technological refinements of CCD (charged-coupled device) created a drastic change in the history of textile computing. The advancements of CCD technology used in input devices of scanners contributed to the introduction of the first textile design computer system in the late 1980s. Prior to this, drum scanners with PMT (photomultiplier tubes) sensors were the dominant input devices for digitizing textile print designs. These drum scanners are massive and expensive devices that have limited access by exclusive end-users. However, flatbed scanners with CCD sensor technology can input the textile design quickly and economically by scanning images in 300 dpi, which is the textile design standard. After scanning the analog paper textile designs into the computer for editing, these digitized designs were further color-separated and manipulated in the computer design systems (Howard, 2010). Moreover, one of the most revolutionary inventions in output devices was the introduction of the Iris continuous flow inkjet printer, which was developed and introduced by Iris Graphics Inc. in the US in 1989 (Wilhelm, 2006). Compared to the previous poor qualities of waxy printed surfaces from output devices, which used electrostatic printing technology, the image quality of the Iris inkjet printer was cleaner and closer to the hand-painted surface of paper. At the same time, it increased optical density and the color gamut. Since the introduction of the Iris continuous inkjet printer, inkjet-printing technology has developed into a dominant paper and fabric-proofing device for the textile design industry.

The computer is used in the design process as:

- an information gathering tool
- a designing tool
- a design editing tool
- a presentation tool.

11.3.1 The computer as an information gathering tool

In general, one of the functions of a successful textile designer is to generate new textile designs. Successful textile design should encompass commercial viability and original creativity, which is largely determined by market trends and designer's esthetic intuitions. The trends of a given time period are established by the symbiosis of political, economical and cultural factors in our society (Ujiie, 2006, p. 343). Before existing Internet information technology had mass appeal, the main source of trend research came from printed hard copy, which was published by trend forecasting companies. These companies gather visual information through social symbiosis, to analyze and create visual representations as trend publications. Textile and apparel designers use the information as a starting point for their design concepts. Since the mid-1990s, as Internet technology became our main source for information gathering, trend forecasting companies have now established online designer recourses. Today, some of the more elite textile design companies also take advantage of this ubiquitous access to Internet technology by researching style trends through blogs, Twitter, and social network sites.

11.3.2 The computer as a designing tool

In the classic textile design workflow, after designers create initial design concepts, they begin to visualize their ideas as samples or croquis, which in turn helps to develop design work for the appropriate market. Generally speaking, the majority of original design work created by textile designers is sold in vast trade shows to the industry. After textile companies purchase and acquire original design samples and croquis from individual designers, designers in the company start to edit and alter the designs depending on their production methods. Textile designers employed by textile companies, which include manufacturers, jobbers, converters and mills, will often take these designs and adjust them to conform to standard repeat sizes, color and scale for a particular market. Computer technologies today function differently for each of the diverse textile markets, which include weave, print and knit. In general, all three markets generate textile design croquis and samples, which are necessary as visual representations of the final design inspiration and fabric quality of the final product.

For constructed textiles, design concept outcomes are actualized by physical weave and knit samples, which consist of materials, structures and colors. Generally, computer technology is utilized as a tool to shorten sample production timelines. Computer software has now replaced manual weave and knit drafting processes, which were traditionally executed on a graph paper with pencil. Additionally, computer software can give designers two- and three-dimensional simulations of constructed structures and finished cloth. However, designers only use these simulations for quick reference, since physical finished samples provide more accurate visual and tactile information for esthetic decision-making. In the designing sample stage for constructed textiles, use of computer technologies provides designers more time to explore creative ideas. Integration of computer technology aids in generating more samples within a shorter timeframe, compared to manual methods. Nonetheless, the fundamental role of computer technology is to assist production, and it is far from a creative medium unto itself.

Conversely, computer technology functions quite differently for printed textile design. Through computer software/hardware, textile print designers can develop textile prints and gain additional new creative applications to their existing conventional manual method. As metamedium, textile CAD (computer aided design) software provides new sets of tools, which can include:

- Step and repeat in real time
- Drawing ability with pressure sensitive stylus
- Changing colors and transparencies – trapping and fall-on
- Filling additional patterns in layers and layering
- Duplicating, scaling and rotating transforming motifs.

Moreover, aside from developments of textile CAD software, technical advancements in computer hardware contribute to the textile designers' creative processes. According to Jill Simmons, former Director of Design Solutions and Business Development at Lectra, 'Introduction of cordless pressure sensitive stylus in the late 1980s was revolutionary. Textile designers,then gained creative liberty by freely drawing with a stylus pen to create textile patterns' (Simmons, 2010).

Introduction to commercial off-the-shelf graphic software also plays a large part in the creative process. Since the early 1990s, after economical commercial off-the-shelf graphic software was first introduced, some textile designers have begun to utilize such software into their textile pattern creations. With the advancements and affordability of personal computer hardware, peripheral devices, input devices and output devices, today commercial off-the-shelf graphic software such as Adobe Illustrator and Photoshop are essential design tools for textile designers, along with proprietary textile CAD software. Moreover, recent development in digital inkjet textile printing technologies has also accelerated original creativity further. Inkjet textile printing does not require image-transfer devices of screens and rollers and can visualize images in 24-bit color spaces. Consequently, new textile design styles have been established, which include:

- Design with millions of colors
- Diminutive images of extreme tonal and fine lines
- Photographic manipulation
- Special digital effects with filters
- Large single engineered image (Ujiie, 2006, p. 345).

Today, all of these esthetic variations can be seen in high-end fashion print and home furnishing printed textiles.

There are two types of graphic systems used in textile CAD: (1) raster graphics and (2) vector graphics. Raster graphics systems represent images as sets of pixels in a grid format and each pixel can be represented in a different hue, value and chroma. Therefore, they are suitable for drawing patterns with continuous tones as well as for assigning weave and knit structures by replacing traditional methods of drafting structures on graph paper. In contrast to a raster graphic system, vector graphics systems utilize mathematical equations to represent geographical primitives, which include points, lines, curves and shapes. One of the advantages of a vector graphics system is that the images are resolution independent and can retain clarity and sharpness in any enlargement or magnification. In the late 1980s, when textile CAD was still in the development stage, some of the design systems were solely developed with vector graphics systems, since the final ideal output resolution was still unclear in the industry (Howard, 2010). However, today most of the proprietary textile CAD software utilizes both raster and vector systems, and designers have choices as to which system is appropriate for which production outcome.

11.3.3 The computer as a design editing tool

Today, the majority of proprietary textile CAD software is developed for utilization in textile design editing. While constructed weave and knit samples are utilized for color reference, tactile quality, and construction, print design croquis are specifically utilized for pattern development. These include graphic patterns for textile print designs, as well as jacquard weave and knit designs. Depending on particular production methods and speci-fications, these croquis are edited and altered with similar processes. The most common workflow usually follows these sequenced steps:

1. digitization of croquis
2. color reduction and cleaning
3. finishing design in step and repeat, and
4. assigning technical production information.

Most print design croquis are originally created by hand, and are usually painted or printed on paper or fabric. At first, these design croquis are digi-tized by using a scanner to input artwork in 24-bit color space. Depending on input images, color corrections are made, and Adobe Photoshop soft-ware is one of the most popular choices. In the early developments of textile CAD software, some of the standards were unclear among software devel-opers, including file formats, working file resolutions, file compression rou-tines, color management and so on. One of the missions for CITDA (the Computer Integrated Textile Design Association) was to create standards for the industry, and in the early 1990s CITDA made several suggestions such as TIFF (tag image file format) as a standard textile file format and JPEG (joint photographic expert group) as a standard file compression route and independent color space device (Keating, 1994). Similarly, the standards for input and output raster design resolutions were also unclear in the early stage of software developments; however, today these are standardized as 254 to 300 dpi as working resolutions.

After the patterns are digitized, the designers start to color-reduce the designs according to the number of screens for printing, and the number of structures for weave and knit jacquard production specifications. Designs that have 24-bit color space are converted to smaller 8-bit color space. Depending on the design, one pixel color (or an average of a group of similar colors) can be selected as the key color for each color separation. This separation can also be depicted as flat color, or as continuous tonal colors. One of the most critical points in this editing process is that the color that has been reduced should visually look close to the original design. The textile designer/stylist's esthetic judgment plays a key role in creating a successful color-reduced design. During their color reduction process at the production mill, they prepare textile print as well as jacquard weave and knit designs for production. After the initial color separation stage, the so-called cleaning process starts, where stray pixels in the separated file are removed and all design motifs are reshaped in CAD software. These outline stages in the print production editing process are mostly executed by dedicated proprietary textile CAD systems.

After the color reduction process, textile designers start to manipulate the color-reduced designs and finalize them into step and repeat. Although this is the typical order for development workflows, depending on types and styles of designs, color reduction can also come after the design manipulation and step in repeat stage. Nonetheless, this is the core of design preparation, and textile CAD software provides designers with various creative tools to finalize designs for production. At the same time, particularly in printed textile design, several colorways of an original design are created as color variations to encourage marketability.

Lastly, in all three textile categories – knit, weave and print – the finished designs are sent to either the screen engravers for engraving raster information, or technical departments for weave and knit structure for the proofing of strike-off samples for production.

11.3.4 The computer as a presentation tool

Computer technologies used for presentations are divided into two categories: (1) soft proofing, and (2) presentation for final textile products. Soft proofing takes place in the design process as a visual simulation, where design information is integrated with production information. At the same time, computer technologies are also used for creation of the final product presentation for potential clients.

Soft proofing for print design simulates the appearance of the final design production, which includes engraving raster information. Proofing samples of weave and knit designs also imitate the final appearance of constructed cloth, which includes construction structure and yarn style. Final editing decisions are made before production, by viewing simulated textile croquis which, once made, can cut unsuccessful sampling costs. In printed textile design, the latest developments in digital inkjet textile printing have provided a more effective proofing process. Instead of a visual simulation on paper, inkjet printers can actually print on fabric, to create a simulation of what the fabric would look like if it was mass produced. Digitally printed pattern design samples on fabric are thus shown in presentation meetings to market clients, prior to the final conventional mill strike-off process. Today, this is still one of the main uses of the inkjet textile printing technologies, and this process saves time and expense, as compared to conventional strike-off printed textile production (Ujiie, 2006, pp. 340–341).

Computer technologies are also used to create two-dimensional presentations in design storyboards. Usually storyboards consist of visual images that reinforce a design concept, including weave and knit simulations, print design, colorway variations, and pattern mapping. In addition, most of the proprietary textile CAD software is equipped with three-dimensional pattern mapping capabilities that include style templates of apparel designs and furniture in interiors. Some of the proprietary CAD software allows for seamless integration of separated design modules, so that if colorways and designs are altered, the alterations are automatically reflected in the presentation modules. Due to the current advancement of computer processing power and data storage capacity, use of multimedia in design presentations has become increasingly popular. These presentations can be assembled together with a variety of media, including still images, movie clips and sound effects, all of which are created by personal computers.

Currently, CAD, CIM and CAM technologies have been wildly diffused in the textile design fields. Simultaneously, only selected proprietary systems have survived and been utilized at this end of the industry. For example, in the 1990s there were numerous textile and apparel CAD manufacturers, including AFSO/CRE8TIV, Athena Design Systems, AVA CAD/CAM, AVL Looms, Barco Graphics, Cadtex, CDI, EAT, InfoDesign, Gerber Technology, JacqCAD, Lectra Systems, Negraphics, Point Carre, Scotweave, Sophia, TCS, Viable Systems, and so on (CITDA, 1993, 1995; Melling, 1998). However, in 2010, only a handful of selected textile apparel computer systems remained, due to consolidation, mergers and acquisitions among the vendors. The computer technologies used for designing textiles have established their own functions.

11.4   Benefits and limitations of computers for textile design

Throughout the maturation of computer textile design technology, especially in the early stages, textile technologies have changed their capacities and functions which, in turn, have changed users' perceptions and attitudes. Proportionally, benefits and limitations related to textile computing have also changed over time. Today, computer technologies provide the benefits of improved quality, productivity and flexibility, while mechanical functionalities and system costs are unfortunately still their main limitation (Yan and Fiorito, 2007).

Interestingly, in the current textile design field today, Walter Benjamin's classic reproduction theory, which states that the quality of mass-produced objects is lessened due to heightened production, no longer exists (Benjamin, 1969). Computer technology has had a dramatic positive impact on mass textile production, by becoming a *metamedium* that is able to function on many levels at once, contributing to many improvements in the quality, productivity and flexibility in the textile designing process. Depending on the position and function of a textile designer in their company, degrees of creative use of computer technology can seamlessly contribute to productivity and flexibility in the design process. For example, as mentioned, textile designers in the industry use computer hardware and software for multiple variations in design esthetics, mapping into storyboards for marketing and sales, and editing and cleaning for end-use and production at the textile mill. Creative and technical design development by computer technologies is common in the current domain of textile design. At the same time, development of digital inkjet textile printing technologies has brought a new aspect of interrelationships of the computer technologies as a production tool and metamedium. Significantly, in the current digital inkjet textile printing workflow, production inkjet textile printers can print any digital textile design created by CAD software directly onto cloth. This process has eliminated the need for engraving and screen-making processes (Ujiie, 2006, pp. 341–343). These digital textile designs can incorporate any two-dimensional visual effects, created and manipulated by any CAD software, and they are printed seamlessly. Moreover, since it is now possible to print directly on fabric as part of the textile computer design process, the boundaries have been obscured between production and design creation. Textile computer technology has truly evolved into a metamedium that reinforces a generation of processes to become one flawless workflow (Treadaway, 2004).

It is important to mention that textile computer technologies still need technical improvements in representation of tactile materiality and color management. Aside from the visual design appearance of compositions and colors, which can be visualized on VDU and printouts through computer systems, it is almost impossible to communicate the physical parameters of the tactile nature of the printed cloth. The final textile fabric product should retain desirable tactile feedback, and this tactile materiality feedback is one of the most critical design parameters. Significantly, tactile feedback is one of the objectives in haptic computer technology, and this technology has been actualized in the human touch response to the interface of mobile devices and video game controllers. Nonetheless, developing a computer interface that integrates tactile feedback in textiles is far more complex and sophisticated (Ruvinsky, 2003). Considering the fact that textile and apparel designers use their entire hand (with their palm and five fingertips) when they analyze the tactile quality of cloth, computer technologies need more advancement in universal input devices for this purpose.

Similarly, color management systems also need universal standards and enhancements. Since the 1990s, textile CAD systems have started to adopt ICC (International Color Consortium) compliance to their proprietary color systems, in order to accommodate device-independent and cross-platform color management. However, the current textile and apparel industries do not always utilize one single CAD system, but often utilize multiple systems. With multiple systems, the standards of color management become too diverse, and it is necessary to establish one universal color and color management standard in the industry (Swedberg, 2004).

In conclusion, although the current cost of proprietary CAD software has become more economical, textile home furnishing and apparel companies still require sizable investments to facilitate their systems. In the late 1980s, when textile CAD systems were initially introduced to the market, CAD system manufacturers were required to provide proprietary hardware and software, because computer technologies were not advanced enough to manage the massive design files. Each system per station, including hardware and software, cost as much as over $150,000 in the US, and only a limited number of large corporations could afford the systems. However, in today's market, a CAD software system per station (without hardware) cost as little as $25,000 (Simmons, 2010). Yet, it is not easy for small- to mid-sized textile and apparel design companies to adopt the proprietary systems (Yan and Fiorito, 2007). Because of this, new services for renting out proprietary CAD systems have emerged. These rental services are ideal solutions to cost reduction, and provide improved quality, productivity and flexibility in textile and apparel design.

11.5   Future trends

In the current textile apparel industry responsibilities for textile and apparel designers have become more diversified and expanded, due to the rapid design and trend cycles and worldwide competition caused by globalization. At the same time, the role of the textile designer has changed to become a far more multi-tasked and dynamic occupation than formerly. Significantly, the latest trend in the textile design field is that designers are more involved with managing and developing the entire product, instead of just designing and selling a single product. Hence, computer technologies have begun to focus on systems to manage complete product development. Wide ranges of PLM (product lifecycle management) systems have been developed; these are systems that are designed to optimize all the processes involved in managing, designing, developing, and manufacturing collections. Computer technologies of the future will develop systems related to efficient workflow and supply chain management in conjunction with CAD, CIM and CAM applications.

Moreover, the impact of emerging digital inkjet textile printing technologies has had a major impact on textile designers. This printing technology is embedded with a conglomerate of computer technologies, from design to production. Digital inkjet printing has been instrumental in creating new role models that represent entrepreneur designers who have complete control over their own design and manufacture. This new evolving self-employed industry has been referred to as the *neo cottage industry* (Ujiie, 2005). In this model, textile and apparel designers can become designers, printers and manufacturers, and create their own brand, if they have access to a designing environment with a digital inkjet printer. For short-run production, designers can design and produce their final production in-house. When a large volume is required, designers can also outsource printed designs by commissioning a digital printing service bureau. This is an exciting benefit in the development of computer technologies for textile designers.

11.6   Sources of further information and advice

Because of the speedy changes in the CAD, CIM and CAM industries, the best source of computer technology information for textile and apparel designers is journals and websites. Some of these are no longer operational; however, there are still good information resources such as *Textile World*, *Apparel Magazine*, CITDA reviews and conference reports, www.techexchange.com, and websites of CAD and CAM vendors. Another helpful avenue for understanding textile production is by researching reports related to each ITMA (International Exhibition of Textile Machinery) event.

11.7   References

Benjamin W (1969), 'The work of art in the age of mechanical reproduction', in *Illuminations*, ed. Hannah Arendt, trans. Harry Zohn, pp. 217–252.

CITDA (1993), *First Annual Computer Integrated Textile Design Association Conference*, Cary, NC.

CITDA (1995), *Third Annual Computer Integrated Textile Design Association Conference*, Cary NC.

Howard M (2010), Principle of design technologies (personal communication, 28 May).

Kay A (1984), 'Computer software', *Science American*, 25, 3, 52–59.

Keating M (1994), 'CITDA Standards Committee Report', *CITDA News*, 3, 4–13.

Lourie J and Lorenzo J (1967), 'Textile graphics applied to textile printing', *AFIPS Joint Computer Conference Proceeding*, 14–15 November, 33–40.

Melling K (1998), 'Print patterns: creativity, precision', *Textile World*, November, 148, 11, 92–100.

Noonan K (n.d.), 'CAD/CAM's effect on the jacquard weaving industry and what can be expected in the future', available online from http://www.techexchange.com/thelibrary/jacquard_weave.html (accessed 12 June 2010).

Ruvinsky J (2003), 'Haptic technology simulates the sense of touch – via computer', *Stanford Report*, available online from http://news.stanford.edu/news/2003/april2/haptics-42.html (accessed 8 June 2010).

Simmons J (2010), Sales, marketing and business development executive (personal communication, 19 May).

Swedberg J (2004), 'Designers embrace CAD differently', *Apparel Magazine*, May, 45, 9, 38–42.

Textile World (1967), 'Now – it's fabric design by electronics', *Textile World*, January, 75–77.

Treadaway C (2004), 'Digital imagination: the impact of digital imaging on printed textiles', *Preview Textile: The Journal of Cloth and Culture*, December, 2, 3, 256–273.

Ujiie H (2000), Integration of computer in textile design field, presented at *Surface Design International Conference*, Kansas City, MO, May 2000.

Ujiie H (2005), Innovative product development in digital fabric printing, presented at *Digital Textile 2005*, Berlin.

Ujiie H (2006), *Digital Printing of Textiles*, Cambridge, UK, Woodhead Publishing.

Wilhelm H (2006), 'A 15-year history of digital printing technology and print performance in the evolution of digital fine art photography – From 1991 to 2006', *Final Program and Proceedings: NIP22: 22nd International Conference on Digital Printing Technologies*, 308–315.

Wilson J (2001), *Handbook of Textile Design*, Cambridge, UK, Woodhead Publishing.

Yan H and Fiorito S (2007), 'CAD/CAM diffusion and infusion in the US apparel industry', *Journal of Fashion Marketing and Management*, 11, 2, 238–245.

12

Digital printing technology for textiles and apparel

D. J. TYLER, Manchester Metropolitan University, UK

Abstract: Digital technologies are prominent at every stage of textile digital printing. This chapter reviews their role in computer-aided design, in the definition and management of colour, and in the control of the machinery that will deposit precise quantities of dyes onto textile substrates. Textile digital printing is a multi-disciplinary field and it is important for product developers to work in a concurrent mode to ensure effective communication across the supply chain.

Key words: textile digital printing, colour management, CAD software, RIP software.

12.1 Introduction

In many sectors of industry, computer-aided design (CAD) systems have been successfully integrated with computer-aided manufacturing (CAM). This chapter is concerned with the digital printing of textiles, which starts with CAD designs and finishes with those designs printed onto textile substrates. The pathway for achieving digital printing on paper has been with us for many years, but it is only during the past decade that the digital printing of textiles has been able to move away from small bureau activities to sizeable commercial ventures.

The emphasis in this chapter is on the use of digital technologies in three areas: developments in CAD technologies to exploit the potential of digital printing, the use of digital technologies to define and manage colour, and the deposition of colour onto substrates. In order to present these developments in context, sections of the chapter will consider the machinery technologies that have been developed, the expansion of this industrial sector, and the ways fabric dyes can function as printing inks.

In conversation with staff in a company offering digital printing services, the following opinion was expressed about the designs submitted for printing: 'The biggest challenge is to change the mindset of designers! They do not know how to take advantage of the freedom to use colour and to expand their options for repeat lengths and drops.' This comment was made after noting that very few of the designs supplied by customers exploited the potential offered by digital technologies.

259

For many years, industrial textile designers knew that their products would be screen or roller printed in four, six, 10 or more colourways. Digital printing was relevant only as a prototyping tool, to run off samples prior to bulk production being authorised. Whilst this reduced development costs, it meant that these designers had no motivation to exploit the design potential offered by digital printing. This situation has changed slowly as fashion designers have explored design possibilities when preparing collections for shows such as London Fashion Week. Significantly, the first textbook written specifically to support digital textile design was published as recently as 2009. The Introduction has this paragraph about the expanded options available to designers:

'The development of digital printing onto fabric is changing printing methods and removing the restrictions that textile designers have traditionally faced: freed from concerns about repeat patterns and colour separation that are key considerations in screen- and roller printing, designers are able to work with thousands of colours and create designs with a high level of detail. There is also greater freedom for experimentation, facilitating one-off production as well as smaller print runs and prints engineered specifically to fit the form of a garment.' (Bowles and Isaac, 2009, p. 7)

We should recognize that designers work to a brief, and the brief is developed by design directors, brand managers and buyers. The will to innovate and to move into new areas of design needs to be accompanied by a willingness to work with new technologies and different supply chain dynamics. This chapter will point out that the elements for making digitally printed products are in place, and new opportunities are available to all who will commit to working in this multidisciplinary field.

A monograph reviewing the state-of-the-art was by Tyler (2005), and a textbook on textile digital printing was edited by Ujiie (2006). Whilst focusing on issues concerned with computer technologies, this chapter revisits and updates material previously presented.

12.2   Review of digital printing technology

The human vision system is equipped with receptor cells that are sensitive to red, green and blue light. Our brains combine the resultant signals to allow the perception of all the colours of the rainbow. A digital image on a CAD screen is made up of pixels, each with a red, green and blue component. Our vision system integrates these components and we are capable of sensing the full spectrum of colours. Printing colour digitally involves printing pixels, each made up of different ink colours. Our eyes do not detect the pixels (except under a magnifying glass) but vision integrates the visual stimulus.

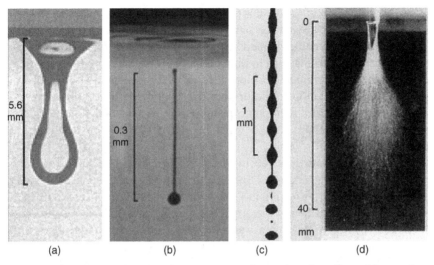

(a)   (b)   (c)   (d)

12.1 Fluids emerging from a nozzle, illustrating the effect of increasing flow rate (images courtesy of Xennia Technology Ltd): (a) dripping faucet; (b) drop on demand (DOD); (c) continuous ink jet (CIJ); (d) atomisation.

The principle of digitising the printing process can be grasped by observing a dripping tap as in Fig. 12.1(a). Each drip allows a drop of water to fall onto the surface below. If the position and movement of the faucet is controlled, the drops of water can be made to wet a substrate in a controlled way. If further controls are introduced so that for each position of the faucet, the drop can either fall or not fall, then the system mirrors the digital printing process reasonably well.

With digital printers, the drop sizes are much smaller than those coming from a dripping tap. As the flow rate is increased, the system changes from drop on demand, illustrated in Fig. 12.1(b), to continuous ink jet (c) to atomisation (d). Each of these modes of operation has been explored by digital print head manufacturers.

The DOD mode shows the drop with a long tail. This is a characteristic feature of forming these tiny drops, and managing the tail so that it does not degrade the image is a major task for both ink and head manufacturers. The CIJ mode shows the continuous jet emerging from a nozzle breaking up into droplets. The challenge for machinery builders is to manage this process and then control where the droplets fall. The atomisation mode represents a higher flow rate and there is no possibility of controlling individual droplets. This latter mode has led to the design on pulsed jet machines, where the heads emit either an atomised spray or they do not.

Using these mechanisms of putting ink onto substrates, machinery builders have developed a large number of options for purchase by printers.

However, as a generalisation, it is possible to say that most companies offering digital print services make significant use of DOD technologies.

The printing machinery needs to produce drops of ink of a precise size (typically 50–60 microns in diameter) and ensure they fall on the substrate in a predetermined position. Printed pixels are made up of several drops of different colours. The printed pixels must be of a predetermined density. For example, a popular print quality is 360/540 dots per inch (dpi), meaning 360 dpi in the weft direction and 540 dpi in the warp direction.

For many applications, a resolution of 360 dpi would be sufficient if the drop sizes were large enough. However, the print heads being used in many bureau machines are working in a range of 3 to 12 pl (picolitres) of droplet size. With such print heads, the application of ink at a 360/360 resolution is inadequate. For this reason, a warp-ways resolution of 540 dpi is used to increase the amount of ink put down, even though this means a reduction in productivity. If this method is still inadequate (as some substrates need much ink to be printed properly) it is possible to overprint: the same design information is printed twice.

An additional capability is that printers can create drops of varying size. Larger drops allow the substrate to be covered quicker, but run the risk of pixellation (so the human eye discerns individual drops making up the image). Smaller drops allow for greater print resolution, and are used wherever pixellation is an issue. When printing with variable drop sizes, printers refer to the greyscale mode of operation (Tyler, 2005).

Printing heads are often made up of banks of 180 nozzles for each colour. Some machines double up on the available printing heads, simply to increase the throughput rate. Consequently, the task of controlling when and how each nozzle fires during each pass of the printing head is very demanding and is put under the control of specialist computing software (RIP software – see Section 12.5).

Inks comprise a colourant, a carrier base and various additives. The carrier base (or vehicle) may be water, solvent, oil, phase change fluid (hot melt), or UV curable fluid. However, to date, nearly all the inks for textile printing are water based because they are designed for heads that require water-based inks. This is convenient, because textile colour chemistry makes use of dyes that are applied using the medium of water.

The usefulness of the carrier base is over once the ink is deposited on the substrate. Water is ideal in that it can evaporate relatively easily and is non-toxic. Other carrier fluids may need to be evaporated in a controlled way and fumes exhausted or reprocessed to ensure a safe working environment. The carrier base may constitute 80% of the ink.

Whilst ease of evaporation is desirable once the colorant is on the substrate, it is not desirable at the nozzle. All digital printers have to be able

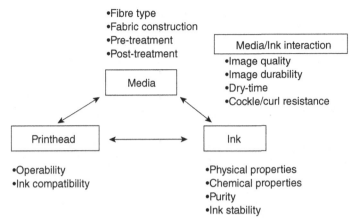

12.2 Interaction of the three key components involved in producing an inkjet image (redrawn from Provost, 2009).

to monitor and correct for blocked nozzles, which can easily occur if heads are inactive and exposed to the air for any length of time.

Colourants are either pigments or dyes. Dyes have to be capable of being chemically bonded to the textile fibres that form the substrate. For example, cellulosic fibres require reactive dyes, polyester requires disperse dyes, and protein fibres require acid dyes. Before printing, fabrics need to be clean and there may be a need for pretreatment. After printing, it is necessary to fix the dye by the application of heat or steam.

The interactions between the ink, the media or substrate, and the printhead are complex. Decisions affecting one component affect the others. The nature of these impacts, with implications for the media–ink interaction, is illustrated in Fig. 12.2. Further discussion of these issues is in Tyler (2005) and Provost and Lavery (2009).

12.3 Global developments in digital printing technology

For the past decade, analysts have been saying that the tipping point for digital printing of textiles is close. The goal has been to mainstream digital printing: to make it as competitive as screen printing for lengths of about 1000 metres. However, despite many innovations and the launching of new, improved machines, the breakthrough has been elusive.

Inkjet printing technologies, however, have witnessed a worldwide annual growth rate of 11% (Herrera, 2009). This has been achieved by the parallel development of inks, printing technologies and new application areas. Herrera provides specific examples of such developments:

- Water-based inks were developed so that they could be used in large-format display graphics.
- Water-based pigment inks were introduced, giving enhanced fade resistance for signage applications.
- Colour gamuts of these new inks were a significant improvement on traditional silver-halide prints, so the digital photo and fine art industry was transformed by high resolution inkjet printers and specially coated substrates.
- Solvent-based pigment inks were developed to print directly on uncoated vinyl to achieve photographic quality for the sign industry.
- UV-curable inks were developed to print uncoated packaging materials.

This buoyancy has stimulated innovation and a readiness to invest in new technologies. Digital textile printing has benefitted because many of these technologies can be adapted and reconfigured for textile applications.

Between 2000 and 2005 digitally printed textile output rose by 300% to 70 m square metres (Milmo, 2007). In May 2008, Tait wrote: 'Despite all the hype about digital printing over the last few years, just 2% of the world's printed fabric is produced using this method. One of the reasons for this slow uptake is that digital printing is far from straightforward.' In May 2009, at the FESPA Digital Textile Conference, it was estimated that the annual print figure was 30 billion square metres, of which only 0.5–1.0% was digital (Holme, 2009). This means that the production was assessed at 1500–3000 m square metres. From these various figures, it appears that the pace of change is increasing.

The past decade has undoubtedly witnessed major investments. A report by Milmo (2007) documents extensive work by ink manufacturers targeting textile digital printing, as well as mentioning the greater availability of 'prepared for print' fabrics, and other innovations. Machinery developments at the May 2009 FESPA Digital exhibition were said by Scrimshaw (2009) to provide 'a rare flash of optimism in an otherwise gloomy season'. New printers displayed included the Nassenger VII by Konica Minolta, the MS-XR48 by MS, the COLARIS® by Zimmer, and the Isis by Osiris – all of which targeted production fabric printing.

It is thought that the annual global figure for printing textile fabrics in 2010 will remain at about 30 billion m^2, most of which (about 80%) is screen printed. A major report produced by Pira (2009) has attempted to document the present situation and to forecast growth in digital print for textiles over a five-year period. The report considers production in four sectors: display, garments, household and technical. The largest of these sectors is display: for retail environments, special events, exhibitions and hospitality applications (with novel textile products like graphic walls, lightboxes, and wayfinding signage). One of the major drivers for this growth is considered to be the display and signage sector, which represents about 75% of textile digital print production and has always been characterised by vigour and innovation (Provost, 2009). Based on the trends in machinery and markets, the report considers that there will be a nine-fold expansion of industrial output, with an expectation that digital textiles will be about 10% of the total digital print volume in 2014.

Whilst fast production speeds are routine in the digital display sector, it is only recently that machinery for garment fabrics and household textiles has reached the stage where market share can be taken from screen printing. The trend has been for the batch size of screen print runs to reduce and it is not unusual to find minimum order quantities of 1000 metres specified by manufacturers. This is important because the costs of making the screens and starting up the print run are substantial. Digital printing, of course, has no screen costs and no fabric or ink waste during the startup.

As an example of a modern machine for textile digital printing, the COLARIS® was launched at the 2010 FESPA show by J. Zimmer from Austria. Printing is possible on a great variety of substrates (up to 320 cm width and 10 mm thickness) with four, six or eight water-based process colours. COLARIS® is a DOD printer with piezoelectric print heads, capable of producing variable drop sizes and a print resolution of 180–1200 dpi. COLARIS® uses Seiko 508 GS print heads with a droplet size between 12 and 36 pl. This allows printing a design at a resolution of 360/360 dpi with an excellent print image. Zimmer offers a one-stop solution: the COLARIS® Inkjet printer includes the fabric unwinder, the printing machine, a dryer unit, plus control software.

Printing capacity is measured in linear metres per hour of operation and, of course, this varies with the resolution selected and fabric width. Data provided by COLARIS® is shown in Fig. 12.3. Four different resolution values are shown. The first number shows the resolution in the weft direction and the second number shows the resolution in the warp direction. The third number shows overprint (2) or no overprint (1).

When compared with screen printing, the numbers in Fig. 12.4 reveal competitive costs for most scenarios for printing a batch of up to 1000 metres. Developments like this bring the breakthrough for garment and household fabrics very close.

Another machinery development that sets out to compete with screen printing is the Isis for Osiris. This printer has a resolution of 144 dpi, which is significantly lower than that of other products targeting this market. The explanation (reported by Scrimshaw, 2009) is that the Isis is designed as a direct competitor to screen printing: it has the same quality but a better price/performance ratio for shorter print runs demanded by the market. It is reputed to print at 1000 square metres per hour, compared with 360 square metres per hour for the COLARIS® at 360 dpi.

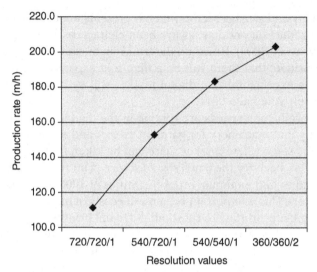

12.3 COLARIS® production rates with 180 cm fabric width (graph courtesy of J. Zimmer GmbH).

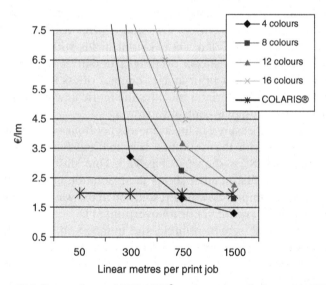

12.4 Comparison of COLARIS® and screen printing with 180 cm fabric width (graph courtesy of J. Zimmer GmbH).

12.4   Colour technology and colour management

12.4.1 Light emitters and light absorbers

People see colour in two ways: from light emitters (such as the Sun and computer colour monitors), and from light absorbers (such as fabric, skin, hair). There are significant differences here that should be understood by users of printing technologies.

The Sun gives us a complete spectrum of visible light, as is apparent when we use an artificial glass prism or look at a rainbow, which is a natural means of refracting light. Other emitters are more restricted in the colours they offer: tungsten filament lights have more yellow and red than sunlight, and low-pressure sodium street lights are almost monochromatic with an orange light (wavelength 589 nm) that leaves everything looking washed out. CAD screens have been cleverly designed to allow users to perceive millions of colours by combining the three primary colours: red, green and blue (RGB). The human vision system combines these colours to give oranges, yellows, white, etc. Our eyes have three receptor cells that are sensitive to these three colours and our brains combine the signals and allow the perception of all the colours of the rainbow. CAD screens work by adding red, green and blue together to create the perception of a continuum of colour. This is referred to as 'additive' colour.

Light absorbers work rather differently. Pigments in a material absorb certain frequencies and allow others to be reflected. Human vision senses the reflected frequencies and registers the colour of the material. The colour that is seen is the result of selective absorption of incident light and is referred to as 'subtractive' colour. Human vision is able to combine the colours cyan, magenta and yellow (CMY) so as to reconstruct the primary colours. Therefore, in principle, CMY colours can be used in printing to reproduce the colours available via a CAD screen. In practice, the colour gamut is more restricted and colour printers often have to use other colours to achieve desired colour effects. Furthermore, to get a good black, printers have traditionally used a black ink (K) rather than a combination of CMY. Hence, the standard office colour printer has four colourways: CMYK. With pastel shades, pixellation may be a problem. To overcome this, light cyan, light magenta and light yellow may be added to the available inks. To increase the colour gamut (the range of colours that can be printed), other colours may be added (e.g. orange).

CIE colour space was developed in 1931 as a way of analyzing vision quantitatively. The intention was to model the vision sensitivity of a human observer and to express the colour by a set of coordinates. CIE space is represented by an x–y plot with green, blue and red primaries occupying different limits of the colour space. Moving around the boundary allows a change of *hue*: moving anticlockwise goes through the spectrum: red, orange, yellow, green, blue, indigo, violet. In Fig. 12.5, numbered locations along the curve are the wavelengths (in nanometres) of the colours. The filled area represents the space of colours that can be perceived by the human eye. As one moves from the outside margin to the centre, there is a dulling of the colour: the hue becomes less saturated.

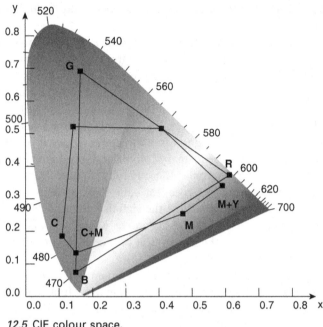

12.5 CIE colour space.

Also marked are the colour gamuts for a CAD screen (RGB colour) and for a four-colour printer (CMYK). Printers, CAD screens and other tools for communicating colour are not able to reproduce all the theoretical colours. For each, a colour gamut must be defined. These colour gamuts, if known, will inform the user about the colours that are within the capability of the office printer, CAD screen, scanner or textile printer.

Different manufacturers produce monitors with slightly different gamuts. Furthermore, as monitors age, the phosphors emitting RGB colours deteriorate and the gamut changes. Clearly, any attempt to standardise the colours seen on CAD screens will require calibrating the monitors. A similar situation applies to printers because inks exhibit signs of ageing.

Black is printed as a 'composite black' using CMYK inks. Minimising the use of black ink overcomes the problems of having dirty-looking shades and pixellation. The colour of black is dependent on the dyes used. For example, with disperse dyes, some blacks are yellower and some are redder.

The CIE colour space allows meaningful relationships to be established between additive colour (e.g. from a CAD screen), subtractive colour (e.g. a printed fabric) and human vision. Using this approach, the colour produced by one set of inks (CMYK process colours) can be matched to the colour produced by another set of inks (spot colours as used in screen printing) by adjusting them to produce the same CIE coordinates. This application is utilized whenever digital printing is used for prototyping a screen-printed textile. The issues involved are discussed by Mikuž *et al.* (2005).

12.4.2 The challenge of colour management

Gamut mapping

The first colour management issue concerns the accurate reproduction of a specified image using the printer and inks available. The problem is that a four-colour printer cannot reproduce all the colours seen by the human eye. It has a reduced colour gamut. Consequently, compromises have to be made. The techniques used are known as gamut mapping and all involve computer processing.

Colorimetric correction matches exactly all the colours within the gamut of the print system being used, and adopts the closest match for all the others. This may deliver acceptable results, but it has the reputation of being unsuitable for photographs with significant out-of-gamut colours. Perceptual (or photometric) correction applies a common correction to all the colours, so that all can be located within the gamut of the print system. This means that the printed image will have less saturated colours than the original, but there will be an internal consistency in the way colours are represented. Other approaches to correction are possible (see, for example, Bae and May-Plumlee, 2005). The technical issues of gamut mapping are surveyed by Morovic and Luo (2001). CAD systems associated with printing software will normally offer the user several of these options.

An alternative strategy is to enhance the technology by increasing the number of inks (by using a six-, seven- or eight-colour printer) or by improving the quality of the inks being used with the four-colour printer.

Colour specification

There is a second, and more significant, colour problem relevant to textile digital printing. Often the starting point for colour selection by designers is not standardised. Some may work from a CAD screen (RGB colour), some will use swatches of fabric from suppliers or a Pantone standard (spot colour), and some will start with a printed or scanned image (CMYK colour). These different starting points have to be addressed by any colour management system subsequent to the design work.

The tools used to process colour at the design stage are CAD systems, scanners, digital camera images and office printers (which have RGB and CMYK gamuts). Thus, in many cases, spot colours are converted to a digital format in order to process them digitally. This is not a problem if the spot colour lies within the gamut of the particular tools being used, but it is a problem if it cannot be accurately reproduced by RGB or CMYK gamuts.

The finished product, if dyed or screen printed, has a spot colour and, if digitally printed, it has a CMYK colour gamut. If the designer starts by specifying a colour outside the range of CMYK printing, then there can be no possibility of reproducing that shade using digital printing technologies.

Colour management solutions

The problems are illustrated by the experience of Sublitech, an Australian company specialising in digital dye sublimation and digital direct printing (Adams, 2010):

> 'We work with the top end of the swimwear industry', Faill [the CEO] says. He acknowledges this form of printing is 'not the cheapest' but the main concern for this Australian operation is on maintaining a high quality rather than competing in a more price sensitive market sector. 'We're always under pressure for delivery, important is short turnaround', he says. 'China is attractive to a lot of people until they open a container and find the "blue" they wanted is "green" or the "green" they asked for is "pink", I've seen a lot of that', he says. 'We work very hard on colour matching and colour consistency which is really important for our larger customers. When they order so many metres of "x, y and z", they want it to be the same as their last order of "x, y and z".'

A significant step towards standardisation is possible using physical colour standards. Companies may create a colour atlas that represents the colour gamut of the printer. This works well as long as the designer/buyer adopts the colour atlas and as long as the inks used by the printer are unchanged. This approach works best with solid colours. However, textile digital printing can make use of images and gradations of colour. With prints of this type, where there are effectively hundreds or thousands of discrete colours, the colour atlas approach is of limited value in the management of colour.

Pantone, Inc. has made a business based on providing standard swatches, linked to CAD data that can be incorporated into colour palates. Designers choose colours and develop palettes using the Pantone colour book. This means that the starting point is standardised, and the supply chain can acquire the same standard swatches and incorporate them within their quality systems. There is no doubt this has had a beneficial effect. However, Pantone colour standards are drawn from a gamut that is broader than that achievable with CMYK printing. Consequently, with digital printing, a subset of Pantone standards has to be used, confined to those shades that are achievable commercially. With this proviso, the Pantone system is workable.

Utilising the concepts of the CIE model, colour control is possible by specifying spectral reflectance data. Ultimately, this is the most meaningful approach for textile dyers and printers. Contemporary CAD software systems are capable of managing colour to link sample, monitor display, paper printing and textile printing. This is done by extensive calibration of components and software routines that ensure the whole system is coordinated. A more detailed overview of some commercial systems is by Henry (2004). A technical presentation relating to one specific colour management solution is by Oulton and Young (2004).

The generic procedure for textile digital printing is as follows:

- A setup routine uses a specified mix of inks with known spectral data.
- A substrate is printed with a range of standards.
- Spectral data of the prints is obtained and compared with the dyestuff database.
- Revised specifications of mix are calculated so that the intended print spectral data are obtained.

Calibration is also extended to the screen display (RGB colour) and the paper printer (CMYK). Tools and software routines for doing this are becoming widespread.

The main difficulty is that the rest of the supply chain lacks familiarity with colour technology and procedures for the quantification of colour standards. In particular, designers and buyers are not used to working with spectral values. It should be noted that this situation is not static, and designers are perfectly capable of using computer tools when these are perceived as enhancing the design function. Developers of colour standards based on spectral reflectance values have sought to provide tools that 'capture' spectral data in the earliest stages of design work, in a way that is unobtrusive and that does not inhibit creativity.

Small, desktop spectrophotometers are available for the use of design personnel. The device can capture the spectral reflectance data of a swatch, magazine image or photograph, and provide an RGB representation of the data on the designer's CAD screen. The designer does not need to know anything about the meaning of this data, only that a 'fingerprint' of the colour has been stored for reference by others downstream in the supply chain.

The principle is essentially very simple. Once the fingerprint of the colour has been defined, an integrated system of colour management through the supply chain is viable. A review of the current situation is given by Xin (2005), with comments on some of the outstanding issues. The AATCC evaluation procedures for instrumental colour measurement, for instrumental assessment of the change in colour of a test specimen, and for the visual assessment of the colour difference of textile samples, are in the

AATCC Technical Manual (2005). Azoulay (2005) quotes one system supplier as saying:

> 'Color standards are becoming as precise as technology can produce. Using the best current technology (spectrophotometry and quality control software) it is possible to produce standards that if matched will ensure a dyer of first lab dip approval. This is being done around the world today.' (p. 10)

Another colour management specialist contrasts the spectral reflectance approach with more traditional practices (cited by Azoulay, 2005):

> 'Electronic standards are the spectral reflectance measurement of a rigorous color standard. If you plan to send physical color standards to several different mills (either to get competitive dips or to match for different component fabrics), then you will be sending out swatches which are not identical. Even if they are all cut from one single piece, they will not be identical after they arrive at the mills and are exposed to handling, different temperature and humidity conditions, etc. So, it is preferred to provide a single measurement of this standard, and treat the physical pieces as examples which represent the standard, but the "real standard" is the electronic measurement. The physical samples are useful to look at and to understand what the color looks like, but lab dips should be compared against the "real standard" (so that all mills are aiming for the same standard).' (p. 11)

There is, therefore, convergence between the colour management systems appropriate for dying solid colours and those appropriate for digital printing. If the output required is a printed textile, the designer needs to start with colours that are obtainable by the digital printing system. The principles of colour science outlined above are relevant here, and the tools offered by CAD software should be capable of handling issues of gamut limitations in a way that is understandable to users.

Within a digital printing CAD system, either an image can be processed as a scanfile or parts of it can be isolated and edited using colour separation techniques. As an example, a flag can be isolated so that corporate colours can be applied. The system will alert the user if the colours go outside the colour gamut of screen or print.

12.5   Three stages of computing for digital printing

Computing power is employed for three distinct stages in the process of printing textiles digitally. First comes the design work, involving a textile print designer; second, processing the image so that it is in a suitable form for printing; and third, controlling the printer so that each pixel is laid down in the right place with the right mix of inks. These are considered below.

12.5.1 CAD software for creating designs

The first stage makes use of standard CAD software and there are no specialist computing issues to note. The comments of Bowles and Isaac (2009) below are representative of feedback received by the writer from bureau digital print companies:

> 'Software programs such as Adobe Photoshop and Illustrator present the perfect platform for textile design. These have become the industry standard tools for textile designers, offering them the freedom to work with both bitmap and vector based imagery, manipulate drawings and photographs, and create accurate details and graphic effects.' (p. 7)

Digital technologies have the capability of creating prints with a much higher resolution, with as many colours as the designer wants to use, and with no restrictions on the way patterns are repeated. Designers who seek to exploit the potential of these technologies will develop specialist skills in these areas. As an example, consider the opportunities opened up by incorporating digital images into textile designs. Previously, this was of no interest to users of screen printing, because there is inadequate resolution and because of the need to use relatively few solid colours. The first attempts to incorporate photography came in the 1960s and 1970s with the heat-transfer printing of polyester fabrics, but this was limited by the capabilities of the CAD software. Today, with enhancements affecting both software and printing machinery, designers are unconstrained by these technologies.

> 'Being able to manipulate and transform an image digitally means that incorporating photography into textile design is now much more sympathetic to the nature of cloth as a material. Cloth comes to life in a way that paper does not: it moves, reflects light, and is often transparent or highly textured. Photographs formatted as if for printing onto paper can make a stark and incongruous statement when translated onto fabric. Designing a textile often involves a very different sensibility from that inherent in pure photography. On paper, photographs are usually intended as narrative documents, whereas the hybrid use of photography in textile design has begun to create a very different style in which the image is subtle or abstracted.' (Bowles and Isaac, 2009, p. 13)

The other major issue to consider as a digital designer is that of colour management. The conceptual framework for this has been presented in Section 12.4. Designers need to ensure that the spectral data of colours they are working with are captured at an early stage and certainly before the decision is made to release the design work to later stages of product development. This applies to swatches, Pantone colours, printed materials, CAD screen colours or any other source of colour inspiration. In addition, it

12.6 The Kaledo approach to colour management and communication (© Lectra SA and used with permission).

should be checked that the selected colours fall within the colour gamut of the supplier's printers.

CAD systems have been available for several years that address these colour management issues. An example is the Kaledo Print solution by Lectra SA, illustrated in Fig. 12.6.

Alongside the CAD software, designers work with a calibrated VDU, a portable spectrophotometer and a light cabinet for swatches. With these tools, there can be confidence that meaningful information is available to be communicated through the supply chain.

12.5.2 Prepare-for-print CAD software

Design images can be printed directly, with no intermediate processing. For example, this may be the route to print a photographic image for a wall hanging or picture. However, in most cases, a specialist CAD interface is needed. Some application areas are as follows:

- To develop the design in repeat
- To ensure an accurate match between the specified colour and production samples
- To separate colours in the original design and set up alternative colour palettes
- To separate flat and tonal designs
- To simulate various textural effects.

It is not unusual for graphic images to leave design studios with stray pixels, unclear boundaries between areas of different colour, tonal effects that grade into flat colour, and a resolution that does not match with the requirements of digital printing. The prepare-for-print process addresses all these issues. The work is usually undertaken by the printing company making the product: their role is not to contribute to the design but to process the image file to best realise the design concept.

To achieve colour management, printer profiles are required to link the colours specified in the CAD image with the printer/inks/substrates in use. This aspect of the process is technically demanding, but essential. The printer will prepare textile samples representative of the colour gamut and these are scanned. Then a colour profile is created as a data file. In principle, this allows any specified colour to be converted to print instructions that will deliver it. This service carries a significant cost, so producing a colour profile is not a routine matter. When suppliers of dyes/inks make changes, this may trigger a reprofiling exercise. Working with new suppliers may also trigger a new profile. However, the hope is that the new dyes will give results within commercial tolerances of the old dyes, so the printer will always try it out to check.

The experience of one print bureau may be of interest. In their experience, most colours received from customers are poorly defined – often based on a CAD screen display. The bureau will work on the colours, picking the nearest match that they know can be achieved using the relevant colour profile. They then seek customer approval for the suggested matches. Sometimes, it is known that the profile used is not perfect. For example, reds may be too orange, so this stimulates some human tweaking of the colours selected. Generally, these tweaks relate to greens and reds.

Over the past few years, software enhancements have been incremental but significant.

- Ease and speed. This affects, for example, the way colour palettes can be adjusted and the rapidity with which different design files can be embedded on the same page layout.
- Quick colourways. This enables multiple colourways to be prepared with ease.
- Vector graphics. This avoids pixellation when manipulating images. According to one print bureau, there used to be a big problem with image rotation because every time the image needed rotation, it had to be redrawn.
- Preparing and splitting large prints. There is a potential loss of clarity when images are enlarged, so the software has special algorithms to maintain the integrity of lines and avoid pixellation effects. Large-format signage and interior wall coverings are made in sections and sewn

together. Splitting the original image so that it is appropriate for each section has now been made easy.
- Ink consumption monitoring. This information has been integrated with reporting.
- Reductions in manual editing. Automated and semi-automated features assist the tasks of initial image processing and colour separation.

12.5.3 RIP software for precision printing

This specialist software provides colour management according to the printer profiles and controls the execution of the print. There are two levels of control: colour and the speed/quality of the printing process. This software can be considered as a print driver.

When an image is scanned, the two-dimensional area is subdivided to the pixel level and, for each pixel, information is gathered about the hue, saturation and luminance. When all has been gathered, these data can be reassembled by the CAD software and viewed on-screen. The resultant image is known as a 'raster image'. It is a grid of x and y coordinates on a display space, with each point linked to colour (or monochrome) values. A widely used synonym is the bitmap image. Examples of raster image file formats in widespread use are BMP, TIFF, GIF and JPG.

Raster images tend to be large (as each pixel is linked to colour data) and they are not easy to modify. Nevertheless, digital applications cannot do without them. Raster images are produced by scanning and by digital photography, and raster images are required to print (which can be considered as reversing the process of scanning). However, manipulation of images, and much design work, is best done using vector graphics.

A vector is an object that combines quantity and direction. A line can be a vector: it can be defined by its length and its orientation. Vector graphics builds up whole images in terms of vector statements, and this is the basis of most CAD drawing packages. Objects can be modified and resized very easily by changing the parameters of the vectors. They can be manipulated by mathematical processing and other instructions given by the designer. An additional benefit is that vector graphic images are smaller than raster graphic images. Popular drawing packages like Adobe Illustrator, CorelDraw and AutoCad all make use of vector graphics. Many fonts are infinitely scalable, which is a sure indication that they are stored as a set of vector statements.

It is relatively easy to convert vector graphics to raster graphics, but it is not at all easy to do it the other way round. Thus, Adobe and CorelDraw software both allow users to save image files in a raster file format. This conversion of vector to raster graphics is a necessary step for printing images. The hardware/software product that converts CAD images to raster graphic format suitable for specific printers is referred to as a Raster Image Processor (RIP). As the technology has developed, RIP software has been produced with additional functionality, and there is likely to be some overlap with prepare-for-print software.

During printing, the image file is rasterised (converted) to a bitmap, with each pixel associated with colour information. Based on the resolution that has been set, the software works out the speed of execution, the movements of the printhead(s) and the nozzles that will fire at each position. If variable drop sizes are possible (the greyscale capability), this also has to be computed. The processing requirements are immense, so it is customary for each printing machine to have its own RIP computer control.

12.6 Future trends

Textile digital printing is a rapidly developing field and there are developments affecting every aspect. Thus far, the trends in computing infrastructure have been incremental. There is already some convergence between prepare-for-print software and RIP software, but the likelihood is that there will continue to be two stages of activity here: one involving a human operator working on the design image and the other automatically driving the printer.

Inks are continuing to be developed. The hidden cost here is that each new range of inks may trigger the preparation of a new print profile. A recent novelty is the introduction of silver and white colours to extend the range of solvent inks for large-format printers (Mimaki, 2010). By combining silver with other colours, Mimaki claim that gold and bronze effects can also be achieved. At present, this will affect the display sector of the industry.

For some years now, pigment inks have been hailed as a possible way forward for textile digital printing, avoiding the pre- and post-treatments necessary when printing silk and cotton. However, although there have been incremental developments, the goal is still elusive. The problem facing printers is the relatively high viscosity of the polymers needed to bind the pigments to the textile fibres. An update on this research is provided by Provost and Lavery (2009).

Nanotechnology is appearing more frequently in the media about digital printing inks. There is some value in making pigment inks using nanoparticles, reducing clogging and enhancing control over drop formation. There is also scope for the micro-encapsulation of nanoparticles to impart antimicrobial, fragrance and other properties to inks. Evidence of interest in such applications is provided by Leelajariyakul et al. (2008). Perhaps the greatest potential for nanotechnology is in the area of printed electronics on flexible substrates. This has been 'blue sky' research for several years, but a recent press release from Xennia Technology (2010a) indicates the availability of a commercial tool: the precision inkjet development dispenser, the XenJet™ 4000, specifically designed for the deposition of printed electronics and other functional materials:

> 'The innovative proprietary Xennia software incorporated in the XenJet™ 4000 is designed to suit a range of demanding functional and nano-material applications. Benefits of the new software include an increased degree of control of drop volume for layer/film thickness optimisation, full control of image processing parameters for multi-layering applications and a fully integrated intuitive user interface allowing simple operation, or detailed access to print parameters, as required.'

Xennia Technology is participating in two EU-funded collaborative projects, with the codenames 'LOTUS' and 'NOVA-CIGS'. The targeted applications are RFID, photovoltaics, OLED displays and low-cost solar cells (Xennia Technology, 2010d).

An emerging area for textile digital printing is the application of coatings to technical textile products. Xennia Technology Ltd and inkjet printhead manufacturer Trident (Xennia Technology, 2010b) have announced an initiative to target this market. 'Trident's printhead technology is important for Xennia as it is extremely robust and enables printing of large drops that spread easily to form uniform layers, ideal for coatings in key industrial applications such as textiles, glass, floor and wall coverings, furniture laminates and decor papers', said Dr Alan Hudd, Managing Director of Xennia.

Printing machinery developments require close collaboration between printhead manufacturers, ink suppliers and machinery builders. Earlier in this chapter, COLARIS® was the subject of comment. Another innovative machine targeting textile applications has been jointly developed by Xennia Technology Ltd and Reggiani Macchine (Xennia Technology, 2010c):

> 'The patented combination of continuous substrate movement and diagonal printhead movement eliminates the banding issues that can occur with a normal step and scan system, makes transport design easier and allows scalable web widths. The system has a wide print carriage giving extraordinarily high throughput even on very wide webs. The system can use printheads from several manufacturers, allowing flexibility in printhead choice depending on application requirements. This first demonstrator has 16 printheads, giving a throughput of 360 m^2 per hour. Xennia are currently developing a second prototype with double the amount of printheads giving twice the throughput. Reggiani will commercially launch the new production systems onto the market in early 2011.'

All these developments, whether in hardware or inks, must go hand in hand with software to ensure the process is controlled and robust. Integration of these technologies must accompany innovation.

12.7   Conclusions

Digital printing provides a multidisciplinary environment for the development and production of innovative textile and apparel products. The sector is characterised by a large number of companies that have been exploiting available technologies and extending the boundaries of commercial application. It is emerging as a significant industry and as the employer of professionals who are competent in their chosen field of technology.

The work of designers is significantly affected. This is not just because of the skills they need to create designs for digital printing. The lead times for digitally printed products can be very short, because production batches can be as long or as short as the market will take. Consequently, designers exploiting the potential of the technology will be working close in time to the point of sale: for garments, this is the domain of 'Fast Fashion'.

Product development also needs to be short and streamlined in order to respond to market demand. The traditional sequential processes will not suffice: concurrent practices are needed (Tyler, 2008) with teamwork across the relevant disciplines. The core skills to manage the production technologies are to be found in the supply chain, and these skills also need to be utilised in the product development process. The opportunity exists for marketing and buying staff to be effective team members, with the goal of responding to market demand and surprising the market with innovative products.

Happily, this integrated supply chain model is exactly what is needed to achieve colour management. With designers capturing the spectral data of the selected colours, the rest of the supply chain has the information it needs to get samples and production right first time. The discipline of working to numerical colour standards allows problems to be anticipated and corrective action can then be taken.

12.8   Sources of further information and advice

Literature

Bowles, M. and Isaac, C. *Digital Textile Design*, Laurence King Publishing, London, May 2009.

Tyler, D.J. Textile digital printing technologies, *Textile Progress*, 37(4), April 2005, 1–65.

Ujiie, H. (ed.). *Digital Printing of Textiles*, Woodhead Textiles Series No. 53, Woodhead Publishing, Cambridge, UK, 2006.

Trade sources

Advanced Digital Textiles: http://www.advdigitaltextiles.com/
AVA CAD/CAM Group Ltd: http://www.avacadcam.com/en/
Citrus Rain Ltd: http://www.citrus-rain.com/
Color Solutions International: http://www.colorsolutionsinternational.com/
Digetex Ltd: http://www.digetex.com/
Dream Fabric Printing: http://www.dreamfabricprinting.com/
FESPA: http://www.fespa.com/
Forest Digital Ltd: http://www.forestdigital.co.uk/
Lectra Kaledo Design: http://www.lectra.com/en/design/index.html
Mat Creations Ltd: http://www.matcreations.com/
Mimaki Engineering Co. Ltd: http://www.mimaki.co.jp/english/top/index.php
RA Smart Ltd: http://www.rasmart.co.uk/services/digital-printing
Stork Prints: http://www.storkprints.com/
The Inkdrop Boutique: http://www.inkdropboutique.com/
X-rite Ltd: http://www.xrite.com/home.aspx
Zimmer – Austria: http://www.zimmer-austria.com/

Websites

Digital Textile (via Inteletex): http://www.inteletex.com/newsindex.asp
Provost Ink Jet Consulting Ltd: http://provost-inkjet.com/1.html
Speciality Graphic Imaging Association: http://www.sgia.org/Publications/journal/index.cfm
TechExchange – Textile digital printing: http://www.techexchange.com/textile-printing.html

12.9 Acknowledgements

Alan Johnson and Andrew Mosley of Citrus Rain, Josef Osl of J. Zimmer Maschinenbau GmbH, and Tim Hoyle of Tim Hoyle Digital Print & Design Ltd are thanked for feedback and advice when preparing this chapter. Xennia Technology Ltd, J. Zimmer GmbH and Lectra SA are thanked for the use of images as credited in the text.

12.10 References

AATCC (2005), Technical Manual, American Association of Textile Chemists and Colorists, North Carolina.
Adams, D. (2010), Sublitech's Consistent Colour, *atf magazine*, Jan–Feb, 24–25.
Azoulay, J.F. (2005), The devil in the details, *AATCC Review*, 5(2), 9–13.

Bae, J. and May-Plumlee, T. (2005), Integrated technologies for the optimum digital textile printing coloration, *Proceedings of the Textile Institute 84th World Conference*, 20–25 March, Raleigh, NC, 1–19.

Bowles, M. and Isaac, C. (2009), *Digital Textile Design*, Laurence King Publishing, London.

Henry, P. (2004), Computer aided design: Its pedigree and future contribution to the success of digital printing, in: Dawson, T.L. and Glover, B. (eds), *Textile Ink Jet Printing*, Society of Dyers and Colourists, Bradford, UK, 38–43.

Herrera, R. (2009), Latex inkjet printing technologies: an evolution in large-format printing, *SGIA Journal*, Third Quarter, 27–31.

Holme, I. (2009), Insight into diverse markets, *Digital Textile*, No. 3, 29–31.

Leelajariyakul, S., Noguchi, H. and Kiatkamjornwong, S. (2008), Surface-modified and micro-encapsulated pigmented inks for ink jet printing on textile fabrics, *Progress in Organic Coatings*, 62(2), 145–161.

Mikuž, M., Šostar-Turk, S. and Pogačar, V. (2005), Transfer of ink-jet printed textiles for home furnishing into production with rotary screen printing method, *Fibres & Textiles in Eastern Europe*, 13(6), 79–84.

Milmo, S. (2007), Developments in textile colorants, *Textile Outlook International*, No. 127, 19–42.

Mimaki (2010), Mimaki introduces new silver ink for JV33 and CJV30 series (20 July 2010), http://www.mimaki.co.jp/english/sg/topics/2010/20100720.html

Morovic, L. and Luo, M.R. (2001), The fundamentals of gamut mapping, *Journal of Imaging Science and Technology*, 45(3), 283–290.

Oulton, D.P. and Young, T. (2004), Colour specification at the design to production interface, *International Journal of Clothing Science and Technology*, 16(1), 274–284.

Pira (2009), *The Future of Digital Printing for Textiles: Market Forecasts to 2014*, Pira International, December.

Provost, J. (2009), Soft signage: a major growth area, *Digital Textile*, No. 3, 6–9.

Provost, J. and Lavery, A. (2009), Interactions of digital inks with textile and paper substrates in ink jet printing. Provost Ink Jet Consulting Ltd, http://provost-inkjet.com/resources/Interactions+of+Ink+and+Media+in+Ink+Jet+Printing.pdf

Scrimshaw, J. (2009), Buoyant mood belies economic downturn, *Digital Textile*, No. 3, 22–28.

Tait, N. (2008), Digital printing in action in Asia, *just-style.com*, 27 May, http://www.just-style.com/analysis/digital-printing-in-action-in-asia_id100908.aspx

Tyler, D.J. (2005), Textile digital printing technologies, *Textile Progress*, 37(4), 1–65.

Tyler, D. (2008), Advances in apparel product development, in: Fairhurst, C. (ed.), *Advances in Apparel Production*, Woodhead Publishing, Cambridge, UK.

Ujiie, H. (ed.) (2006), *Digital Printing of Textiles*, Woodhead Textiles Series No. 53, Woodhead Publishing, Cambridge, UK.

Xennia Technology Ltd (2010a), Xennia introduces latest generation inkjet development dispenser at Printed Electronics Europe 2010, Press Release, 12 April, http://www.xennia.com/newsandevents/news-detail.asp?ItemID=60

Xennia Technology Ltd (2010b), Xennia signs supply agreement with Trident for industrial and textile inkjet coatings, Press Release, 24 May, http://www.xennia.com/newsandevents/news-detail.asp?ItemID=93

Xennia Technology Ltd (2010c), Xennia and Reggiani announce their jointly developed industrial digital textile printer, Press Release, 25 June, http://www.xennia.com/uploads/100625-PR-Reggiani.pdf

Xennia Technology Ltd (2010d), Xennia starts two major new EU printed electronics collaborative projects, Press Release, 5 January, http://www.xennia.com/news-andevents/news-detail.asp?ItemID=56

Xin, J.H. (2005), Colour management for the globalised textile and clothing industry, *Textile Asia*, 36(1), 30–35.

13

Approaches to teaching computer-aided design (CAD) to fashion and textiles students

P. C. L. HUI, The Hong Kong Polytechnic University, Hong Kong

Abstract: This chapter aims to review the current pedagogic approach including constructivism, problem-based learning, experiential learning and global-scope learning in teaching computer-aided design (CAD) to fashion and textiles students. The historical development and challenge to each pedagogic approach are addressed. The main challenges include rapid changes of industry needs and wants, rapid changes of computer technology, global working environment for fashion and textiles, and non-standardization of jargons used in fashion and textiles. Comparison of each pedagogic approach in terms of principles, learning outcomes, materials/technology required, role of teacher, and assessment method has been made. Through such comparison and case studies in two various academic environments, three areas of improvement on pedagogic approach are proposed.

Key words: CAD, pedagogic approach, fashion and textiles.

13.1 Introduction

Pedagogy or teaching and learning involves knowledge and skills, classroom management, and effective teaching practices. It is a complex blend of professional knowledge and practitioner skills (Lovat, 2003). There are various pedagogic approaches (i.e. teaching and learning strategies) used in teaching computer-aided design (CAD) to fashion and textiles. This chapter reviews each approach and its challenges. Based upon the challenges of each approach, a case study is used to identify improvements in teaching CAD.

13.2 Review of approaches to teaching computer-aided design (CAD)

13.2.1 Constructivism

In constructivism, constructive processes operate and learners form, elaborate, and test candidate mental structures until a satisfactory one emerges (Perkins, 1991). This implies that experience is created by learners rather than existing in the world independently. Based on the influence of the

283

Swiss psychologist and epistemologist Jean Piaget (1896–1980), the nature of knowledge should be studied empirically where it is actually constructed and developed. This can be done either through the historical development of knowledge or by studying the growth and development of an individual. Constructivism has developed from such a Piagetian perspective, and has drawn on other theorists who put more stress on social and cultural conditions for learning. This may explain why there are many varieties of constructivism. The other main contributor to this development is the Russian Lev Vygotsky (1896–1934). Both of them have some fundamental similarities. Piaget was interested in knowledge creation but Vygotsky was more interested in understanding the social and cultural conditions for human learning.

Constructivism is concerned with how learners construct knowledge. The construction of knowledge is a function of the prior experience, mental structures and beliefs that a learner uses to interpret objects and events. Constructivism does not preclude the existence of an external reality. Learners construct their own knowledge for their use. Constructivism will be discussed based on the following assumptions (Uden, 2004):

1. All knowledge is constructed and not transmitted.
2. Knowledge and meanings result from activity and are embedded in activity systems.
3. Knowledge is distributed in persons, tools, and other cultural artefacts.
4. Experience arises out of interpretation and thus multiple perspectives are recognized.
5. Knowledge construction is prompted by problems, questions, issues, and authentic tasks.

Hence, the basic premise is that people are active learners and must construct knowledge themselves (Geary, 1995). Another premise is that teachers should not teach in the traditional way of delivering knowledge to a group of students. Teachers should define situations in a manner that learners become actively involved with the content through some activities such as observing phenomena, collecting data, and working collaboratively with others (Schunk, 2004). Application of constructivism includes self-regulation and instructional scaffolding. Self-regulation refers to planning, checking, and evaluating (Bruning *et al.*, 1999). Instructional scaffolding refers to the process of controlling task elements that are beyond the learners' capabilities so that learners could focus on and master those features of the task the learners can grasp more quickly (Bruning *et al.*, 1999). In teaching CAD to fashion and textiles students using constructivism, the student is urged to construct questions and seek out possible answers rather than feeding information to the students through direct instruction. Using this approach, the teacher is maintaining the role of facilitator in the learning process.

13.2.2 Problem-based learning

Problem-based learning (PBL) is a curriculum development and delivery system that recognizes the need to develop problem-solving skills as well as the necessity of helping students to acquire necessary knowledge and skills.

The first application of PBL, and perhaps the most strict and pure form of PBL, was in medical schools which rigorously test the knowledge base of graduates. Medical professionals need to keep up with new information in their field, and the skill of life-long learning is particularly important for them. Hence, PBL was thought to be well suited for this area.

According to Vernon and Blake (1993), problem-based learning (PBL) is an instructional approach that uses problems as a context for students to acquire problem-solving skills and knowledge. It is the study of real or hypothetical cases in small discussion groups engaged in student-to-student and/or student-to-teacher collaborative study using hypothetical-deductive reasoning with a style of direction that concentrates on group process rather than the provision of information (Vernon and Blake, 1993). The main focus of PBL is to confront students with a problem to solve as a stimulus for learning (Boud and Feletti, 1991). The problem does not test skills, but assists in development of the skills. Problems in PBL have no single solution. Problems are ill-structured, messy, and complex in nature and require problem-solving and critical thinking skills to resolve. By working through the problems, students learn the necessary skills to deal with real-life problems that they have to face in the real world (Uden and Beaumont, 2006). Students in PBL acquire many important skills that are essential in life. PBL encourages students to take a detailed view of all issues including concepts and problems. It also enables students to develop skills such as literature review and information processing.

PBL may be useful for educating computing students because this subject undergoes rapid changes every day. There is a large amount of material to be learned and it is conceptually complex. In real life, most computing problems are ill-structured and thus decisions must be made with incomplete information. Meanwhile, computer technology is advancing at a rapid rate, so students must be able to keep pace with the advancements and acquire the necessary knowledge on demand. PBL offers a teaching tool that lets students learn the knowledge through problem solving (Uden and Beaumont, 2006).

13.2.3 Experiential learning

Experiential learning provides a holistic model of the learning process and a multilinear model of adult development. Both of them are consistent with what we know about how people learn, grow and develop. It emphasizes the central role that experience plays in the learning process and how experiential learning is differentiated from both cognitive learning, which tends to emphasize cognition over effect, and behavioral learning, which denies any role for subjective experience in the learning process.

Experiential learning is the process of making meaning from direct experience. It focuses on the learning process for the individual so that learners make discoveries and experiments with knowledge first-hand instead of hearing or reading about others' experiences. Experiential learning requires no teacher and relates solely to the meaning-making process of the individual's direct experience. Such experiential learning originated in the experiential works of Dewey, Lewin, and Piaget. Taken together, Dewey's philosophical pragmatism, Lewin's social psychology, and Piaget's cognitive–developmental genetic epistemology form a unique perspective on learning and development (Kolb, 1984). According to David Kolb, an American educational theorist, the following abilities are required for gaining knowledge from experience (Merrian et al., 2007):

- The learner must be willing to be actively involved in the experience.
- The learner must be able to reflect on the experience.
- The learner must possess and use analytical skills to conceptualize the experience.
- The learner must possess decision-making and problem-solving skills in order to use the new ideas gained from the experience.

Experiential learning can be a highly effective educational method. It engages the learner at a more personal level by addressing the needs and wants of the individual. Experiential learning requires qualities such as self-initiative and self-evaluation. For experiential learning to be truly effective, it should employ the whole learning wheel from goal setting, to experimenting and observing, to reviewing, and finally to action planning. This complete process allows learners to gain new skills, new attitudes or even new ways of thinking. It is vital that individuals are encouraged to directly involve themselves in experience to gain a better understanding of the new knowledge and retain the information for a longer time.

In teaching CAD to fashion and textiles students using this approach, students must return to the lesson more responsibly and with a greater insight into the nature of the design problem. The teacher acts as an experiential facilitator who provides only sufficient encouragement to ensure that students continue to confront the challenge of the activity, and ample feedback so that students visualize the steps to the outcome and understand their own progress through the activities.

13.2.4 Global-scope learning

This is an innovative learning strategy which integrates an international intercultural dimension into teaching, research and community service to enhance its academic excellence and relevance of its contribution to society. It has evolved into formal exchange programs, virtual universities, and double degrees between global institutions (Ankerson and Pable, 2008). This approach allows learners to gain knowledge and skills a cross borders for enriching their global views. It is well adopted by many interior design programs to broaden the perspective of their learners. Some institutions, such as Texas Christian University, are taking the step of participating in consortiums of institutions in teaching interior design that cross national borders.

Interdisciplinary approaches into opportunities to study abroad are necessary. While studying abroad provides meaningful experiences, technology plays an integral role in embedding international activities into the classroom, particularly for students who cannot go abroad. To implement such pedagogical approaches, technology such as WEB 2.0 plays an integral role in embedding international activities into the classroom. Such technology could facilitate classroom learning and online discussion to understand the global issues of a particular study area. It is particularly useful for sharing cultural issues of fashion design work among different countries. Collaboration between different academic institutions is a key issue in implementing such approaches successfully. Teachers at various academic institutions act as supporters for sharing their knowledge in order to give learners a global view of a particular subject (Ankerson and Pable, 2008).

In order to have a better understanding of each pedagogic approach used in teaching CAD of fashion and textiles, comparison of each pedagogic approach in terms of principles, learning outcomes, materials/technology required, the role of the teacher, and assessment method is illustrated in Table 13.1.

13.3   Challenges presented by each approach

The main challenges of teaching CAD to fashion and textiles students are directly related to the industry needs and wants. In the past, the industry expected graduates to possess the essential skills of mastering state-of-the-art CAD software in order to produce the marker plan, grading plan and production sketches with respect to a particular fashion design and/or style.

Table 13.1 Comparison of four pedagogic approaches

	Constructivism	Problem-based learning	Experiential learning	Global-scope learning
Working principles	Experience is created by learners rather than existing in the world. It mainly concerns with how learners construct knowledge.	An instructional approach that uses problems as a context for students to acquire problem-solving skills and knowledge. It does not test skills but assists in development of the skills.	Learners make discoveries and experiments with knowledge first-hand instead of hearing or reading about others' experiences.	An innovative learning strategy that integrates an international intercultural dimension into teaching, research and community service to enhance academic excellence.
Learning outcomes	Learners construct their own knowledge through active learning	Learners learn the necessary skill (such as literature review and information processing) to deal with real-life problems. It encourages learners to take a detailed view of all issues.	Learners gain new skills, new attitudes or new ways of thinking.	Learners gain knowledge and skills across borders for enriching their global views.
Materials/technology required	Activities such as observing phenomena, collecting data, and working collaboratively with others are included.	An ill-structured problem, computing technology, and some teaching tools are used.	The whole learning wheel is employed, from goal setting to experimenting and observing, reviewing, and finally action planning.	Technology such as WEB 2.0 could be employed to facilitate classroom learning and online discussion. Collaboration between different academic institutions is necessary.
Role of teacher	Facilitator	Facilitator	Experiential facilitator to encourage their learning.	Supporter for sharing their global knowledge.
Assessment method	Learners could be assessed by direct and indirect measures such as project assessment and external feedback.			

Basically, CAD is a computer tool that allows a fashion designer to visualize his or her design work. If his or her design work is accepted by the market, CAD becomes the communication tool that allows a producer to prepare a production plan such as marker and grading for mass production. With the rapid advances in computer technology, in order to increase the productivity of designers and to improve communication between designers and producers, the features provided by state-of-the-art CAD software are updated from time to time. In order to meet such quick response concepts, the teaching materials of CAD for fashion and textiles students are similarly kept up to date. Meanwhile, designers and producers cannot work in isolation; they need to work globally and share their views of fashion trends as they occur. Standardizing the jargon used in fashion and textiles is necessary in CAD systems and thus learners are kept up to date with such knowledge before they work in the industry.

Numerous researchers (Barker, 1999; Goodyer, 1999; Hannifin, 1999; Hayes *et al.*, 2001) identify the following changes of pedagogy in teaching computer-based technologies:

- A shift from 'instructivist' to constructivist education philosophies
- A move from teacher-centered to student-centered learning activities
- A shift from focus on local resources to global resources
- An increased complexity of tasks and use of multi-modal information.

Such proposals may not fully meet the above challenges. For example, in constructivism, learners should construct their knowledge through some activities such as observing phenomena, collecting data and working collaboratively with others. However, in real life, an improper learning environment exists because it is difficult to make observations directly in the field, and collaboration with industry is not adequate for their learning. In addition, there is no clear mechanism to standardize the jargon used in fashion and textiles at this moment. This will greatly affect the implementation of each pedagogic approach.

13.4   Case study

In order to analyze the current pedagogic approach used in teaching CAD to fashion and textiles students, two case studies have been carried out. The first is from the Higher Diploma program and Bachelor of Arts degree program in Fashion and Textiles offered by the Institute of Textiles and Clothing, the Hong Kong Polytechnic University (ITC, HKPolyU). The other case study is from the Bachelor of Arts degree program in Fashion Design offered by the School of Design and Manufacture, De Montfort University, Leicester, UK (SDM, DeMontfortU).

13.4.1 Teaching and learning facilities

For the former case, all teaching classes are conducted in two computer laboratories. Each laboratory has been well equipped with high-end CAD workstations with the following software and optical scanner.

- PrimaVision for Knit: This CAD software is used for designing knitted fabric with various types of yarn materials.
- PrimaVision for Weave: This CAD software is used for designing woven fabric with various types of yarn materials. Both PrimaVision for knit and PrimaVision for weave could simulate the finished fabric design and would prepare the fabric sample via digital printing.
- Photoshop/Illustrator: This software is used for preparation of design work on scanning any background work. It supports the CAD software for knit and weave to develop a complicated fabric design.
- Gerber for preparation of marker and grading: This software is used for preparation of marker layout and grading in apparel production based on the outcomes of the apparel CAD software.

Two color printers and a plotter are connected with each workstation via a local area network. Each laboratory could accommodate more than 20 fashion and textiles students for learning.

For the latter case, at De Montfort University, all practical classes are conducted in two CAD laboratories. One CAD laboratory has 48 IBM PCs, nine monochrome printers, two color printers, two A3 color thermal wax printers, an A4 color dye sublimation printer, an A0 plotter, two A3 plotters and two A3 color scanners. In this laboratory, software of MS Word, Corel Paint and Corel Draw, Aldus Pagemaker, Autodesk AutoCAD and AEC, AutoCAD 3D Studio, Ormus Fashion, Tex-Design, Tex-Dress and Clarkes Shoemaster are installed in each PC. The other CAD laboratory houses 39 Apple Macintosh computers with two monochrome printers, two color printers and a color scanner; software of Aldus Freehand, Quark Express, Adobe Photoshop, Macromedia Director and Adobe Premier are installed in each Apple Macintosh.

13.4.2 Curriculum design

The Institute of Textiles and Clothing (first case study) adopts the learning outcomes approach. This refers to what graduates are expected to be able to do or demonstrate on completing the subject. In the curriculum design, the learning outcomes are as below:

1. To describe the application and benefits of the CAD systems in the fashion industry

2. To apply the principles of CAD to fashion and textile design and product development
3. To integrate the CAD system with other computer systems to achieve quick response and other benefits in the fashion industry.

The subject is covered in two parts: 14 hours of lectures and 42 hours of hands-on practice within 14 weeks. The teacher introduces the core principles of the subject and demonstrates the problem-solving skills in the CAD software in all lectures. Hands-on practice allows students to concentrate on their skills development and encourages collaborative experiences among peers of developing fabric design work including fabric construction, color combinations and texture effects. Problem-based projects may be assigned to each team, which comprises four or five students working together to strengthen their basic team/communication skills. Through this curriculum design, students are expected to achieve the above-mentioned learning outcomes.

The School of Design and Manufacture (second case study) adopts the modules-based approach. The curriculum of teaching CAD consists of three modules: Elementary, Intermediate and Advanced. The modules have been designed to complement each other and to progressively develop the CAD knowledge and skills of all the design students. The elementary module aims to provide all the design students with an introduction to the basic theory of computing and a broad appreciation of how computers are used in design and manufacture in order to build a foundation for a more detailed study of CAD. The elementary module is delivered through a series of 12 lectures for 100 students and weekly practical classes for 40 students. The intermediate module aims to develop the students' understanding of how CAD is used within their own design specialization and how it is linked to the manufacturing process. The intermediate module is delivered through a weekly practical or seminar for 20 students. The advanced module aims to apply their newly acquired CAD knowledge and skills to either a major design project or a major investigation into some aspect of computing and design. The advanced module is run in the form of a weekly supervised practical workshop and individual tutorials. These are complemented by independent student CAD projects. Through this curriculum design, the learners strengthen their CAD knowledge and practical skills progressively.

13.4.3 Classroom management

In order to develop the students' learning on CAD in fashion and textiles, in the former case all lectures and workshops are conducted in the CAD laboratory. This environment should encourage the students' learning attitude. However, in the latter case, only practical classes are conducted in the CAD laboratory. To ensure all students attend each lecture and workshop, the teacher requires each student to submit a small assignment at the end of each class in both cases. These assignments are used to reflect the degree of students' learning in order to provide appropriate feedback from the teacher in the next class.

Meanwhile, the teacher adopts time management to optimize student learning in both cases. According to Kauchak and Eggen (2008), the class time is divided into three overlapping categories, namely instructional time, engaged time, and academic learning time. Instructional time is the time allotted for teaching the concepts and basic skills of CAD in fashion and textiles. Engaged time is the time allotted for students' active learning such as asking and responding to questions, completing worksheets and exercises, preparing report and presentations, etc. Academic learning time occurs when students participate actively and are successful in learning activities. According to past experience, the proportion of time allocation for these three categories is 2:2:1. For effective classroom management, the proportion of academic learning time should be maximized.

13.4.4 Assessment of effectiveness of teaching and learning

In both cases, there are direct and indirect measures of the effectiveness of teaching and learning. Direct measures are based on direct assessment of a student's work, performance or behavior. Some common examples of direct measures include course-embedded assessment, capstone project, portfolio assessment, performance assessment in work-integrated education/placements, and tests and examinations. Indirect measures usually involve stakeholders' perceptions of how well the students have attained the learning outcomes and thus are relatively more subjective in nature. Some common examples of indirect measures include alumni surveys/interviews, employer surveys/interviews, student surveys/interviews and external reviews.

As the direct measures are more costly to collect, it is necessary to prioritize the most essential outcomes for assessment. In the first case study, learning outcomes 2 and 3 are more important than learning outcome 1. Thus, the direct measures focus on these two most important outcomes. However, indirect measures are easier to collect but are less objective and credible. They mainly provide the overall subject outcomes according to stakeholders' perceptions.

In the first case study, the teaching approach was moving from 'instructivist' to constructivist education philosophies and a learning outcomes approach was being adopted. Nowadays, it is observed from direct measures that student technology skills increase and more interaction between peers occurs as students are more concerned with how to solve immediate CAD problems and are more engaged and more responsible for their own learning. From indirect measures, the responses made by employers and alumni are positive and encouraging, and employers feel that graduates are more active in learning and have good problem-solving skills. In the second case study, the teaching approach was moving to a 'modular scheme' and is currently affected by three important factors: (a) the rapidly changing nature of CAD technology; (b) the changing CAD expertise of the students; and (c) the changing requirements and changes in the nature of the industry. This implies that continuing development of the curriculum design and teaching approach are necessary for all CAD courses in order to adapt to these environmental changes.

13.5 Areas for improvement in teaching computer-aided design (CAD)

Based on the above reviews of four pedagogic approaches and two case studies, the following areas need to be improved in order to strengthen the effectiveness of teaching CAD to fashion and textiles students:

1. The 'constructivist' approach is a core approach in teaching CAD with the support of three other approaches – problem-based learning (PBL), experiential learning (EL), and global-scope learning (GSL). For example, PBL enables students to develop skills and learn the knowledge of CAD through problem solving; EL encourages students to involve their experience directly to gain a better understanding of this new knowledge of CAD and retain it for a longer period; and GSL provides a way of sharing their knowledge of CAD with other academic institutions in order to give them a global view of CAD in fashion and textiles. GSL is important for design students to strengthen their creativity.

2. Standardization of facilities used in teaching CAD of fashion and textiles is necessary for improving communication among students and industrialists. For example, jargon used in fashion and textiles such as color and fabric construction and the architectures of hardware and software (including coding and operating systems) used for CAD systems needs to be standardized, which could improve students' learning in various environments. This is important for speeding up communication among practitioners in the fashion and textiles industry.

3. Life-long learning needs to be encouraged to allow students to keep their CAD knowledge up to date with the rapid changes in computer technology, especially in 3D and 4D technology development and image

processing. This is critical for students in maintaining state-of-the-art CAD knowledge in fashion and textiles to strengthen their competitive advantage.

13.6   Conclusions

There are four pedagogic approaches in teaching CAD to students of fashion and textiles: constructivism, problem-based learning, experiential learning, and global-scope learning. These are compared in terms of principles, learning outcomes, materials required, role of the teacher, and assessment method. The historical development and main challenges to each pedagogic approach in teaching CAD to fashion and textiles students are addressed. The main challenges include rapid changes of the industry needs and wants, rapid changes of computer technology, the global working environment for fashion and textiles, and non-standardization of jargon used in fashion and textiles. Two case studies were carried out and the results were analyzed based on teaching and learning facilities, curriculum design, classroom management, and assessment of the effectiveness of teaching and learning used in each case. Based on the main challenges and the results of the case studies, there are three areas for improvement in teaching CAD to fashion and textiles students, including 'constructivist' as core among existing approaches, standardization of facilities, and encouragement of life-long learning. Fashion and textiles courses will continue to be popular, but it is essential that they adapt their curriculum to provide students with a portfolio of skills that will enable them to work in a rapidly changing industry. It is unlikely that this will be achieved without some innovation in the way that fashion and textiles are delivered within the curriculum.

13.7   References

Ankerson, K.S. and Pable, J. (2008). *Interior Design: Practical Strategies for Teaching and Learning*. New York: Fairchild Books.

Barker, P. (1999). Using intranets to support teaching and learning. *Innovations in Education and Training International*, 36(1), 3–10.

Boud, D.J. and Feletti, G. (1991). *The Challenge of Problem-based Learning*. New York: St Martin's Press.

Bruning, R.H., Schraw, G.J. and Ronning, R.R. (1999). *Cognitive Psychology and Instruction* (3rd edition). Upper Saddle River, NJ: Merrill/Prentice Hall.

Geary, D.C. (1995). Reflections of evolution and culture in children's cognition: Implications for mathematical development and instruction. *American Psychologist*, 50, 24–37.

Goodyer, A. (1999). *Workshop on Information and Communication Technologies and the Curriculum*. Office of the Board, www.boardofstudies.nsw.edu.au/docs_general/occasionalp2_ict.html#heading2

Hannifin, R.D. (1999). Can teacher attitudes about learning be changed? *Journal of Computers in Teacher Education*, Winter, 15(2), 7–13.

Hayes, D., Schuck, S., Segal, G., Dwyer, J. and McEwen, C. (2001). *Net Gain?: The integration of computer-based learning in six NSW government schools 2000.* The University of Technology, Sydney Faculty of Education Change and Education Research Group.

Kauchak, D. and Eggen, P. (2008). *Introduction to Teaching: Becoming a Professional* (3rd edition). Upper Saddle River, NJ: Pearson Education.

Kolb, D.A. (1984). *Experiential Learning: Experience as the Source of Learning and Development.* Englewood Cliffs, NJ: Prentice-Hall.

Lovat, T.J. (2003). *The Role of the 'Teacher' Coming of Age?* Australian Council Deans of education, discussion paper, available at http://www.acde.edu.au/assets/pdf/role%20of%20eds%20-%2024%20june.pdf

Merrian, S.B., Caffarella, R.S. and Baumgartner, L.M. (2007). *Learning in Adulthood: A Comprehensive Guide.* San Francisco: John Wiley & Sons.

Perkins, D. (1991). Technology meets constructivism: Do they make a marriage? *Educational Technology*, 35(5), 18–33.

Schunk, D.H. (2004). *Learning Theories: An Educational Perspective* (4th edition). Upper Saddle River, NJ: Pearson Education.

Uden, L. (2004). Editorial. *International Journal of Learning Technology*, 1(1), 1–15.

Uden L. and Beaumont, C. (2006). *Technology and Problem-based Learning.* Hershey, PA, and London: Information Science Publishing.

Vernon, D.T.A. and Blake, R.L. (1993). Does problem-based learning work? A meta-analysis of evaluative research. *Academic Medicine*, 68(7), 550–563.

14

Three-dimensional (3D) technologies for apparel and textile design

C. L. ISTOOK, North Carolina State University, USA,
E. A. NEWCOMB, North Carolina Agricultural & Technical State
University, USA and H. LIM, Konkuk University, South Korea

Abstract: This chapter provides an overview of the technologies used for
3D apparel CAD systems. The apparel industry can benefit immensely
from the development of integrated 3D CAD systems, in which body
models play a critical role. While significant advancements have been
made in computer technologies and approaches that improve 3D
modeling, research continues in the development of more realistic and
mechanically accurate models. Completed models may be integrated
with 3D garment design and visualization and 3D fabric modeling
systems to create a truly interactive 3D product development and
marketing tool.

Key words: 3D body model, avatar, virtual simulation, 3D draping, 3D
garment design.

14.1 Introduction

In the last few years, the apparel industry has become increasingly interested in the development of three-dimensional (3D) CAD systems.
Technological advancements in the areas of 3D human body, fabric, and
garment modeling have enabled researchers to develop integrated design
and development systems that benefit from 3D interaction. Many analysts
are predicting that the interactive 3D design and visualization of garments
will have dramatic effects on apparel product development and the way
consumers purchase clothing in the future.

Three-dimensional CAD systems have widespread potential benefits for
the apparel industry and its consumers. For instance, they can significantly
affect the product development process. Traditional apparel CAD systems
are centered on 2D pattern design, requiring the development of physical
prototypes to determine the effectiveness of a garment style or fit. This
iterative process contributes to an apparel product development stage that
is almost three times the consumption stage (Rodel *et al.*, 2001) and constitutes 4–6% of total garment costs (Farber, 2002). However, the fashion
industry works on very short cycle times, requiring fast product adjustments
and responses to market demands. Three-dimensional visualization, in

296

which products are designed and tested on a virtual human body, would speed the product development process and help firms respond to the market faster – given that the body, fabric, and garment systems are handled realistically (Chittaro & Corvaglia, 2003; Desmarteau & Speer, 2004; Stylios *et al.*, 2001; Xu *et al.*, 2002).

In addition, firms are considering the use of 3D CAD systems to create competitive opportunities in made-to-measure applications for custom fit garments (Cho *et al.*, 2003; Dieval *et al.*, 2001; Xu *et al.*, 2002). Research shows that consumers cite a general dissatisfaction with the fit of ready-to-wear apparel as one of the primary reasons for returns (Azoulay, 2004; Desmarteau, 2000). Companies have started to recognize the competitive advantages of customization over mass production and have begun to demand CAD applications with 3D apparel fit evaluation and patternmaking (Cho *et al.*, 2005).

Interest in 3D systems for the apparel industry has also been fueled by the increased importance of electronic retailing for apparel retailers. However, apparel consumers often cite the lack of ability to physically see, feel, and try on fashion products as a hindrance in purchasing and the reason behind dissatisfaction with the garments they do purchase (Cordier *et al.*, 2003). The ability to virtually try on products in 3D environments has the potential to revolutionize online apparel shopping by simplifying product evaluation (Gurzki *et al.*, 2001; Protopsaltou *et al.*, n.d.).

While there are clear benefits in the development of 3D CAD systems, ultimate success in the apparel industry depends on the inclusion of several critical components in the system. First, the system should accurately and realistically model the human body in size, shape, texture, and movement. In addition, the system should enable garment design and visualization, allowing for the assessment and adjustment of garment styling and fit in a virtual environment. Last, the system should be able to realistically model the properties and behavior of a wide range of fabrics in a variety of environments. These three areas should be incorporated appropriately to create the most useful and interactive 3D CAD system for apparel (Gong *et al.*, 2001; Hardaker & Fozzard, 1998; Imaoka, 1996; Inui, 2001; Ng *et al.*, 1993; Protopsaltou *et al.*, n.d.; Rodel *et al.*, 2001).

14.2   Applications of three-dimensional (3D) human body modeling

A realistic representation of the body in shape, size, and appearance for use in 3D virtual environments is known as an avatar (Hilton *et al.*, 2000). The design and development of avatars, referred to in this chapter as 3D body models, is a critical component in the development of complete 3D CAD environments for the apparel industry (Jones *et al.*, 1995).

14.2.1 Apparel

Three-dimensional body modeling for apparel has benefited from contributions from entertainment, medicine, and the military, but also has much to offer to the field of 3D body modeling itself. For one, it can be viewed as an exercise in combining the expertise from two (sometimes very) different fields – computer science and textile and apparel science. Historical approaches to 3D modeling by these two disciplines have been different in the sense that computer graphics scientists generally focus on visual realism and animation efficiency, while textile and apparel scientists are more concerned with the mechanical accuracy of the body, fabric, and garment representations (Imaoka, 1996; Choi & Ko, 2005). The 3D modeling field as a whole benefits from both approaches (Hardaker & Fozzard, 1998).

The balance between visual realism and technical accuracy is difficult, especially considering the computational and system requirements for realistic 3D modeling. In the search for the most realistic models, computational design is often very complicated and time consuming. To determine the most appropriate model, it is important to consider its intended application, including the primary users, system requirements and capabilities, and required functions. These considerations can be very different for those in computer graphics and textile and apparel fields. At a minimum, any 3D modeling system should balance scope and realism with robustness and speed of computation (Volino & Magnenat-Thalmann, 2000).

The importance of the body model in 3D CAD environments cannot be underestimated, representing the foundation of the system, upon which garment designs can be created and assessed virtually (Jones *et al.*, 1995). As such, there are specific considerations and requirements in the development of 3D body models for the apparel industry. Primarily, the body should accurately represent real human body shapes, accommodate individual variability in body morphology, and have a realistic appearance and texture (Gurzki *et al.*, 2001). This includes the development of models across a wide range of sizes, nationalities, and ethnic groups (Cichocka *et al.*, 2005), and may necessitate the use of actual human measurement data (Paquette, 1996). In addition, if the body is animated, movements should be realistic and portray actual interaction of the body with garments and fabric (Protopsaltou *et al.*, n.d.; Hilton *et al.*, 2000). Visual realism of the body has consistently been the biggest challenge to 3D developers, but it is also the most critical for widespread use and acceptance.

Realistic and accurate body shape representations rely on a deep understanding of the composition of the human body. The body is composed of a rigid, articulated bone structure that forms the basis of the shape, and deformable tissues (i.e. muscle and fat) that cover the bones. The outermost visible layer of the body is skin, a deformable tissue that seamlessly and smoothly covers the body's bone segments and joints (Azuola *et al.*, 1994). An accurate body model will realistically illustrate these components in a computationally stable environment.

As discussed, accurate human body models will have dramatic implications for the apparel industry by enabling design and evaluation of products directly in 3D. However, despite the benefits, research in this area has not been as widespread as in 3D garment and fabric modeling. Ng *et al.* (1993) cited the lack of an accurate mathematical model to express human body variability as the most significant hindrance to 3D pattern design. Okabe *et al.* (1992) also predicted that the development of more precise mechanical body models would contribute significantly to custom 3D tailoring. Stylios (2005) cited the 3D modeling of real humans for garment visualization as a major challenge for the industry. Thus, research continues into the development of 3D human body modeling technologies.

14.3 Technologies of human body modeling in three dimensions (3D)

Three-dimensional human body model development begins by capturing an image and data for a given body, followed by processes to translate these images into 3D space. Depending on the type of body model being created, the model may then be surfaced with or without animation capabilities. The most simple body model is a mannequin represented only by its surface, which is rigid and static. A more advanced model is one that represents the individuality of the human form, and may be constructed with a skeletal base to allow animation, or a soft surface that will yield under certain types of garments, fabrics, or fits (Imaoka, 1996). This section will cover the steps involved in creating simple and sophisticated 3D body models, including imaging, body surfacing, application of a skeletal structure, and animation of the finished model. The differences between generic and individualized body models will also be discussed.

14.3.1 Imaging

Imaging is the first stage of body model development and involves the capture of actual body data from a specific form. Many methods have been researched, from manual methods to the incorporation of 2D photography and 3D whole body scanning.

Manual methods

One of the most basic ways to obtain an initial 3D body image is 3D digitizing (Hardaker & Fozzard, 1998). In this process, the body is manually defined as the intersection of numerous latitudinal and longitudinal vertices in the coordinate plane. This process can capture individual detail, but storage of images containing the many vertices needed is time consuming and sometimes inconvenient (Volino *et al.*, 1996).

Similar to 3D digitizing, Kang and Kim (2000) reported the creation of a 3D body model by manual measurement and translation into 3D space. In this method, the human torso is divided into 20 parallel cross-sections and measured using a sliding gauge. Then, each cross-section is divided into 60 sectors, and the radius of each section is calculated. From these calculations, the body model is reconstructed and smoothed in 3D using the cylindrical coordinate plane (Kang & Kim, 2000). Manual methods have largely been bypassed in recent years in favor of 2D photography and 3D whole body scanning techniques.

Two-dimensional photography manipulation

Reconstruction of a 3D form from 2D images is a frequently used method of 3D body model development. While researchers may employ slightly different methods, the basic process of using 2D photographs to create a 3D image is very similar. First, images are captured from different orthogonal viewpoints, most often corresponding to front, back, and side views (Furukawa *et al.*, 2000; Guerlain & Durand, 2006; Hilton *et al.*, 2000; Lee *et al.*, 2000; Lee & Magnenat-Thalmann, 2001; Stylios *et al.*, 2001; Stylios, 2005). Once the images are captured, the silhouette is extracted from 2D photographs and 'feature points' are identified. These points are located in specific regions of the body, generally corresponding to anatomical landmarks, skeletal features, or key body locations and extremities (Furukawa *et al.*, 2000; Hilton *et al.*, 2000).

Feature point identification provides the direction necessary to translate the body from the 2D to 3D environment. In this process, feature points defined on the 2D silhouettes are matched with corresponding feature points on a generic 3D model (surface and/or skeleton) (Furukawa *et al.*, 2000; Hilton *et al.*, 2000; Lee *et al.*, 2000; Lee & Magnenat-Thalmann, 2001). Once in 3D, the control points defining the generic model's skin surface can be moved individually to follow the silhouette of the subject from the 2D image.

Reconstruction of the human body using 2D photography provides a low-cost, basic alternative to some of the more advanced 3D techniques that exist (Stylios, 2005). However, the reconstruction process can be time-consuming and complex, and may not be the best choice for highly detailed body models, or for individuals with shape anomalies (Linney *et al.*, 1997).

14.3.2 Three-dimensional body scanning

Three-dimensional body scanning is a method used to rapidly and accurately acquire a 3D image and measurements (Yu, 2004). Initial scanning attempts focused on specific body parts such as the head, face, hands, feet, and torso. This eventually moved into 3D whole body scanning, which is now the most advanced method of obtaining the body data needed to create a 3D model (Paquette, 1996). Some researchers use extracted measurements to geometrically create a body model, while others utilize the actual 3D output and enhance it to create the model.

Three-dimensional body scanning systems utilize optical devices to obtain 3D data and create a 3D surface image. Using structured light, laser, infrared, or photogrammetry as a light source, these systems deliver a collection of 3D points, known as a point cloud, that approximates the surface of the body (see Fig. 14.1) (Hilton *et al.*, 2000; Yu, 2004).

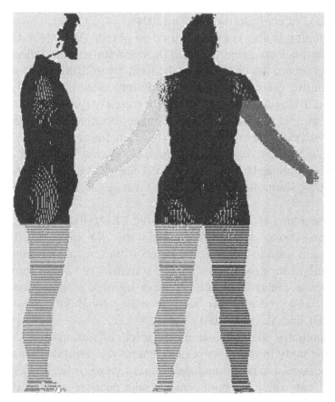

14.1 3D point cloud of body surface (from [TC]², retrieved 11 November 2010 from http://www.tc2.com/pdf/nx16.pdf).

The collection of points are in a common coordinate frame, corresponding to width, height, and depth (Pargas *et al.*, 1997), and are estimated based on the angle of light source and the reflection of that light from the body (Buxton *et al.*, 2000). Some methods utilize static heads that capture images from different views, while others use rotating heads (Xu & Sreenivasan, 1999). A greater number of cameras increases coverage and processing time. Subjects are most often scanned in close-fitting garments to expose the body.

Structured light body scanning methods typically involve the projection of a light pattern or grating onto the body surface. As the light is reflected from the body surface, a series of images captures the distortion of the pattern as it comes in contact with the 3D body. The distortion reveals information about body contours and surfaces, while the synthesis of the multiple images provides the 3D point cloud data. Information about the white light methods developed by companies such as [TC]^2, Wicks & Wilson, and Telmat has been presented by Azoulay (2004), Buxton *et al.* (2000), Istook (2008), Istook and Hwang (2001), Paquette (1996), Simmons and Istook (2003), Yu *et al.* (2001), and Yu (2004).

Laser scanning is also commonly used to obtain 3D body data. In this method, laser lines are projected onto the body surface and reflected onto cameras that record the distortion of the lines. By utilizing an arrangement of mirrors during the scanning process, different views of the body surface are obtained. Triangulation methods are then used to synthesize these views and obtain the 3D point cloud data. A range of information exists about laser body scanning technologies, for both commercial (i.e. Cyberware, TecMath, and Vitronic) and non-commercial systems (Buxton *et al.*, 2000; Istook, 2008; Istook & Hwang, 2001; Kim & Kang, 2002; Linney *et al.*, 1997; Paquette, 1996; Simmons & Istook, 2003; Xu *et al.*, 2002; Yu *et al.*, 2001; Yu, 2004).

Infrared scanning systems employ infrared LEDs (light emitting diodes) and position-sensing detectors to obtain the 3D shape of the body. Triangulation is used to calculate 3D points from several viewpoints, much like the method used in laser scanning systems (Yu, 2004). Information regarding the commercial development of infrared scanning systems by Hamamatsu and Cousette has been presented by Buxton *et al.* (2000), Istook (2008), and Yu *et al.* (2001).

Photogrammetric methods use the principle of stereovision to capture images of the body. In this process, two cameras are calibrated for a specific range of the body and positioned for different views of the body. As a grid is projected onto the body, both cameras take pictures of the body region at the same time, resulting in slightly different images. By analyzing the distances between the points on the object and the base plane, points captured with both cameras are matched and translated into 3D (Kang & Kim,

2000; Yu, 2004; Yu *et al.*, 2001). Reconstruction of the model based on only a few views of the body can be difficult, especially in the more inaccessible parts of the body (e.g. armpits) (Guerlain & Durand, 2006).

The use of low-frequency radio waves in 3D body scanning has garnered attention in recent years. However, the accuracy of surface data in this method has been questionable, so its use in 3D modeling is not likely with the current technology. It has generally only been used for size projection in apparel retail outlets or in airport terminals for security purposes.

Image processing and measurement extraction

Capturing surface information is only the first step in creating a 3D image from 3D body scans. Since most body scanners capture multiple views of 3D data, the next step involves the registration and merging of these views into a point cloud (Buxton *et al.*, 2000). These point clouds must then be cleaned and refined to remove outliers or artifacts that often result when cameras pick up additional information from the environment during scanning. Nearest-neighbor analysis and evaluation of point intensity can be used to eliminate points beyond the body surface (Jones & Rioux, 1997; Buxton *et al.*, 2000).

Once the images have been cleaned and refined, body landmarks are then defined, enabling measurement extraction from 3D body models. This process is very important because landmarks and measurements are commonly used as feature points or lines to guide the surfacing of 3D models and contribute to their use in automated patternmaking systems.

The process of measurement extraction begins with the definition of body landmarks. In manual measurement, this is done by experts who touch the body to feel for specific bony protrusions under the skin or a joint that connects two bones. In 3D measurement, the process is more difficult because the landmarks can be very subtle, there is high variation in body shape for different people, and the landmarks may be hidden from view (Buxton *et al.*, 2000). Researchers have used various methods to locate landmarks, such as searching for the intersection of angles at points of extremities or vertical locations, or changes in curvature along body cross-sections or profiles (Dekker *et al.*, 1999; Kim & Kang, 2002) Researchers have also utilized belts or other markers during image capture to help predefine a body landmark that can be easily located by a computer after scanning (Buxton *et al.*, 2000; Dekker *et al.*, 1999; Pargas *et al.*, 1997; Wang *et al.*, 2007; Zhong & Xu, 2006). Some laser scanning systems can record color values, so that landmarks identified with color markers can be extracted after scanning (Paquette, 1996; Wang *et al.*, 2007). While virtual landmark location is difficult (and sometimes different in different scanning systems) (Bye *et al.*, 2006; Carrere *et al.*, 2001; Simmons & Istook, 2003), accurate identification is critical for measurement extraction in addition to 3D surfacing, animation, and body adjustment.

Once landmarks are located, measurements are then extracted. Rapid automatic measurement extraction is one of the primary advantages of 3D body scanning systems over manual measurement techniques (Pargas *et al.*, 1997; Simmons & Istook, 2003; Xu & Sreenivasan, 1999). Length, width, depth, and angle measurements not obtainable through manual measurement can be obtained through automatic 3D extraction. This has enormous possibilities for made-to-measure patternmaking, development of sizing strategies, and size prediction in 3D visualization systems (Bye *et al.*, 2006; Pargas *et al.*, 1997).

14.3.3 Selecting the best imaging technique

The two major categories of imaging include image capture using traditional camera techniques, which results in a 2D photograph, and image capture using lighting and cameras for 3D images. Available resources and the use of resulting body models dictate the best technique. Three-dimensional models created from 2D images are often more photo-realistic in conveying skin color, facial features, and basic shape information. They can also be acquired with less financial investment. Those created from 3D scanning technologies lack photo-realism in the face and skin, but give considerably more accurate size and shape information in a faster reconstruction approach (Hilton *et al.*, 2000; Lee *et al.*, 2000).

14.4   Development of the body surface

Output for the imaging techniques is rough, and while possibly serving as a good representation of the human form, it is not sufficient for use in virtual environments. Surfacing the body improves the realism of the 3D body and its use in 3D CAD and patternmaking systems (Kim & Kang, 2002; Kim & Park, 2004; Zhong & Xu, 2006).

The simplest representation of the body is a wire frame, consisting only of points, lines, and curves that describe the edges of objects. To create a wire frame, the body image is acquired and the body is reconstructed as a series of circumferential cross-sections connected to each other using parametric curves (Hinds & McCartney, 1990; Jones *et al.*, 1995; Ng *et al.*, 1993; Stylios, 2005; Xu *et al.*, 2002). This wire frame is a good basic body model, but it is really only a representation of the skeleton, without surface information. More surface definition can be achieved for enhanced visual realism with the use of explicit, implicit, and advanced methods of surfacing described in this section.

14.4.1 Explicit methods

Explicit methods of body surfacing involve the use of mathematical approaches to define the body surface with parametric or polygonal patches. Polygons are commonly used to surface the body because they are the simplest geometric form (Volino & Magnenat-Thalmann, 2000). To define a surface using polygons, vertices making up the 3D body model (such as those defining cross-sections or 3D point clouds) are connected to form a mesh. These vertices may be moved independently to deform the mesh and provide new surface shape. The most common shapes are triangles or quadrangles, as these shapes easily lend themselves to geometrical computations used in mechanical modeling.

An extension of polygonal surfacing, subdivision surfacing also works well to represent fine human body details. Used by Pixar to model clothing, this process begins by modeling the body as a rough collection of large polygons. These polygons are repeatedly subdivided into smaller collections of polygons until the surface is visually smooth. The faceted surface that results in traditional polygonal surfacing at high magnification can be minimized by the use of subdivision. The subdivided surfaces also lend themselves to the application of hierarchical collision detection and response rules in 3D animation, making them a good choice for modeling (House & Breen, 2000).

High-order surfaces defined by parametric polynomial functions represent another type of body surfacing. These functions are easy to apply and provide adequate surface smoothness and continuity without significant hindrance to computational time (Volino & Magnenat-Thalmann, 2000). Examples of these types of curves include Bézier curves, B-splines, and non-uniform rational B-splines (NURBS), each of which are similar in their application, but differ slightly in their mathematical definition (Buxton et al., 2000; Volino & Magnenat-Thalmann, 1997).

14.4.2 Implicit methods

Implicit surfacing methods typically have smaller computational time and storage requirements than the explicit methods described above. Implicit models begin with simple building blocks known as primitives, such as meta-balls, which are blended or fused together to create the body surface. The primitives can then be moved or deformed to change the surface shape, or attached to joints to aid in animation (Shen & Thalmann, 1995; Volino et al., 1996).

As already discussed, explicit and implicit surfaces have advantages and disadvantages for 3D body modeling. Researchers in 2003 presented a method that combined both explicit and implicit surfacing to take advantage of the benefits of both methods. In this approach, researchers start modeling with explicit surfaces, which are more easily deformed manually, and transform to implicit surfaces, which are best for modeling noisy 3D data (Ilic & Fua, 2003).

14.4.3 Advanced surfacing techniques

Surfacing 3D scan data

The explicit and implicit surfacing methods described above can be used to surface 3D body scan data, though not without some difficulties. Point cloud surface data may be inconsistent, the location of points representing the surface can be inaccurate, and point cloud information can be entirely missing in some areas. This causes problems when trying to surface the body, in addition to the problems already discussed related to measurement extraction (Allen *et al.*, 2003; Buxton *et al.*, 2000; Guerlain & Durand, 2006).

In some cases, 3D datasets may have substantial missing data, occluded areas, or outliers that render cleaning and subsequently smooth surfacing difficult. Since the advent of 3D point cloud data collection, researchers have been investigating a variety of methods to surface these 'unclean' datasets. Two early methods researched involve the principles of inflation and deflation (or sculpting) to surface 3D data. In 1988, Choi, Shin, Yoon, and Lee discussed the use of inflation to surface a point scatter. In this approach, a surfaced template mannequin is created in 3D using triangulation (polygonal mesh surfacing). This mannequin is then placed inside 3D scatter plots and after simple transformations the mannequin is 'inflated' until it forms to the shape of the outer scatter plot (Choi *et al.*, 1988). A disadvantage of this system is that the original virtual mannequin may not be of the same basic shape or distribution as the data being fitted, which complicates the surfacing process. In an opposite method described by Boissonat (1984) and Veltkamp (1991) a convex polygonal mesh is placed around a point cloud. Specific polygons are then removed according to Delaunay tetrahedrization until the surface conforms to the point cloud (Boissonat, 1984; Veltkamp, 1991).

While these inflation and deflation methods appear to be well suited for filling in missing data, there are more advanced methods of surfacing 3D point cloud data using generic 3D template models. This approach involves the application of a generic template model to an uncleaned dataset using landmark or feature point extraction and matching. Once superimposed, the generic model is deformed to take on the individual data's shape, filling in any missing data (Allen *et al.*, 2003; Bruner, 2010; Seo & Magnenat-Thalmann, 2003).

Texture mapping

Surfaced body models represent the dimensions of the human form adequately, but often do not provide the photo-realism wanted for many virtual environments. For photo-realism, texture mapping can be used to increase fine surface detail and visualization of colors, textures, and surface contours (Buxton *et al.*, 2000). Texture mapping involves the transfer of surface properties from a given object to a 3D geometrical model (Volino & Magnenat-Thalmann, 2000). To obtain the most photo-realistic surface information, many researchers turn to 2D photographs, which provide more color and texture surface information than 3D body scanning (Hilton *et al.*, 2000; Lee *et al.*, 2000). When attempting to texture-map from a 2D image to a polygonal mesh, the 2D texture coordinates are associated with each mesh vertex, while barycentric coordinate interpolation helps to transfer surface information to points inside the triangles within the mesh (Hilton *et al.*, 2000; Lee & Magnenat-Thalmann, 2001).

Pressure-sensitive surfacing

Body models that yield to pressure would be beneficial for the fit evaluation of close-fitting garments, such as knitted garments (Hardaker & Fozzard, 1998; Inui, 2001). Pressure-sensitive surfacing involves the definition of the external body shape and internal structure, and the application of mechanical properties to these entities that allow deformation. Inui (2001) created pressure-sensitive models by first defining a triangular mesh body surface. Strain is applied to the vertices of the body, in much the same way as in methods used to model fabric movement. However, the mechanical properties of fabric can be estimated more easily using measurement methods (i.e. Kawabata fabric evaluation system or the FAST system) than properties of the body. To estimate the mechanical properties of the body, Inui calculates strains on an object in a specific loading condition and compares them to strains on a non-deformed object. Each loading condition that could affect the body must be estimated in order for the mechanical properties to be accurately simulated. Once the mechanical properties have been determined and applied to the surface, collision detection and response algorithms are used to register contact between the body and fabric and respond to it by contraction of the body. An actual body model was not developed by Inui, though the application of mechanical properties to simulate deformation of a solid object was achieved.

Commercial software providers such as OptiTex have been researching the development of a body model that can simulate soft tissues for the lingerie and swimwear market for years (Desmarteau & Speer, 2004), though the process is difficult.

14.5   Animation

Initial attempts at body modeling centered on static bodies (Ng *et al.*, 1993). Even today, static 3D body models are useful in 3D visualization for the apparel industry. However, to be the most useful and realistic for product developers and consumers, full 3D visualization of apparel cannot be realized without animation of the human form. Through animation, users can evaluate how the garment interacts with the body as it moves, how the fabric drapes, and how apparel fit may be affected during certain movements. Thus, successful animation not only involves rendering the body during movement, but also rendering the movement of garment and fabric in relation to the body during movement (Yang *et al.*, 1992). This section will contain a discussion of commonly used animation methods, as well as a brief acknowledgment of the importance of realistic interaction between the garment, fabric, and body during animation in 3D CAD systems.

14.5.1 Skeletal-driven deformation

A widely used and reported method of body model animation is the skeleton-driven animation and deformation of the skin. This involves surfacing the body and the deformation of this surface related to movement of joints. The skin surface is often attached at specific joint locations, such that during movement, the surface moves according to joint movement (Allen *et al.*, 2003). The H-Anim skeleton standard contains rules for the attachment of skin parts to skeleton joints (Humanoid Animation Working Group, n.d.; Lee *et al.*, 2000).

14.5.2 Animation without skeletal definition

While not quite as popular as skeleton-driven deformation, researchers have also been exploring the option of animating body models without first defining a skeleton. In this approach, regions of articulation are approximated by analyzing transformation of the body surface mesh during movement. During movement, some triangles within the body mesh rotate, while others do not; clustering of triangles with similar rotation sequences reveals locations of articulation. Results of the clustering can also be synthesized to estimate the number of bones needed to produce the rotation seen within the mesh. Using this approach, researchers have been successful at estimating the bone segments and locations of articulations to animate living objects, such as horses, birds, and the body without prior application of a skeletal structure (James & Twigg, 2005).

14.5.3 Animation in a 3D CAD system

Realism of body model animation relies significantly on the completeness with which the body skeleton and surface are defined and modeled. Greater definition (such as smaller polygons or more sophisticated polynomial functions) in the joints and areas of articulation will allow for a more continuous range of motion, without the discontinuities that often appear when using a rough approximation of the body surface and skeleton (Azuola *et al.*, 1994). Also, use of a seamless mesh to define the body surface, rather than segmented meshes based on the connection of body segments, would result in a smoother, more realistic animation sequence (Hilton *et al.*, 2000). Furthermore, capturing a broader range of motion in initial motion capture sequences would allow for the creation of more realistic human movement in 3D. However, it must be noted that with greater definition in surfacing, skeletal structure, and motion, there is a trade-off between computational time and accuracy.

Realistic body animation also relies heavily on accurate fabric and garment models that include collision detection and response. Collision detection is the act of finding points of contact between 3D objects, while collision response involves the application of some repulsive force or energy to remove the collision (Volino & Magnenat-Thalmann, 2000; Yang *et al.*, 1992). The process of collision detection and response is critical in a 3D CAD system to ensure that the components making up the system (body, garment, and fabric) do not penetrate each other or themselves during animation, yet behave as they would in real-life environments (Imaoka, 1996; Magnenat-Thalmann & Volino, 2005; Volino & Magnenat-Thalmann, 1997).

The importance of realistic 3D fabric modeling in the development of 3D CAD systems cannot be overestimated. For specific information regarding fabric modeling, two useful resources are *Cloth Modeling and Animation* edited by Donald House and David Breen (2000) and *Three Dimensional Modeling of Garment Drape* by Narahari Kenkare (2005). The book by House and Breen covers a variety of topics related to fabric modeling, including the different methods used to model fabrics (geometric and physical), approaches for collision detection and response, and application of fabric mechanical parameters for realistic woven and knit fabric animation. The book not only covers methods for rendering fabric in 3D, but also stresses the importance of understanding the factors affecting fabric behavior in general, such as the type of fabric, its construction, fabric treatments, surface-to-surface friction, and garment construction details (House & Breen, 2000). Kenkare also presents a thorough investigation into current best practices in 3D modeling of garment drape, including a history of fabric drape and fabric modeling, as well as the development of a framework for capturing, evaluating, and characterizing 3D drape in a virtual environment (Kenkare, 2005). Other useful resources for fabric modeling include papers by Cordier & Magnenat-Thalmann (2002), Hardaker & Fozzard (1998), Magnenat-Thalmann & Volino (2005), and Volino & Magnenat-Thalmann (1997).

Combining expertise in the areas of body, fabric, and garment modeling, researchers at the University of Geneva's MIRALab have developed a 3D CAD system for garment and fabric visualization using animated body models. Called MIRACloth, their system features 2D pattern design, virtual garment assembly, and animation of the fabric and body based on the application of mechanical properties to the fabric and application of skeletal structure to the body. For a detailed overview of the design and functionalities of the MIRACloth system, refer to the book *Virtual Clothing: Theory and Practice* (Volino & Magnenat-Thalmann, 2000). In addition, the MIRALab website contains a wealth of information about the research projects conducted by the group, including projects on advanced 3D body, fabric, and garment modeling that are very relevant to the topics of this chapter (MIRALab, 2006).

14.6   Generic vs individualized body models

For the best human body representation, body measurements and proportions should be simulated as realistically as the body surface and skeleton. In the search for realism, there are basically two kinds of body model generation. In the first, generic bodies are created by utilizing one body image (possibly an averaged image) which may then be adjusted at specific locations to accommodate some of the individual variability in the human body. In the second, body models are directly created from individualized body images (such as those acquired through body scanning). The second method most closely reproduces the individual human form, but there are drawbacks to this process, such as the difficult integration of often very large body scan data files with software programs and increased time required for cleaning, surfacing, and animation (Cordier *et al.*, 2003; Kim & Park, 2004). Thus, many studies use generic body models to reduce the large costs and storage in acquiring, processing, and surfacing of individual 3D body geometry (Protopsaltou *et al.*, n.d.).

14.6.1 Generic body models

Much of the literature in the area of 3D body model development discusses the use of generic body models created using standard sizing systems (Chittaro & Corvaglia, 2003). Researchers also commonly use mannequins or dress forms, which are then digitized or scanned, to create 3D models.

These forms have historically been created using standard sizing information as well (McCartney *et al.*, 2004; Okabe *et al.*, 1992). While the use of standardized data to create models results in statistically smooth body models (Azuola *et al.*, 1994), users may not be able to identify with the models if they do not belong to the same group from which the model was created or if the data is not current. This supports the need to refine current sizing standards to better represent today's populations (Newcomb & Istook, 2004), so that when standard information is used, the model created will be more representative.

While generic models may not represent the complete individuality of human bodies, they can have adjustable body measurement contours or skeletal systems. Some models only feature adjustable horizontal cross-sections (Protopsaltou *et al.*, n.d.), while others add in vertical adjustment parameters (Cichocka *et al.*, 2005; Kim & Park, 2004; Volino *et al.*, 1996). The contours chosen for adjustment should represent important parameters for shape definition, so that adjustment of the body will accommodate shape variation (Cichocka *et al.*, 2005).

However, adjustment of this type, where only a few dimensions are altered on a single generic model, often does not accommodate unique proportions completely. To improve results, researchers have increased the number of generic models to levels from 5 to 5000, depending on the researcher (Protopsaltou *et al.*, n.d.; Wang, 2005). In addition, 3D body scan data can be used to create a range of generic models based on a more accurate representation of the body (Jones *et al.*, 1995; Wang, 2005). The more variety in shapes and sizes offered in generic body models, the less adjustment must be done to result in a closer, more accurate representation of the human body – even if exact replication cannot be achieved (Wang, 2005).

Generic models are commonly used because of their ease of adjustment and fairly good representation of body dimensions. Another advantage of generic body models is the fact that surface and skeletal information can be perfected for generic models and then transferred to 'non-perfect' individual models. Cordier *et al.* (2003) illustrated this in their research combining an adjustable generic model and a perfected generic model with body surface and skeleton definition. Once the generic model is 'fitted' onto the adjusted model, surface contours and joint hierarchy can be transferred to the adjusted model and exploited for animation of individual body models.

14.6.2 Individualized body models

In made-to-measure applications, generic body models do not provide the full range of information needed to create garments with good fit, or to communicate fit in 3D. For instance, the smoothing of circumferences, lengths, and breadths that occurs in generic body model development does not fully represent the individuality in the human form that naturally exists. This makes the use of individualized body models, such as those based on 3D body scans, the best choice for automatic pattern generation, made-to-measure applications, and the representation of unique body shapes and sizes in 3D (Azoulay, 2004).

While individualized models are best for some applications, the process of surfacing individual body scans involves several labor-intensive steps such as cleaning and refining the point cloud data. To overcome these issues and quickly and smoothly create surfaced body models, a variety of methods have been researched. One method, presented in Section 14.4.3 of this chapter, combines a template 3D model with individual 3D body scans to compensate for any missing information in body scans. Various template models can be used to accommodate a wide range of scan data and to ensure that very little (if any) of the original individual scan data is lost. A range of body models can be created using a generic model database and this morphing technology.

Generic template models are often used to create individualized body models. They improve the look of final rendering, as well as decrease the time and effort involved in body model development. Generic models can also be used in combination with individualized body scan data to create animatable body models seamlessly, without the effort required to individually assign surface and skeletal information to each body scan. As described by Seo and Magnenat-Thalmann (2003), generic models containing skeletal and skin attachment information are combined with individual body scan data using feature point and landmark matching. Skin refinement using polygonal mesh surface relaxation techniques assures that the final body model is an almost exact match of the original 3D scan, while the template ensures a cleanly surfaced, immediately animatable model (Seo & Magnenat-Thalmann, 2003). To improve the simulation and provide additional functionality, subsequent work incorporated weight, hip-to-waist ratio, and posture adjustments. Adjustment of these additional features allows for the visualization of body changes as it gains or loses weight, redistributes weight, or changes posture.

Examples of individualized body models

The example below demonstrates how an individual body scanned image can be used to create an avatar or body model for use in apparel design, virtual fit, or e-tailing situations. In this case, a subject was scanned and the resulting point cloud data was used in two ways. One way was to use an existing template (avatar) provided from a company such as OptiTex, Lectra, Browzwear, or Tukatech (to be discussed shortly) and 'morph' the

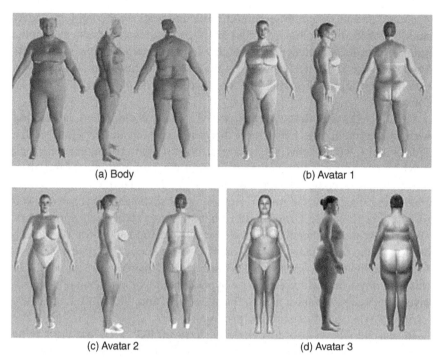

(a) Body   (b) Avatar 1

(c) Avatar 2   (d) Avatar 3

14.2 Examples of directly morphed and manually created avatars.

two images together. Figure 14.2(b) is an example of this method. The original scanned image that has not been smoothed or cleaned is shown in Fig. 14.2(a). The second way in which the scanned data was used was to extract measurement data from it based on the specific measurements required by the 3D design systems being used. This information was then entered manually into a dialog box, inside the software, so that the size and shape of the avatar could be adjusted to reflect the measurements of the subject. Examples (c) and (d) in Fig. 14.2 show the results of this effort using two different software systems. The difference between the two is the difference in the number and location of measurements available for control. It is easy to see that the creation of the more 'real' simulated person comes from the directly 'morphed' avatar inside the body scan software and not from manually input measurement data (Lim, 2009).

14.7 Virtual try-on technologies

The development of realistic 3D body models for textile and apparel applications involves the creation of bodies that are not only visually appealing but also accurate in their representation of the human body surface, skeletal structure, and movement. Successful modeling rests on a complete understanding of the technologies behind 3D body modeling, but also a commitment to the development of new technologies. Accurate 3D body models have widespread applications within the apparel industry, in areas such as product development, marketing, and virtual garment try-on for apparel consumers.

In 3D CAD systems, the body model is a critical component, interacting with the garment and fabric, and directing garment design (Inui, 2001). Traditional CAD systems for apparel utilize 2D patternmaking functions, in which patterns are designed, then cut and sewn for evaluative (and often iterative) prototyping (Chittaro & Corvaglia, 2003; Hardaker & Fozzard, 1998; Yang *et al.*, 1992). However, technological advancements have allowed researchers to create *virtual* patternmaking and prototyping approaches. For instance, researchers have developed systems that allow for 2D pattern definition and subsequent 'sewing' on a 3D virtual body (Cordier *et al.*, 2003; Imaoka, 1996; Okabe *et al.*, 1992; Volino *et al.*, 1996; Yang *et al.*, 1992). In this way, physical prototypes can be bypassed in favor of virtual prototypes, given that accurate fabric models, body models, and the ability to evaluate garment style and fit exist in the program.

Researchers have also developed customized 3D to 2D patternmaking systems, which could not be effective without accurate human body models. In the 3D to 2D process, virtual garments are created by drawing lines (or specifying them) that follow the shape of the body to create pattern pieces. Patterns are then 'unwrapped' and flattened to create a 2D pattern (Cho *et al.*, 2003; Griffey & Ashdown, 2006; Heisey *et al.*, 1990; Hinds & McCartney, 1990; Kang & Kim, 2000; Kim & Kang, 2002; McCartney *et al.*, 2004; Outling, 2007; Xu *et al.*, 2002). These 2D to 3D and 3D to 2D systems to create, evaluate, and modify garment patterns in 3D can significantly reduce time spent in patternmaking and physical prototyping. For the advantages of 3D body modeling to be realized in product development applications, research must continue towards integrating all components of a 3D CAD system, including fabric and garment models. The evaluation of apparel fit in 3D is another critical area that must be improved for product development processes to truly benefit from 3D CAD systems.

Several CAD systems including Browzwear, OptiTex, Lectra, and others have developed software for the apparel industry enabling visualization of garments on three-dimensional avatars (Virtual Fashion Technology, 2007). These avatars can be adjusted to generally mimic a specific set of fit measurements by fine-tuning points of measure, such as the bust, waist, hips, abdomen, etc. Two-dimensional garment patterns can then be set to 'sew' together on the body demonstrating the drape of the fabric and the potential fit of the garment. These 3D virtual try-on technologies allow consumers to try on or 'wear' the garment with their body measurements and other specifications such as silhouette, fabric, color, and embellishments on their 3D avatar and evaluate clothing fit over the Internet (Istook, 2008).

14.7.1 Browzwear

Browzwear International, founded in 1999, develops and manufactures advanced 3D fashion design and communication software for the garment and textile industries. Browzwear's 'V-Stitcher' is based on 3D visualization that enables users to create a virtual garment from a 2D pattern over a 3D body image (Techexchange.com, 2008, March 2). V-Stitcher™ integrates conventional design techniques with real-time, true-to-life 3D technology. The software enables users to create a virtual 3D garment and model it on a customized 3D human avatar. It is also a communication solution allowing a real-time 3D representation of the garment design at a remote location (*Apparel Magazine*, 2008, January 15). V-Stitcher's virtual stitching technology transforms 2D patterns into lifelike 3D simulations of completed garments with multisize grading over a virtual body. Texture mapping capabilities enable photo-quality representation of fabric, seams, prints, and colors. The 3D model can be viewed from 360° and zoomed in on the garments to see fabric or stitching detail (Hwang, 2004).

Browzwear developed 'C-me' for virtual fit application in the Business-to-Consumer (B-to-C) environment by allowing Internet shoppers to dress their own virtual 3D image. In other words, C-me™ is an online 3D viewer that enables remote, lifelike presentations of collections for real-time collaborative fitting and design sessions. It enables customers to share collections with buyers, suppliers and retailers at any time during the pre-production, production and merchandising process. It can offer an exciting new dimension to e-shopping, allowing retailers to invite their customers to step into a virtual fitting room. With C-me™, shoppers are able to share the experience with friends for instant feedback, or even access the opinion of professional fashion consultants, transforming e-shopping from a solitary experience into a socially interactive one (Browzwear International, 2008).

14.7.2 OptiTex

Founded in 1988, OptiTex specializes in 2D–3D CAD/CAM solutions and offers an integrated CAD package, including software solutions, digitizer, and pen or inkjet plotter (OptiTex, 2008). Like other CAD suppliers, OptiTex USA, Inc. also focuses a 3D virtual design and information management via the Internet. One of its core software products, 3D Runway

Designer 10, has tools to enhance remote collaboration between designers and production professionals. In 2001, OptiTex introduced 3D4B2B that is a web-enabled 3D virtual fit modeling application that can accept 2D flat patterns from a variety of CAD systems. The 2D patterns can be draped onto a user-definable 3D model that is displayed on a B2B website (Techexchange.com, 2008, May 27). The 3D product can be flattened out to 2D patterns for production (OptiTex, 2008). 3D4B2B can be plugged into an existing B2B website, and it can be used in conjunction with a variety of PDM systems and body scanner systems.

3D Runway Designer is a realistic cloth simulation and cloth modeling software system based on accurate CAD patterns and real fabric characteristics. The OptiTex™ 3D Garment Draping and 3D Visualization software system allows designers, patternmakers, and retailers to visualize any pattern modifications instantly in full 3D. It uses 3D models with adjustable body measurements and posture positions that move with the garments. In the OptiTex Runway™ 3D, different fabrics and stitch properties can be entered, so that cloth simulation and stitch properties such as gathering, shrinkage, strength, width and texture can be applied. A tension map illustrates clearly in red where the cloth is under tension and in green where the cloth is relaxed.

Virtual draping modules developed by OptiTex based on physical cloth modeling techniques are a recent addition to CAD systems. Typically, this application uses a powerful simulation engine with a built-in or customizable pattern and mannequin. The particle model algorithm is used in most of the commercial software due to its advantage in speed and reliable accuracy (Kenkare, 2005).

14.7.3 Lectra Systems, Inc

Founded in 1973, Lectra Systems, Inc., one of the leading CAD/CAM suppliers to the apparel industry, provides made-to-measure solutions that offer a wide range of pattern, color, fabric, and fit (Lectra Systems, 2008). Lectra's virtual 3D prototyping solution, Modaris 3D Fit, enables patternmakers, designers, developers, and sales and marketing teams to simulate and visualize their models in 3D on a virtual mannequin, including the colors, motifs, and fabrics originally created in 2D. With Modaris 3D Fit, the look and fit of a garment can be verified, and its style and that of entire collections can be validated. Virtual 3D prototyping ensures the quality of a garment and its look and fit in all graded sizes, reduces the number of physical prototypes necessary to finalize a model, and makes communication more fluid among the actors in product development (Techexchange.com, 2008, June 12). Modaris 3D Fit can also provide a tension map.

14.7.4 Tukatech

Tukatech was founded in California in 1997 and has launched a CAD product that can help reduce sample approval and production time by allowing the entire supply chain to visualize garments via the Internet and share input simultaneously (Tukatech, 2008). In 2004, Tukatech began work on a new software product that it dubbed the 'e-fit Simulator' by TUKA. The e-fit Simulator software allows designers, patternmakers and manufacturers to make comments on fit and suggest alterations more quickly. It creates a virtual garment from a 2D pattern and drapes this over a 3D body image with lifelike movement capabilities. The same patterns used to cut actual samples are used by e-fit Simulator, and the virtual-fit models are created using body scan data of real-fit models. Dozens of preset cloth types can be blended together to test garments using the intended fabric, and additional presets can be created by the user. The e-fit Simulator creates virtual prototypes of garments, reducing time to market and increasing efficiency in the apparel product development process (Just-style.com, 2004, September 7).

14.7.5 Virtual try-on example

Fabric properties

An accurate simulation depends on several factors including cloth parameters, stitch attributes, and so forth. To get the accurate simulation fitting with the actual garment, it is essential to acquire accurate input values converted from the real fabric mechanical properties into the virtual fabric parameters. Figure 14.3 shows how the real fabric mechanical properties obtained from the FAST (Fabric Analysis by Simple Testing) System were input into the Fabric Converter (or Fabric Editor) of the OptiTex software.

Garment design and stitching

In most of the 3D systems, it is possible to import design work from one system to another, enabling collaboration more easily. In this example, the final garment patterns were developed in the Gerber AccuMark PDS system and exported using the DXF AAMA format files. Then, these .dxf format files were imported into OptiTex software. To simulate clothing in a virtual environment, the imported patterns were sewn through a 'stitch' tool. Figure 14.4(a) shows the 2D patterns for a tank top and the stitch box that describes the matched stitch lines (e.g., the right side seam line of the front piece matched with the side seam of the back piece). Figure 14.4(b) shows the stitch line (circled) of the 3D garments placed on the 3D avatar.

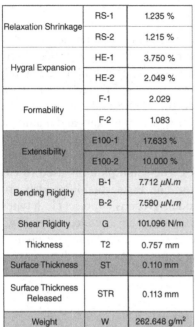

Relaxation Shrinkage	RS-1	1.235 %
	RS-2	1.215 %
Hygral Expansion	HE-1	3.750 %
	HE-2	2.049 %
Formability	F-1	2.029
	F-2	1.083
Extensibility	E100-1	17.633 %
	E100-2	10.000 %
Bending Rigidity	B-1	7.712 $\mu N.m$
	B-2	7.580 $\mu N.m$
Shear Rigidity	G	101.096 N/m
Thickness	T2	0.757 mm
Surface Thickness	ST	0.110 mm
Surface Thickness Released	STR	0.113 mm
Weight	W	262.648 g/m²

14.3 Fabric properties input.

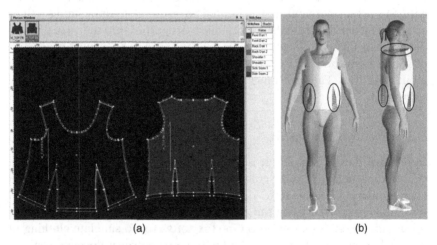

(a)   (b)

14.4 Virtual stitching for a tank top garment.

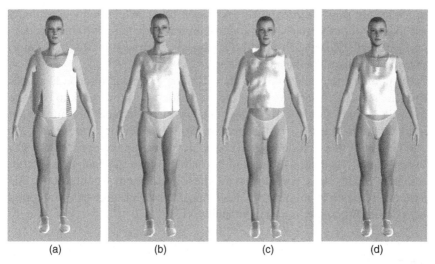

(a)   (b)   (c)   (d)

14.5 Draping simulation for tank top garment using OptiTex software.

Garment simulation

The stitched virtual patterns are placed on a virtual avatar by using the 'Place Cloth' button. Figure 14.5(a) shows the placing process of tank top pattern pieces around the avatar; (b) shows the particle model simulation process in progress; and (c) shows the final completed garment on the avatar. The actual stitching connections marked in the 3D garment are seen as shown in Fig. 14.4(a). This 3D virtual garment was simulated by using the 'Simulate Draping' tool, and the duration of simulation time varied depending on the size of the clothing items and the simulation resolution.

14.8   Conclusions

This chapter has focused on the development of human body modeling and fabric simulation for 3D CAD applications. Three-dimensional body, garment, and fabric modeling have significant implications not only for the way designers design products and businesses communicate product designs, but also for how consumers buy apparel products. The Internet is becoming a very important shopping outlet for apparel products (and other consumer goods). Realistic 3D body models can be used to improve the online shopping experience for consumers as part of a virtual try-on environment. In this application, consumers create personalized virtual body models (using either adjustable generic models or individualized 3D body scan data) who can actually try on products in a virtual dressing room. Companies utilizing this type of technology have said their customers are able to visualize product offerings and assortments better and evaluate product attributes in a way not possible previously (Azoulay, 2004). While many consumers are excited about virtual try-on environments, body models, fabric models, and garment styles and fit must be accurately portrayed in 3D for them to provide the most beneficial experience.

In summary, the field of 3D body modeling has made incredible strides over the last 20 years. However, there is considerably more work to be done in order to fully simulate the range of size, shape, surface, and movement individualities that exist in the human form. In addition, widespread use and benefit of these models cannot be achieved within the apparel industry unless equally accurate fabric and garment models are included. Future research in this area will center on the development of interactive 3D systems, seamlessly integrating mechanical, physical, and environmental information to create 'real life' virtual worlds.

14.9   References

Allen, B., Curless, B., & Popovic, Z. (2003). The space of human body shapes: reconstruction and parameterization from range scans. *International Conference on Computer Graphics and Interactive Techniques*, 587–594.

Apparel Magazine (2008, January 15). *Browzwear develops new software features, promoting better collaboration in the fashion industry* [online], retrieved 19 July 2008 from http://www.apparelmag.com/ME2/dirmod.asp?sid=&nm=&type=news &mod=News&mid=9A02E3B96F2A415ABC72CB5F516B4C10&tier=3&nid=8 D02411DB8D340F19AAD0C9C274D15E1

Azoulay, J. F. (2004). Body scans and consumer demands move into the future together. *AATCC Review*, 4, 36–40.

Azuola, F., Badler, N. I., Ho, P., Kakadiaris, I., Metaxas, D., & Ting, B. (1994). Building anthropometry-based virtual human models. In *Proceedings of the IMAGE VII Conference*, retrieved 10 June 2007 from University of Pennsylvania website: http://repository.upenn.edu/cgi/viewcontent.cgi?article=1066&context=hms

Boissonat, J. D. (1984). Geometric structures for three-dimensional shape representation. *ACM Transactions on Graphics*, 3, 266–286.

Browzwear International Ltd (2008). *Company Profile* [online], retrieved 21 July 2008 from http://www.browzwear.com/profile.htm

Bruner, D. (2010). Personal communication, 22 November.

Buxton, B., Dekker, L., Douros, I., & Vassilev, T. (2000). Reconstruction and interpretation of 3D whole body surface images. In *Proceedings of Scanning 2000*, retrieved 17 August 2007 from http://www.somavision.co.uk/i.douros/publications/buxton_dekker_douros_vassilev_scan2000.pdf

Bye, E., Labat, K. L., & Delong, M. R. (2006). Analysis of body measurement systems for apparel. *Clothing and Textiles Research Journal*, 24, 63–79.

Carrere, C., Istook, C., Little, T., Hong, H., & Plumlee, T. (2001, November). In *Automated Garment Development from Body Scan Data – Annual Report I00-S15*, retrieved 18 January 2007 and available from National Textile Center website: http://www.ntcresearch.org

Chittaro, L., & Corvaglia, D. (2003). 3D virtual clothing: from garment design to Web3D visualization and simulation. In *Proceedings of the 8th International Conference on 3D Web Technology*, 73–84.

Cho, Y., Park, H., Takatera, M., Kamijo, M., Hosoya, S., & Shimizu, Y. (2003). *Pattern remaking system of dress shirt using 3D shape measurement*. Paper presented at the 6th Asian Design Conference, retrieved 27 July 2007 from http://www.idemployee.id.tue.nl/g.w.m.rauterberg/conferences/CD_doNotOpen/ADC/final_paper/453.pdf

Cho, Y., Okada, N., Park, H., Takatera, M., Inui, S., & Shimizu, Y. (2005). An interactive body model for individual pattern making. *International Journal of Clothing Science and Technology*, 17, 91–99.

Choi, B. K., Shin, H. Y., Yoon, Y. I., & Lee, J. W. (1988). Triangulation of scattered data in 3D space. *Computer-Aided Design*, 20, 239–248.

Choi, K., & Ko, H. (2005). Research problems in clothing simulation. *Computer-Aided Design*, 37, 585–592.

Cichocka, A., Bruniaux, P., Koncar, V., & Frydrych, I. (2005). Parametric model of 3D virtual mannequin: Methodology of creation. *Ambience 05, International Scientific Conference on Intelligent Ambience and Well-Being*.

Cordier, F., & Magnenat-Thalmann, N. (2002). Real-time animation of dressed virtual humans. *Computer Graphics Forum*, 21(3), 327–335.

Cordier, F., Seo, H., & Magnenat-Thalmann, N. (2003). Made-to-measure technologies for an online clothing store. *IEEE Computer Graphics and Applications*, 23(1), 38–48.

Dekker, L., Douros, I., Buxton, B. F., & Treleaven, P. (1999). Building symbolic information for 3D human body modeling from range data. In *2nd International Conference on 3D Digital Imaging and Modeling*, 388–397, retrieved 20 May 2007 from http://www.somavision.co.uk/i.douros/publications/dekker_douros_buxton_treleaven_3dim99.pdf

Desmarteau, K. (2000, October). Let the fit revolution begin. *Bobbin*, retrieved 25 September 2007 from http://findarticles.com/p/articles/mi_m3638/is_2_42/ai_66936829

Desmarteau, K., & Speer, J. K. (2004). Entering the third dimension. *Apparel Magazine*, 45, January, 28–33.

Dieval, F., Mathieu, D., Herve, K., & Durand, B. (2001). Voluminal reconstruction of the bodies applied to the cloth trade. *International Journal of Clothing Science and Technology*, 13, 208–216.

Farber, Y. (2002). Improving design cycle time and bottom lines by using 3D visualisation technology. *World Sports Activewear*, Summer, 57–58.

Furukawa, T., Gu, J., Lee, W., & Magnenat-Thalmann, N. (2000). 3D clothes modeling from photo cloned human body [electronic version]. In *Virtual Worlds: Second International Conference, VW 2000*, Paris, 1834/2000, 159–170.

Gong, D. X., Hinds, B. K., & McCartney, J. (2001). Progress towards effective garment CAD. *International Journal of Clothing Science and Technology*, 13, 12–22.

Griffey, J. V., & Ashdown, S. P. (2006). Development of an automated process for the creation of a basic skirt block pattern from 3D body scan data. *Clothing and Textiles Research Journal*, 24, 112–120.

Guerlain, P., & Durand, B. (2006). Digitizing and measurement of the human body for the clothing industry. *International Journal of Clothing Science and Technology*, 18, 151.

Gurzki, T., Hinderer, H., & Rotter, U. (2001). A platform for fashion shopping with individualized avatars and personalized customer consulting. In *Proceedings of the World Congress on Mass Customization and Personalization*, retrieved 1 September 2007 from http://www.e-business.iao.fraunhofer.de/Publikationen/FashionMeAdvice.pdf

Hardaker, C. H. M., & Fozzard, G. J. W. (1998). Towards the virtual garment: three-dimensional computer environments for garment design. *International Journal of Clothing Science and Technology*, 10, 114–127.

Heisey, F., Brown, P., & Johnson, R. F. (1990). Three-dimensional pattern drafting: Part II: Garment modeling. *Textile Research Journal*, 60, 731–737.

Hilton, A., Beresford, D., Gentils, T., Smith, R., Sun, W., & Illingworth, J. (2000). Whole-body modelling of people from multiview images to populate virtual worlds [electronic version]. *The Visual Computer*, 16, 411–436.

Hinds, B. K., & McCartney, J. (1990). Interactive garment design. *The Visual Computer*, 6, 53–61.

House, D. H., & Breen, D. E. (eds) (2000). *Cloth Modeling and Animation*. Natick, MA: A K Peters.

Humanoid Animation Working Group, H-Anim (n.d.). *Specification for a Standard Humanoid*, retrieved 25 July 2007 from http://www.h-anim.org/

Hwang, S. J. (2004). Standardization and integration of body scan data for use in the apparel industry, Unpublished doctoral dissertation, North Carolina State University, Raleigh, NC.

Ilic, S., & Fua, P. (2003). From explicit to implicit surfaces for visualization, animation, and modeling. In *Proceedings of the International Workshop on Visualization and Animation of Reality Based 3D Models*, 34, retrieved 15 September 2007 from http://www.photogrammetry.ethz.ch/tarasp_workshop/papers/ilic_fua.pdf

Imaoka, H. (1996). Three models for garment simulation. *International Journal of Clothing Science and Technology*, 8, 10–21.

Inui, S. (2001). A preliminary study of a deformable body model for clothing simulation. *International Journal of Clothing Science and Technology*, 13, 339–350.

Istook, C. L. (2008). Three-dimensional body scanning to improve apparel fit, in Fairhurst, C., *Advances in Apparel Production*, Cambridge, UK: Woodhead Publishing, 94–114.

Istook, C. L., & Hwang, S. (2001). 3D body scanning systems with application to the apparel industry. *Journal of Fashion Marketing and Management*, 5, 120–132.

James, D. L., & Twigg, C. D. (2005). Skinning mesh animations. In *International Conference on Computer Graphics and Interactive Techniques, ACM Transactions on Graphics (SIGGRAPH 2005)*, 24, 399–407, retrieved 20 September 2007 from Carnegie Mellon Graphics website: http://graphics.cs.cmu.edu/projects/sma/jamesTwigg_SMA_lores.pdf

Jones, P., & Rioux, M. (1997). Three-dimensional surface anthropometry: Applications for the human body. *Optics and Lasers in Engineering*, 28, 89–117.

Jones, P. R. M., Li, P., Brooke-Wavell, K., & West, G. M. (1995). Format for human body modelling from 3-D body scanning. *International Journal of Clothing Science and Technology*, 7, 7–16.

Just-style.com (2004, September 7). USA: Tukatech launches e-fit, 3D motion simulators [online], retrieved 15 November 2010 from http://www.just-style.com/news/tukatech-launches-e-fit-3d-motion-simulators_id71248.aspx

Kang, T. J., & Kim, S. M. (2000). Optimized garment pattern generation based on three-dimensional anthropometric measurement. *International Journal of Clothing Science and Technology*, 12, 240–254.

Kenkare, N. S. (2005). Three dimensional modeling of garment drape. Unpublished doctoral dissertation, North Carolina State University, Raleigh, NC.

Kim, S. M., & Kang, T. J. (2002). Garment pattern generation from body scan data. *Computer-Aided Design*, 35, 611–618.

Kim, S., & Park, C. K. (2004). Parametric body model generation for garment drape simulation. *Fibers and Polymers*, 5(1), 12–18.

Lectra Systems (2008). *Modaris 3D fit: 3D virtual prototyping* [online], retrieved 14 July 2008 from http://www.lectra.com/en/cao/modaris_3d_fit/solution_performante.html

Lee, W., & Magnenat-Thalmann, N. (2001). Virtual body morphing. *Computer Animation*, 14, 158–166.

Lee, W., Gu, J., & Magnenat-Thalmann, N. (2000). Generating animatable 3D virtual humans from photographs. *Computer Graphics Forum*, 19(3), 1–10, retrieved 26 May 2007 from MIRALab website: http://www.miralab.unige.ch/papers/27.pdf

Lim, H. (2009). Three dimensional virtual try-on technologies in the achievement and testing of fit for mass customization. Unpublished doctoral dissertation, North Carolina State University, Raleigh, NC.

Linney, A. D., Campos, J., & Richards, R. (1997). Non-contact anthropometry using projected laser line distortion: Three-dimensional graphic visualisation and applications. *Optics and Lasers in Engineering*, 28, 137–155.

Magnenat-Thalmann, N., & Volino, P. (2005). From early draping to haute couture models: 20 years of research. *The Visual Computer*, 21, 506–519.

McCartney, J., Chong, K. W., & Hinds, B. K. (2004). An energy-based flattening technique for woven fabrics. *Journal of the Textile Institute*, 95, 217–228.

MIRALab (2006). *MIRALab: Where research means creativity*. Retrieved 25 January 2007 from University of Geneva website: http://www.miralab.unige.ch/

Newcomb, E., & Istook, C. (2004). A case for the revision of U.S. sizing standards. *Journal of Textile and Apparel, Technology and Management*, 4(1), retrieved 15 May 2007 from http://www.tx.ncsu.edu/jtatm/volume4issue1/articles/Newcomb/newcomb_full_111_04.pdf

Ng, R., Chan, C. K., Au, R., & Pong, T. Y. (1993). Computational technique for 3D pattern design. *Textile Asia*, 24, 62.

Okabe, H., Imaoka, H., Tomiha, T., & Niwaya, H. (1992). Three dimensional apparel CAD system. *Computer Graphics*, 26, 105–110.

OptiTex (2008). Profile [online], retrieved 13 May 2008 from http://www.optitex.com/en/About_Us/Profile

Outling, C. (2007). Process, fit, and appearance analysis of three-dimensional to two-dimensional automatic pattern unwrapping technology. Unpublished master's thesis, North Carolina State University, Raleigh, NC.

Paquette, S. (1996). 3D scanning in apparel design and human engineering. *IEEE Computer Graphics and Applications*, 16, 11–15.

Pargas, R., Staples, N. J., & Davis, J. S. (1997). Automatic measurement extraction for apparel from a three-dimensional body scan. *Optics and Lasers in Engineering*, 28, 157–172.

Protopsaltou, D., Luible, C., Arevalo, M., & Magnenat-Thalmann, N. (n.d.). *A body and garment creation method for an internet based virtual fitting room.* Retrieved 1 June 2007 from MIRALab CUI, University of Geneva website: http://miralab. unige.ch/papers/125.pdf

Rodel, H., Schenk, A., Herzberg, C., & Krzywinski, S. (2001). Links between design, pattern development, and fabric behaviours for clothes and technical textiles. *International Journal of Clothing Science and Technology*, 13, 217–227.

Seo, H., & Magnenat-Thalmann, N. (2003). An automatic modeling of human bodies from sizing parameters. In *Proceedings of the 2003 Symposium on Interactive 3D Graphics*, 19–26, retrieved 18 August 2007 from http://www.miralab.unige.ch/papers/151.pdf

Shen, J., & Thalmann, D. (1995). Interactive shape design using metaballs and splines. In *Proceedings Implicit Surfaces*, 187–196, retrieved 10 August 2007 from http://ligwww.epfl.ch/Publications/pdf/Shen_Thalmann_IS_95.pdf

Simmons, K. P., & Istook, C. L. (2003). Body measurement techniques: Comparing 3D body-scanning and anthropometric methods for apparel applications. *Journal of Fashion Marketing and Management*, 7, 306–332.

Stylios, G. K. (2005). New measurement technologies for textiles and clothing. *International Journal of Clothing Science and Technology*, 17, 135–149.

Stylios, G. K., Han, F., & Wan, T. R. (2001). A remote, on-line 3-D human measurement and reconstruction approach for virtual wearer trials in global retailing. *International Journal of Clothing Science and Technology*, 13, 65–75.

Techexchange.com (2008, March 2). *V-Stitcher™ gets upgraded again: New link to body scanner output* [online], retrieved 20 November 2010 from http://www.techexchange.com/BreakingNews/arc/0306.html

Techexchange.com (2008, May 27). *Always in touch: OptiTex packs even more features into its 3D runway designer 10* [online], retrieved 20 November 2010 from Techexchange.com: http://www.techexchange.com/BreakingNews/arc/0608.html

Techexchange.com (2008, June 12). *With newly extended Modaris 3D fit simulation capabilities, Lectra promotes the use of virtual 3D prototyping in the apparel industry* [online], retrieved 20 November 2010 from Techexchange.com: http://www.techexchange.com/BreakingNews/arc/0608.html

Tukatech (2008). *e-fit simulator™ by TUKA overview* [online], retrieved 20 November 2010 from http://www.tukatech.com/products/efit/index.html

Veltkamp, R. C. (1991). *2D and 3D object reconstruction with the y-neighborhood graph* (CS-R9116).

Virtual Fashion Technology (2007, April 30). *Technology day at FIT part V: MVM Louise Guay* [online], retrieved 23 September 2007 from http://fashiontech.wordpress.com/category/mass-customization/

Volino, P., & Magnenat-Thalmann, N. (1997). Interactive cloth simulation: Problems and solutions. *JWS97-B Geneva, Switzerland*, retrieved 19 August 2007 from MIRALab, University of Geneva website: http://www.miralab.unige.ch/papers/52.pdf

Volino, P., & Magnenat-Thalmann, N. (2000). *Virtual Clothing: Theory and Practice.* Heidelberg, Germany: Springer.

Volino, P., Magnenat-Thalmann, N., Jianhua, S., & Thalmann, D. (1996). An evolving system for simulating clothes on virtual actors. *Computer Graphics in Textiles and Apparel*, 16(5), 42–51.

Wang, C. (2005). Parameterization and parametric design of mannequins. *Computer-Aided Design*, 37, 83–98.

Wang, M., Wu, W., Lin, K., Yang, S., & Lu, J. (2007). Automated anthropometric data collection from three-dimensional digital human models. *International Journal of Advanced Manufacturing Technology*, 32, 109–115.

Xu, B., & Sreenivasan, S. V. (1999). A 3-D body imaging and measurement system for apparel customization. *Journal of the Textile Institute*, 90(2), 104–120.

Xu, B., Huang, Y., Yu, W., & Chen, T. (2002). Three-dimensional body scanning system for apparel mass-customization. *Optic Engineering*, 41, 1475–1479.

Yang, Y., Magnenat-Thalmann, N., & Thalmann, D. (1992). 3D garment design and animation. *Computers in Industry*, 19(2), 185–191.

Yu, W. (2004). 3D body scanning. In J. Fan, W. Yu, & L. Hunter (eds), *Clothing Appearance and Fit: Science and Technology*, 135–168, Boca Raton, FL: Woodhead Publishing.

Yu, W., Ng, R., & Yan, S. (2001). A new approach to 3D body scanning. *Textile Asia*, 32(10), 23–26.

Zhong, Y., & Xu, B. (2006). Automatic segmenting and measurement on scanned human body. *International Journal of Clothing Science and Technology*, 18, 19–30.

15

Integrated digital processes for design and development of apparel

T. A. M. LAMAR, North Carolina State University, USA

Abstract: This chapter examines the role of digital technology in enabling a design and development process integrating the functions of both textile and apparel creation. First, digitally enabled processes integrating design of textile and design of apparel simultaneously are discussed, including examples of garments developed through integrated processes. The focus of the chapter then shifts specifically to the role of computer-aided design (CAD) and visualization technologies in enabling an integrated digital process, and concludes with discussion of future directions.

Key words: integral and seamless knitting, digital textile printing, engineered design, visualization of apparel.

15.1 Introduction

How garments, and many other textile products, are designed and developed has evolved tremendously in the last few decades. Apparel producers and marketers have responded to intense competition from low cost alternatives through adoption of technology and implementation of new business strategies. Specific strategies, such as mass customization and fast fashion, rely on flexibility, even to the level of delivering one-of-a-kind products, and speed in the apparel design and development process. Process flexibility and speed are enabled by technological innovations in many stages of the design and development process. These innovations and changing strategies have led to evolving demands on the design and development process for textiles and apparel. Colchester (1996, p. 21), as cited in Moxey (1998), noted that designers sourced the most innovative fabrics in order to set themselves and their products apart. It is important to note that such a product differentiation strategy relies on innovative textiles and the expertise of the textile designer as well as the apparel designer.

This chapter examines the role of digital technology in enabling a design and development process integrating both the textile and apparel functions. In that context, conventional design and development processes for textiles and garments are introduced with a review of existing models of textile and apparel design. Next, examples of digitally enabled processes integrating design of textile and design of apparel simultaneously are discussed,

326

including how integrated processes compare with conventional processes in terms of design aesthetics, functionality, and production and challenges that surround adopting such processes. The chapter also presents examples of products developed with integrated processes and highlights the unique attributes achieved through utilization of an integrated process. The focus of the chapter then moves specifically to the role of computer-aided design (CAD) and visualization technologies in enabling an integrated digital process. The chapter concludes with a discussion of future developments and their potential for enhancing an integrated process.

15.2 Conventional design, development and production processes for apparel

Aside from handcrafted garments, traditional approaches to producing both garments and fabrics have largely involved separate business entities and processes. Historically, the textile and apparel industry comprised textile producers, who designed, developed and marketed collections of textile fabrics to producers of final products such as apparel. Apparel producers then acquired selected fabrics to be cut and sewn into garments which they then marketed. For printed textiles, little variation existed in this process unless an apparel designer ordered a custom fabric for the apparel line, thus becoming involved in a limited way with designing the textile. In contrast, the approach to creating and producing knitted garments (other than handcrafted) traditionally has been accomplished by one of four methods: cut and sew, knitting garment lengths, full fashioned knitting and integral knitting, depending on the type of apparel being produced (Parrillo-Chapman, 2008). Considering the production of the garment, the methods vary in terms of the amount of garment shaping that occurs at the knitting machine during the knitting process. The first two methods represent apparel production methods similar to those historically used to assemble garments from woven fabrics. Cut and sew methods utilize yardage of knit fabric cut into specific shapes and assembled with apparel production, i.e. usually sewing, equipment. Knitting garment lengths involves production using equipment that knits sequentially lengths of fabric appropriate for a finished garment rather than continuous yardage (Spencer, 2001). Depending on the knitting machine employed and its setup, the output may be either tubular or open width. When such tubes are knitted to size for a specific garment, the garment may be labeled seamless in practice. However, this is only indicative of a lack of side seams, not of no sewn assembly, and should be noted as distinctly different from knit-to-garment processes that may also be referred to as seamless. Knit-to-garment and full fashioning represent the potential for application of integrated processes, so will be discussed in later sections.

In considering the range of producible apparel items and the fabrications in which they are executed, there are two components of the field where digital technologies offer the greatest opportunities for integrated digital concurrent design and development of textile and garment. For this reason, these components will be discussed in detail in the following sections. One of those components is printed apparel where CAD systems and ink-jet printing technologies offer the potential for integrating the design of garment shape and surface coloration. The second component is integral knitting processes where CAD driven electronic knitting machines are able to produce finished garments. In these cases, the knit fabric structure is produced simultaneously with the garment.

15.2.1 Approach and stages for design of textiles

The field of printed textile design is diverse, encompassing products ranging from textiles for specific end uses driven by market forces, to art textiles. Printed textile design as a domain has been described as a 'body of disciplined knowledge ... culturally structured ... that can be acquired, mastered, practiced, and then advanced through the act of creating' (Li, 1997, p. 109). Moxey (1998) explains that as successful innovations are introduced into the domain, the domain expands like tree rings (p. 36). So, the domain of printed textile design is not stagnant, but is fluid and expands with evolutions in the ways that designers design. Envisioned in this way, the impact of historical innovations such as mechanized roller and rotary screen printing can be seen to expand the printed textile design domain. Likewise, the impact of the more recent introduction of digital textile printing on the domain of printed textile design can be envisioned as influencing the ways that designers design and expanding the domain. Digital textile printing continues to influence the process of designing in ways such as providing the ability to immediately view samples of printed designs rather than waiting on strike-offs from the plant, providing capability of producing textile prints ranging from photorealistic imagery to color reduced prints easily adaptable to traditional printing methods with a single printer, and streamlining the design and sample production process, providing opportunity for added experimentation with potential designs.

Textile design has been discussed from the perspective of creativity by authors such as Moxey (1998) and Li (1997), but these authors focus primarily on the creative element rather than the process of designing textiles. Bruce and Cook (1998) pointed out that the industrial textile design process allows little time and space for design creativity, but did not detail that process. Traditional processes for creating a textile design are detailed in practitioner-focused handbooks and texts for designers, such as the ones by Wilson (2001), McNamara and Snelling (1995), Joyce (1993), and Yates (1986).

In their book *Digital Textile Design*, Bowles and Isaac (2009) provide both an overview of the field of digital textile design and hands-on tutorials for specific techniques, but do not provide much sense of the process of designing. When the design of textiles is undertaken with digital tools, the most effective approaches take advantage of the strengths of textile design software – strengths such as the ability to copy, transform and paste design elements and the ability to quickly and easily explore many repeating design structures. An approach from the designer's perspective is summarized in Fig. 15.1. The process reflects conceptually only the designer's

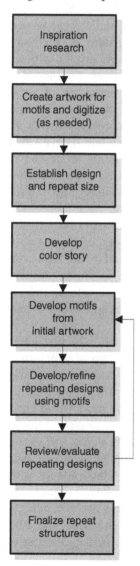

15.1 Textile design process in a digital environment.

creation process, not the business processes that would be integrated in a commercial environment.

Studd (2002) mapped the textile design process more extensively into a generic framework for textile design using data gathered through case study research. In reviewing Studd's framework in relation to models of apparel product development, it is apparent that there are many similarities in the design process for textiles and for apparel. Among these are the use of concept presentations, design review and color development, use of CAD technology for design and visualization processes, and the involvement of many functional areas in the process and in review of designs.

Processes for designing knitted fabrics for cut and sew can be very similar to the process for printed textiles, or even the same. Knitted yard goods, like woven yardage, may be used as base cloth and printed with traditional or digital printing methods. However, in textile design for knitted structures that achieve their design aesthetic through integration of varying yarn colors or specific combinations of knit stitches, the textile designer creates not only the surface design but also the structure of the textile itself. Knit production machinery may be driven with design files from a CAD system. For example, in knitting multicolored tubular fabric on an electronic jacquard knitting machine, a designer may create a colored pattern in the knitted fabric through a digital pattern that controls how the differently colored yarns are knitted into the fabric. So the visual appearance of the textile surface, and the structure that supports it, are created by the knitted fabric designer. As shown in this example, design of many knit fabrics is more integrated with the production of the actual textile than is the typical case for printing, and the designer must have a greater technical understanding of the textile structure. With knit-to-garment and full fashioning, which will be discussed in Section 15.3, integration with garment design processes also occurs.

Computer technologies for design of textiles influence not only the broader process of textile design, but also the ways in which the designer works from day to day. Functions commonly found in industrial textile design systems allow designers to build complex repeat structures in a short time, and to efficiently create and compare more variations of textile designs than is possible when designing by traditional methods. CAD features allow designers to quickly and easily create, adjust and visualize color combinations in the pattern. Options for printed or electronic output from CAD systems facilitate designers working collaboratively with distant designers or producers. Simulation capabilities allow designers to visualize the structure of textiles, such as knits, while they are still being designed. Taken as a whole, such computer technologies provide opportunities for integrating garment design with textile design in the creation of unique products.

15.2.2 Approach and stages for design of garments

In contrast to the textile design field, a number of models for apparel design and development can be found in the literature. The models present researchers' unique perspectives on the increasingly complex process of developing garments. As the industry has evolved to be characterized by huge corporations that design and develop apparel and by retailers who develop their own store brands, both relying on contractors to manufacture the products, processes of design and development have also evolved as reflected in the models. Examples of apparel design and development models include the Taxonomy of Apparel Merchandising Systems (Kunz, 2005), the Revised Retail Product Development Model (Wickett et al., 1999), the No-Interval Coherently Phased Product Development (NICPPD) Model for Apparel (May-Plumlee & Little, 1998) and the Apparel Design Framework (Lamb & Kallal, 1992). Wickett, Gaskill and Damhorst's (1999) Revised Apparel Retail Product Development Model presents the process in phases and sub-phases, providing insight into some of the specific activities that occur at each stage of the process. Likewise, an even more in-depth view of the design and development activities occurring within each phase of the model is shown in the NICPPD Model for Apparel (May-Plumlee & Little, 1998). Kunz's taxonomy as well as the latter two models include fabric selection or design within process stages and recognize the opportunity for the garment designer to become involved in fabric development. However, it is important to note that these models assume traditional fabric development methods, i.e. that the fabrics will be produced as yardage with a repeating pattern, and a cut-and-sew approach to the production of garments from those fabrics.

As was discussed for textile design, digital technologies adopted to facilitate the apparel development process can be envisioned as expanding the domain of apparel design and development by influencing the ways that designers design. Computer aided design programs influence how designers create seasonal apparel product lines through functions that allow creation of garment renderings and overlaying scanned images of selected textile fabrics onto the renderings. Functions to create virtual representations of garments designed in the CAD systems have progressed toward visualizations that allow the designer to see the appearance of finished goods in a virtual environment without the costs associated with traditional physical prototypes. Garments can be fitted to models, the models turned to view designs from various angles, and in some systems 'modeled' by virtual models on a virtual runway so the designer can see the garment in movement. Some proprietary industry CAD systems have extended their software to encompass both textile design and garment design functions, allowing visualization of a garment that doesn't yet exist in a fabric that may not yet exist. These innovations in digital technologies create opportunities for more integrated textile and apparel design processes in practice.

15.3   Simultaneous design of textile and garment utilizing digital technology

I would propose that the expansion of textile and apparel design domains in response to innovations in the fields could be most accurately envisioned as ripples growing around two stones thrown in the water. As the two sets of rippling rings begin to meet, the two domains interface in an ever growing segment where textile design and apparel design occur simultaneously in creating a product, as in Fig. 15.2. In this representation, we would envision the two domains not just as overlapping, as might be suggested by the tree ring analogy, but with each domain in fact informing and shaping the other in the discipline areas where they interface. This interaction and mutual informing of the disciplines is consistent with Black's (2002) observation that the 'Convergence of disciplines is taking place across many areas of design and art practice, including graphic, product, interior and fashion design ...' (p. 9).

I would further conjecture that expansion of a design domain is driven not only by the introduction of innovations in technology but also by innovations in design. More specifically, the expansion of a design domain comes not only through introduction of technologies and accompanying

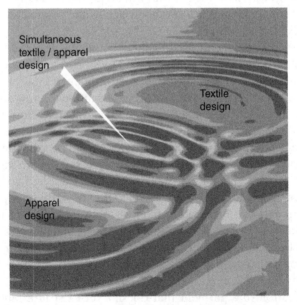

15.2 Interface of textile design and apparel design domains.

exploration through new methods, but also through innovative application of traditional technologies and methods. As a case in point, technology for knitting full garments has been available for centuries, such garments originally being fashioned by hand-knitting methods (Powell & Black, 2005). The technology for knitting full garments with powered knitting equipment has been available since the 1970s, but only more recently has it been applied to creating innovative fashion and performance apparel. This example represents innovation in product design, and how technology is applied – designer rather than technology driven advances.

Some methods of creating textile designs for specific garments represent steps toward integrated processes. The term 'engineered design' is often used to describe textile designs which have been developed, adapted or customized for specific garments. Engineered design, as discussed by Braddock and O'Mahony (1998), comprises a graphic design printed onto a fashion garment in a specific location specified by a textile designer, typically for a runway fashion exhibition. The authors note that if such engineered designs for the runway are adapted for production, the textile designs are generally reworked so that they can be printed continuously on fabric yardage. When the reworked design as it is printed on fabric yardage is then used to produce garments, the garment pattern must be laid out for cutting onto the fabric so that the reworked textile design falls in the desired arrangement when the garment is constructed. In addition to arranging the garment pattern pieces in specific positions on the fabric with respect to the repeat of the reworked continuous textile design, the pieces must also be aligned in a particular direction and orientation in relation to the lengthwise grain of the fabric. The lengthwise grain is a term used to refer to the direction parallel to the selvages of a woven fabric, or the wales in the case of a knit which has no selvage. Figure 15.3(b), (d), and (f) shows a simple garment with a pattern that would be engineered by arranging the pattern pieces on fabric yardage. To achieve this aesthetic in the finished garment, the pattern pieces must be marked to indicate placement on the patterned fabric. Then, the marker, or cutting layout, must be created so that the markings on the pattern align with the design repeat on the fabric. Figure 15.4 illustrates the appearance of a cutting marker for cutting the fabric yardage to engineer the textile design for this particular product. The garment pattern pieces are laid out on the fabric yardage in such a way that the plaid pattern will align on the finished garment. Such engineering is typically feasible only for high-end garments that have a generous enough profit margin to absorb the cost of the fabric wasted in cutting the garment.

Figure 15.5 shows an example of a digitally printed marker for cutting a garment with a similar plaid design. Although the grain orientation must still be accommodated, there are no restrictions on the layout of the pattern pieces regarding orientation and location on the fabric. This is because the graphic pattern is applied to the garment parts rather than continuous

(a) Front: Digitally printed garment

(b) Front: Traditional garment

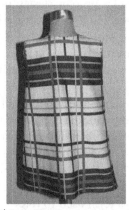

(c) Back: Digitally printed garment

(d) Back: Traditional garment

(e) Side: Digitally printed garment

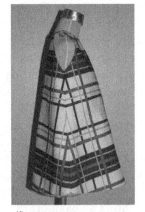

(f) Side: Traditional garment

15.3 Comparison of garment engineered via cutting layout with one digitally engineered.

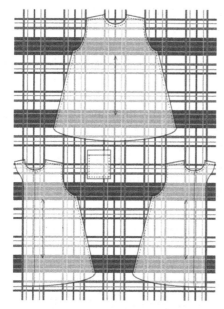

15.4 Cutting marker for garment with traditionally engineered print.

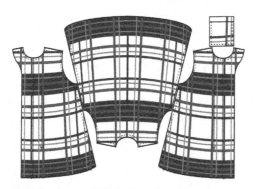

15.5 Engineered digitally printed marker.

yardage, and then only the garment parts need to be printed. It should be noted that this process relies on digital textile printing technology.

15.3.1 Overview of integrated digital processes

Parrillo-Chapman (2004) describes the process of engineering a printed textile design for a specific garment as follows:

> To engineer a print, a product developer brings the garment marker into a textile design software program and creates the textile design within the marker. This design engineering process requires the collaborative efforts of

both a product and textile designer, allowing a better marriage of shape and pattern design. For instance, a panoramic scene could continue across a product, or around a three-dimensional form. The textile design, once created on a two-dimensional surface, now is manipulated across the shape of the body so that it becomes a three-dimensional design.

Chenemilla (2001) explored integrating digitally printed designs into products for mass customization by utilizing a digital archive of previous designs. In this type of approach, the textile and garment designs are not simultaneously created; rather an existing textile design is adapted to an existing garment design. So, in such a model the approach is integrated only in a limited way; the process is more one of modifying a textile design to fit an existing garment design. Integrated design occurs when creation of garment and textile pattern develop in parallel with each development in one component influencing developments in the other component.

Authors including Campbell and Parsons (2005), Gordon (1997), and Bunce (2005) have documented processes for creating art to wear garments using integrated methods. These processes document development of one-of-a-kind garments, and are best suited to similar circumstances. Parrillo-Chapman's (2008) model is one notable framework for understanding integrated processes in the broader context of a production environment. Parrillo-Chapman's model illustrates a holistic process in which the requirements of the end product determine the manufacturing processes, shape or silhouette, and the fabric design and properties. She describes the process as engineering textile design properties within the garment shape and observed that integrated processes for prototyping and for manufacturing the finished product can be the same when creating digitally printed or integrally knitted garments.

Spencer (2001, p. 225) observed that 'Over the past thirty years, many innovations and refinements in knitting technology have gradually evolved and combined to transform the mechanically controlled V-bed machine into a computer-controlled, highly efficient and versatile knitting machine, not only for cut and sew knitwear but also for integrally-shaped panels and whole garments.' Parrillo-Chapman (2008) describes integrally knitted garments as those where traditional cut-and-sew operations are replaced by knitting processes. With integral knitting, apparel products can be knit to shape on a V-bed knitting machine. Complete garment knitting represents integral knitting taken to the furthest application process. Garments produced using complete garment knitting methods after leaving are completed at the knitting machine and thus require no post-knitting assembly. Yielding less finished results, fully fashioned garments are created by knitting garment pieces which are then assembled by sewing or linking into complete garments. Because appropriately shaped and structured garment parts result from full fashioning, it also represents an integrated process in which the garment and the textile must be designed and produced simultaneously. Though some final assembly is required to complete the garment, all necessary shaping of the parts is accomplished during the process of knitting the fabric.

Integrated textile and apparel design processes can encompass a range of textile design applications. Textile design processes can be applied in somewhat limited ways to enhance the aesthetics of the textile surface through processes such as digital printing on existing textiles. At the other end of the continuum, textile design processes can determine the structure of the textile as well as the garment it comprises. Finer-gauge knit garments described as seamless are common in intimate apparel. Growth in the shapewear and activewear market has driven introduction of more complex combinations of structures beyond the 'no or minimal side seams' construction at these finer gauges. Seamless outerwear development has been primarily focused on heavier-gauge sweaters for men and women.

Innovations in knitting technology, such as integral and seamless knitting, have intrinsically connected the creative designer and technical designer with the production of knitted garments. Likewise, availability of digital textile printing has connected the roles of textile designer and garment designer in the creation of engineered printed garments. These innovations create tremendous opportunities for novel design, but also yield unique challenges as discussed in the following sections.

15.3.2 Analysis of integrated digital processes compared to conventional processes

Integrated processes offer the opportunity to create unique aesthetics in finished garments through avenues such as special printed designs on sewn garments and structural designs in knitted ones. The ability to digitally print fabrics is not only changing the way garments made of printed textiles are developed, it has opened a new world of aesthetic design possibilities. Figure 15.3(a), (c), and (e) illustrate how even a simple plaid pattern can be engineered to complement the product design in a digitally printed product. Note the different aesthetics of the two garments shown in Fig. 15.3. In the garment achieved by careful arrangement of the garment pattern pieces on continuous yardage, a chevron effect occurs where the stripes of the plaid meet at the side seam of the garment. However, on the garment cut from the digitally printed marker, the stripes of the plaid flow smoothly and continuously around the garment, following its contours. The difference can be seen most easily near the hemline by comparing the front and back views of the two garment examples. In the garment created from the digitally printed marker, the stripes of the plaid have been engineered to mirror the curve of the hemline, creating a much more pleasing appearance than with the garment cut from repeating yardage. The garment engineered with a precise layout still contains straight lines that, due to their inability to conform to the garment shape, create the illusion of drooping at the outside edges when viewed from front or back.

These examples using simple textile designs illustrate the potential impact on garment aesthetics of creating a textile design for a specific garment design. When design of the textile pattern and structure are developed simultaneously with the design of the garment structure, unique aesthetics can be achieved. The garment shown in Fig. 15.6 was achieved through

15.6 Example of one-of-a-kind engineered garment designed by Anne Porterfield.

designing garment and textile pattern at the same time, digitally printing the textile pattern within the shapes of the garment parts and then assembling the garment. The floral design is placed to complement the garment structure, and, as shown in the detail, continues uninterrupted across structural lines.

Like the printed example, integrated processes can yield knit garments with unique design aesthetics. Whole garment knitting can result in garments with no seams and the same finished appearance on inside and out. A variety of knit stitch structures can be incorporated in a garment to provide shaping and visual design interest. Even some garments of woven fabric are being approached in an integrated fashion, using a computer-driven manufacturing process. Issey Miyake, an apparel, designer, and Dai Fujiwara, a textile designer, invented a process called A-POC (for A Piece of Cloth) for knitting and weaving whole pieces of apparel with no sewing needed (Jana, 2006). Each piece of clothing thus produced can simply be cut out of the fabric and it is complete. Gavenas (2007) describes machinery that takes in a piece of thread at one end and dispenses finished jeans at the other, garments that can be cut but won't ravel so are completely customizable, and garment features such as three-dimensional padded decoration. In addition to the impact on the aesthetics of apparel, the A-POC process provides production benefits, including shortening the fabric development cycle, reducing production time and minimizing fabric waste (Gavenas, 2007).

Like the A-POC process, many integrated textile and apparel creation processes offer benefits in production as well as novel aesthetics. The positive impact of digital printing on material utilization is best seen by comparing the two cutting layouts shown in Figs 15.4 and 15.5. As can be seen in Fig. 15.4, material utilization is far from optimum when textile design is engineered by placement of garment pattern pieces on continuous yardage. Much fabric is wasted in the process of engineering the design to match when the garment is constructed, and the waste is in small pieces of little value for further production. Consequently, manufacturers offering apparel products at a lower price often cannot afford to produce designs engineered this way. It is noteworthy that the larger the repeat of the textile design, the more fabric is required to engineer a design by precise laying out of the garment pattern on fabric yardage. Figure 15.5 illustrates that the amount of fabric yardage required to produce the engineered digitally printed garment parts is substantially less.

In addition to the direct benefits described above, when digital printing is used to add surface design, an apparel producer would need only to stock inventory of white fabrics for printing. When printed with any number of different surface designs, a single fabric can take on the appearance of completely different fabrics. Use of engineered design by digital textile printing allows a reduction in raw material inventory, the flexibility to more fully use fabrics that are in stock, the ability to respond more quickly to changing consumer desires, and the elimination of waste associated with investing in fabrics that ultimately do not sell well in the marketplace.

If we consider alternative methods of constructing knitted sweaters, both integral knitting and full fashioned methods significantly reduce the number of separate steps required for completing a garment as compared to cut-and-sew methods (Brackenbury, 1992). Though the reduction in steps significantly reduces post-knitting assembly time, fully fashioned sweater parts and integrally knitted garments are produced much more slowly than knit yardage. So although the steps are fewer with some methods, the speed of knitting is substantially reduced. It has not been documented whether the fewer steps in fact translate to reduced overall production time.

Processes that integrate fabric and garment production may also have implications for the functionality of the product. Products produced by integral knitting, for example, differ from those produced by cut-and-sew and fully fashioned methods in that production is complete without seaming. Because seam failure significantly reduces quality in knitted products (Clapp *et al.*, 1994), garment attributes such as comfort, fit and durability can be improved by eliminating seams. Strategic placement of different knit stitches on the knitted garment is also possible with integral knitting. Use of integral knitting where the product is actually knitted as a 3D shape offers advantages over seaming of flat panels to provide garment shape (Powell, 2003; Hunter, 2005) in that knit stitch structures can be varied to control shaping and stretch for improved fit and compression. The ability to create areas of compression in garments is useful for specialty applications such as for athletic, medical and shapewear end uses.

15.3.3 Challenges of adopting integrated processes

Despite the appeal of integrated digital processes in terms of potential product features, there are challenges to adoption of these technologies. The most critical challenges are associated with the product design and development skill requirements and knowledge base, and production challenges. Implementation of integrated processes requires a unique and extensive range of design and development skills, and a knowledge base that extends beyond what is typical for both garment and textile designers and developers. For one-of-a-kind garments, the solution is often a collaboration melding the skills and knowledge of textile designer and garment designer and supported by a machine technician with whom the designers work closely. For larger-scale adoption, the solution is more challenging.

Integrating textile design input into cut-and-sew apparel design processes would require a paradigm shift in how designers think about both apparel and textile design. Textile designers rarely have expertise in shaping and assembling cut 2D fabric shapes into the 3D form of an apparel garment. When developing layouts for textile print design, for example, one aspect of the process that has traditionally been important is the arranging of motifs based on their relationship with one another throughout continuous yardage rather than within the boundaries of a specific composition. The logic of the approach is that the motifs will be put into a continuous repeat and the compositional boundaries of the original layout will no longer exist. In an integrated apparel/textile design process, the garment itself defines the compositional boundaries and the textile design can be engineered onto the composition. The textile designer must understand not only how the design will fall on the 3D form, but also how the individual shaped garment parts are assembled to create the form in order to create a design that can be continuous across the sewn lines assembling the garment. Similarly, an apparel designer rarely has depth of understanding in terms of how their flat garment parts can be best designed to accommodate engineered textile design and efficient layout on unprinted fabrics. As observed previously, in an integrated process, the garment pattern pieces can be nested as efficiently as possible across a fabric width without concern for the textile motif layout, and the textile design then engineered to align with the garment pieces.

Brackenbury (1992) observed that the rewards in terms of raw material and labor savings of adopting integral knitting could be substantial. He also acknowledged that a lack of designers and development technicians with the appropriate competencies was a factor limiting such adoptions. Like designers of garments created by full fashion techniques, designers and development technicians for integral garments need to understand the construction of knit stitches and their associated impact on the aesthetics, fit and function of a final garment. They also share with traditional knit garment designers knowledge of raw materials such as fibers and yarns in terms of their impact on the end result. However, to translate ideas into successful, integrally knit garments, these personnel must also have extensive knowledge of machinery technology and operation, which has traditionally been the realm of production personnel. Integrally knit garment designers and development technicians may use CAD software that combines the creative design, technical development and knitting machine operation control in a single application. They also must be able to discriminate among prototype defects and undesirable results that are the result of design shortcomings, raw material issues, and equipment or production defects. Traditionally, the necessary knowledge set belonged partially to the design and development technicians, but partially to knitting machine technicians who may have been located in plants far removed from the design personnel.

Lamar, Powell and Parrillo-Chapman (2009) observe that there is currently no recognized system integrating yarns, knit process and structure, visualization, and finishing, with the aesthetic and performance attributes required to produce integrally knitted garments for varied end uses. Such a system would support designers and development technicians in addressing the unique demands of integrally knit garment design, development and production. Practitioners such as Shima Seiki (http://www.shimaseikiusa. com/shima7102_001.htm or http://www.shimaseiki.com/wholegarment) and Santoni (http://www.santoni.com/en-azienda-seamlessstory.asp) provide rare examples of integrated processes for producing uniquely designed, even custom, integrally knit garments. However, it is notable that it is these machine developers themselves who possess sufficient breadth and depth of knowledge and production expertise to implement a successful, integrated process in a real-world scenario (Lamar *et al.*, 2009).

Implementing integrated digital processes presents production challenges as well. Parrillo-Chapman (2008) notes that the methods of apparel production currently in practice rarely support completely integrated processes, even though digital textile printing technologies would be easily adaptable to process integration. Optimal integrated digital processes would yield unique designs, as seen in the previous examples, printed one at a time in the shapes of the garment parts. Each garment printed could be different from the one printed before and individual cutting and assembly of each garment would then follow for a continuous flow. However, successful models of apparel production have traditionally utilized economies of scale to produce garments profitably. Fifty or more identical garments may be cut simultaneously from inches-deep piles of yard goods spread on a cutting table using a cutting plan, or marker, developed to optimize fabric utilization and minimize waste. Following this, the garments are sent in bulk to production where they are assembled by sewing operators who rely on the repetition of assembling identical products, one after another, to maintain efficiency. Established production facilities laid out and equipped to support mass production processes are not set up to efficiently assemble products one at a time. Substantial modifications in equipment, plant layout and assembly processes are required.

Use of a computer-driven cutter can facilitate efficient cutting of a single ply of fabric, but cutting of printed garment parts presents alignment challenges. For an integrated process, the fabric would need to be carefully placed so that the computer cutting lines aligned with the edges of the digitally printed garment parts. Not only does this require extra time in cutting layout, but poor alignment could easily result in wasted fabric or products of second quality. In assembly, unique, digitally printed garments require particularly careful attention to matching of the textile pattern at seam lines, limiting the speed with which garments can be assembled. Despite the challenges, adoption of an integrated process would then create opportunities for printing and assembling custom garments. Though appealing in principle, providing this level of customization on a mass scale often demands technological capability that is not yet reliably available. For example, digital printing that allows printing of garment parts for engineered design is limited by raw material challenges (May-Plumlee & Bae, 2005) and color management issues (Bae & May-Plumlee, 2005). Commercial firms are making strides in addressing these issues (Rimslow Pty Ltd, 2004; Yuikawa, 2001).

As is the case with digital textile printing, production of engineered, integrally knit garments requires specialized production facilities and dedicated equipment. Integrally knit seamless garments are produced individually on seamless knitting machines at production speeds much slower than typical knitting machines can produce yardage. If each garment to be knit is unique, then yarn may need to be changed and machine settings may need to be adjusted between garments. So, economies of scale are lost and production costs increase to the point at which it may be difficult to produce such garments profitably except at the highest price points. In some regards, it is not a true comparison to compare garment production speed to speed in producing knit yardage. With integral knitting, both fabric production and garment assembly occur at the knitting machine where a complete garment is output. To understand production speed, it would be ideal to compare the time required to produce an integrally knit garment to the sum of the time required for fabric production plus the time required to cut and assemble a similar knit garment. Because these processes typically occur in different segments of the supply chain, such a comparison is difficult.

15.4   Integrated processes in practice

Integrated digital design technologies for garment production can be implemented to support business strategies ranging from entrepreneurial shops producing art and one-of-a-kind garments to competitive practices such as mass customization. Mass customization is defined by Anderson as 'the ability to design and manufacture customized products at mass production efficiency and speed' (Anderson, 2003, p. 271). In their review of the literature, Silveira, Borenstein and Fogliatto (2001) synthesized a number of approaches to mass customization, as presented by other researchers, into a framework reflecting eight levels of customization ranging from collaborative design to pure standardization. Each of the eight levels reflects an increased level of customization and a greater level of customer involvement in the design process when compared with the level preceding it. Each of the levels also reflects an increasing level of integration among various functions of the design process and with the customer, and requires an increasing integration of digital design and development processes. With garments, customization can deal not only with the aesthetics and functional features of the garment, as described previously, but with sizing and fit as well. Integration of non-contact body measurement technology into the development process provides opportunity for customized fit of garments. Brooks Brothers, for example, offers customized men's garments individually fitted based on measurements acquired through a digital body measurement system located in one of their retail stores and transmitted electronically to their factory (Brooks Brothers, 2011).

15.5 Role of computer-aided design (CAD) and visualization technologies in integrated textile product design

Integrated digital processes rely upon integration of enabling computer-aided design technologies, which can itself present challenges. In the case of integrally knit garments, machine manufacturers offer software for designing the garment and knit structures, and translating the resulting pattern to a format to drive the knitting machine within the same software package. Components for visualization are increasingly integrated into such packages in the form of simulations to illustrate specific knit structures and even entire knit garments. As a design tool, these visualization technologies can facilitate more cost-effective production in that, based on simulations, errors can be identified and design revisions can be made prior to the production of physical samples. A reduced need for physical sampling in turn reduces product development time and sampling costs and can begin to offset the costs of slower production.

In the case of digitally printed engineered designs, apparel pattern-making software is optimally used to create shapes of garment parts that will be cut from fabric. Those shapes must then be output to textile print or graphic design software for development of the printed design. The cross-application file transfer must occur without distortion of size or shape of the garment parts, and is complicated by the fact that pattern-making software uses vector-based graphics while textile print design software is primarily raster based. Visualization of an engineered print design is more complex than with integral knitting as well. Many apparel pattern design applications are components of larger software packages which may also include garment simulation capability for visualization of garments. Many include a component for overlaying textile designs onto the simulation as well, but they are commonly programmed to fill with repeating patterns rather than to accept a strategically placed engineered design. Visualizing a finished engineered, digitally printed garment may require exporting the simulated garment back to a graphics program where the surface design can be mapped onto the simulated garment.

15.6 The future of integrated digital apparel design and development processes

Printed textiles are an important component of the fashion market for garments, accessories and home. The dominant technology currently for printing on textiles is rotary screen printing. Of the annual production of printed textiles, ink-jet printing production accounted for only about 0.01% in 2002 (Teunissen *et al.*, 2002) and was projected to have an annual worldwide growth estimated at 13% (Cahill & Ujiie, 2004). Technology advancements have increased printing speeds, enhanced the variety of inks and pigments available for ink-jet printers, enhanced the usability and reduced the production challenges. It is expected that these improvements will continue as the technology matures. Ink-jet printed textiles were determined to have the same or nearly the same color saturation and gamut as screen-printed textiles in a study by Canon Inc., Japan, as cited in (Parrillo-Chapman, 2008, p. 18), and were even found to offer superior performance in some areas such as stiffness, air permeability and crock fastness.

Some of the earliest applications to digital textile printing were for signage and banners, and prototyping in-house for textile designs. Now ink-jet textile printing is increasingly used to produce textile goods available to the consumer. Entrepreneurial printing businesses such as Spoonflower (Scelfo, 2009) and Karma Kraft (http://www.karmakraft.com/) have found successful market niches in printing custom fabric designs for the consumers who created them. [TC]²'s Inkdrop Printing Service (http://www.inkdropprinting.com/) produces custom-printed products by combining digital printing services with cut-and-sew assembly of some basic products such as totes and scarves. Digital printing equipment for printing T-shirts makes garments with custom printing available on the market (Sexton, undated). These steps represent progression toward a time when affordable, customized printed garments are widely available.

Researchers are developing support systems to support use of integral knitting technology. Lamar, Powell and Parrillo-Chapman (2009) are developing an integrated system for engineered knit garment creation that will enable production of a unique, knitted garment engineered to a specific purpose or end-use. The developed system is intended to support creating and modifying knit garment design attributes in a virtual environment, and to support addition of technical knit production data. Yang and Love (2009) observed that seamless and integral knitting technology is not well developed from the designer's perspective. Two recent papers discuss importance of integrated processes, including software applications, to successful implementation of customized knitting by an equipment developer and knitting practitioner (Peterson *et al.*, 2009; Lam & Zhou, 2009).

These trends suggest that the technologies to support integrated digital design and development of apparel will continue to evolve. From knit and weave structure to cut-and-sew garment simulations, virtual models are increasingly useful for communicating the appearance of an apparel product, and its component fabric structures, when a physical product has not yet been fabricated. As such technologies continue to improve, and researchers delve deeper into understanding how digital tools can be used to facilitate integrated processes, application of integrated processes will face fewer challenges to adoption and become more widespread.

15.7   Conclusions

Integrated digital processes for garment design present opportunities for creation of unique garments with great synergy between garment elements and textile structure and aesthetics. Simultaneous creation of the textile design aesthetic along with the garment silhouette and structure allows development of visual design elements that complement the garment in ways not achievable when fabric and garment are completed in separate steps. Further, an integrated process allows development of structures to support specific garment functional requirements as with the case of engineering knitted features for performance garments. As the supporting technologies for integrated digital design continue to advance, production of engineered, even one-of-a-kind garments becomes more viable in the market. Combined with business strategies such as mass customization, such products may become more widely available.

Integration of digital technologies is not seamless. Challenges to integration exist both in terms of product design and technical development expertise and in ability to integrate various digital applications. Garment production methods which historically have involved simultaneous production of fabric and garment, such as knitting, present fewer challenges in terms of digital integration but perhaps more in terms of expertise required. As these challenges continue to be addressed by researchers and practitioners, the potential of integrated digital technologies will be realized.

15.8   Sources of further information

The book *Digital Textile Design* by Bowles and Isaac (2009) provides an overview of the opportunities in digital textile design from craft applications to commercial ones, from input sources to output techniques, and profiles practicing designers. Black's *Knitwear in Fashion* (2002) features visuals and information for many innovative knit garments. TechExchange.

com features a library of articles providing greater depth about technologies for textile and garment development, industry news and events.

For more technical approaches to the subjects, Brackenbury's *Knitted Clothing Technology* (1992) includes a chapter on integral knitting, and Spencer's *Knitting Technology* (2001) provides in-depth discussion of techniques for knit shaping. *Textile Design Engineering Within the Product Shape*, by Parrillo-Chapman (2008), provides greater depth in engineered designing with both digital printing technologies and integral knitting.

Several practitioners of integrated and digital processes provide opportunities to learn more about these processes in practice. For the A-POC process, the article Seamless (Scanlon, 2004) provides a great overview (http://www.wired.com/wired/archive/12.04/miyake.html?pg=1&topic=& topic_set=) and the book *A-POC Making: Issey Miyake & Dai Fujiwara* documents an A-POC exhibition. To learn more about companies that print custom textile designs, Spoonflower (http://www.spoonflower.com/welcome) and Karma Kraft (http://www.karmakraft.com/) provide current examples. To learn more about a service that does customized, digitally printed products and the process, see [TC]²'s Inkdrop Printing Service (http://www.inkdropprinting.com/). The Shima Seiki (http://www.shimaseikiusa.com/), Santoni (http://www.santoni.com) and Stoll (http://www.stoll.com/) websites feature galleries of sample integrally knitted products as well as discussion of their integral knitting technologies.

15.9   References

Anderson, D. M. (2003). *Build-to-Order and Mass Customization: The Ultimate Supply Chain Management and Lean Manufacturing Strategy for Low-Cost on-Demand Production Without Forecasts.* Cambria, CA: CIM Press.

Bae, J. & May-Plumlee, T. (2005). Integrated technologies for optimum digital textile printing coloration. *Textile Institute 84th World Conference Papers* (on CD-ROM), Session 3 (Thursday), Paper 3, presented 24 March 2005, Raleigh, NC.

Black, S. (2002). *Knitwear in Fashion.* New York: Thames & Hudson.

Bowles, M. & Isaac, C. (2009). *Digital Textile Design.* London: Laurence King Publishing in association with Central St Martins College of Art and Design.

Brackenbury, T. (1992). *Knitted Clothing Technology.* Oxford and Boston, MA: Blackwell Scientific.

Braddock, S. & O'Mahony, M. (1998). *Techno Textiles : Revolutionary Fabrics for Fashion and Design.* London: Thames & Hudson, or inventory CIM Press.

Brooks Brothers (2011). About made-to-measure [online], retrieved 7 February 2011 from http://www.brooksbrothers.com/specialorder/mtm_dress_shirts.tem

Bruce, M. & Cooke, B. (1998). Textile design – the right approach. *Textile Horizons*, July/August, pp. 10–13.

Bunce, G. (2005). Digital print: To repeat or not repeat. Designer Meets Technology Europe Conference, hosted by KrIDT, Creative Institute for Design and

Technology and The Center for Excellence of Digital Ink Jet Printing of Textiles at Philadelphia University, 27–29 September 2005, Copenhagen.

Cahill, V. & Ujiie, H. (2004). Digital textile printing 2004. Retrieved 7 February 2011 from http://vcesolutions.com/www/library/Digital%20Textile%20Printing%20 2004.pdf

Campbell, J. & Parsons, J. (2005). Taking advantage of the design potential of digital printing technology for apparel. *Journal of Textile and Apparel, Technology and Management*, 4(3), 1–10.

Chenemilla, P. (2001). Integrating digitally printed designs for mass customization. Masters thesis, North Carolina State University, Raleigh, NC.

Clapp, T., Gunner, M., Dorrity, J. & Olson, L. (1994). The online inspection of sewn seams, National Textile Center Project S94-4 Annual Report, retrieved 7 February, 2011 from http://www.ntcresearch.org/pdf-rpts/AnRp94/A94S9404.pdf

Colchester, C. (1996). *The New Textiles*. London: Thames & Hudson.

Gavenas, M. (2007). A-POC making. *Daily News Record*, 21 May, p. 34.

Gordon, A. S. (1997). Digital printing of textiles in 1997. Unpublished doctoral dissertation, North Carolina State University, Raleigh, NC.

Hunter, B. (2005). Novel seamless sweater technique. *Knitting International*, 112(1324), 26–27.

Jana, R. (2006). Case study: Issey Miyake, the dream weaver. *Bloomberg Businessweek*, Special Report, 25 April 2006. Retrieved 7 February 2011 from http://www. businessweek.com/innovate/content/apr2006/id20060425_361290.htm

Joyce, C. (1993). *Textile Design: The Complete Guide to Printed Textiles for Apparel and Home Furnishing*. New York: Watson-Guptill Publications.

Karma Kraft (2001). Retrieved 7 February 2011 from http://www.karmakraft.com/ AboutUs.aspx

Kunz, G. (2005). *Merchandising: Theory, Principles, and Practice* (2nd edn). New York: Fairchild Publications.

Lam, J. & Zhou, J. (2009). Digital knitting, a new perspective for knitting industry. Presented at AUTEX 2009 World Textile Conference, 26–28 May 2009, Izmir, Turkey.

Lamar, T., Powell, N. & Parrillo-Chapman, L. (2009). *NTC project S09-NS02 annual report*. Retrieved 7 February 2011 from http://www.ntcresearch.org/pdf-rpts/ AnRp09/S09-NS02-A9.pdf

Lamb, J. M. & Kallal, M. J. (1992). A conceptual framework for apparel design. *Clothing and Textiles Research Journal*, 10(2), 42–47.

Li, J. (1997). Creativity in horizontal and vertical domains. *Creativity Research Journal*, 10(2–3), 107–132.

May-Plumlee, T. & Bae, J. (2005). Behavior of prepared-for-print fabrics in digital printing. *Journal of Textile and Apparel, Technology and Management*, 4(3).

May-Plumlee, T. & Little, T. J. (1998). No-interval coherently phased product development model for apparel. *International Journal of Clothing Science and Technology*, 10(5), 342–364.

McNamara, A. & Snelling, P. (1995). *Design and Practice for Printed Textiles*. New York: Oxford University Press.

Miyaki, I. & Fujiwara, D. (2001). *A-POC Making: Issey Miyake & Dai Fujiwara* (Kries, M. & von Vegesack, A., eds), Weil, Germany: Vitra Design Museum; and Tokyo: Miyake Design Studio. Exhibition held at Vitra Design Museum, Berlin, 1 June to 1 July 2001.

Moxey, J. (1998). A creative methodology for idea generation in printed textile design. *Journal of the Textile Institute*, 89(3), 35–43.

Parrillo-Chapman, L. (2004). Engineering whole garment designs. Paper presented at the 6th Annual IFFTI Conference: Best Practices in Fashion Education, New Delhi, India.

Parrillo-Chapman, L. (2008). Textile design engineering within the product shape. Unpublished dissertation, North Carolina State University, Raleigh, NC.

Peterson, J., Larsson, J. & Pal, R. (2009). Co-design tool for customized knitwear. Presented at AUTEX 2009 World Textile Conference, 26–28 May 2009, Izmir, Turkey.

Powell, N. B. (2003). Mass customization in transportation textiles through shaped three dimensional knitting. In *Proceedings: International Textile Design and Engineering Conference*, Edinburgh: Heriot-Watt University, September 2003.

Powell, N. & Black, S. (2005). Evaluation of complete garment and seamless weft knitting technologies from a fashion and design perspective. Presented 4 November 2005 at the International Textile and Apparel Association Annual Conference, Alexandria, VA.

Rimslow Pty Ltd (2004). Steam-X, retrieved 7 February 2011 from http://www.rimslow.com/steamx.html

Santoni (2011). Photo and video, retrieved 7 February 2011 from http://www.santoni.com/en-videofoto.asp

Scanlon, J. (2004). Seamless. *Wired Magazine*, Issue 12.04, April, retrieved 7 February 2011 from http://www.wired.com/wired/archive/12.04/miyake.html?pg=1&topic=&topic_set=

Scelfo, J. (2009). You design it, they print it. *New York Times*, New York edition, 8 January, D3.

Sexton, D. (undated). The dawn of direct to T-shirt printing, retrieved 7 February 2011 from http://www.imachinegroup.com/t-jet_the_dawn_of_direct_to_t-shirt_printing.php

Shima Seiki (2011). Knit sample collection. Retrieved 7 February 2011 from http://www.shimaseiki.com/support/samples/

Silveira, G., Borenstein, D. & Fogliatto, F. (2001). Mass customization: literature review and research directions. *International Journal of Production Economics*, 72(1), 1–13.

Spencer, D. J. (2001). *Knitting technology: A Comprehensive Handbook and Practical Guide* (3rd edn). Cambridge, UK: Woodhead Publishing; Lancaster, PA: Technomic Publishing.

Spoonflower, http://www.spoonflower.com/welcome

Stoll (2011). Online trend collection. Retrieved 7 February 2011 from http://195.145.71.72/epaper/Stoll-TC-SS-12/index.html

Studd, R. (2002). The textile design process. *The Design Journal*, 5(1), 35–49.

[TC]² Inkdrop Printing Service. Retrieved 7 February 2011 from http://www.inkdropprinting.com/

TechExchange.com (2011). The Library, retrieved 7 February 2011 from http://www.techexchange.com/thelibrary.html

Teunissen, A., Kruize, M. & Tillmanns, M. (2002). *Developments in the Textile Printing Industry 2002.* Boxmeer, Netherlands: Stork Textile Printing Group.

Wickett, J. L., Gaskill, L. R. & Damhorst, M. L. (1999). Apparel retail product development: Model testing and expansion. *Clothing and Textiles Research Journal.* 17(1), 21–35.

Wilson, J. (2001). *Handbook of Textile Design: Principles, Processes and Practice.* Cambridge, UK: Woodhead Publishing in association with The Textile Institute.

Yang, S. & Love, T. (2009). Designing shape-shifting of knitwear by stitch shaping combinatorics: A simple mathematical approach to developing knitwear silhouettes efficaciously. IASDR Conference 2009: Design/Rigor & Relevance, Seoul: International Association of Societies of Design Research and the Korean Society for Design.

Yates, M. (1986). *Textiles: A Handbook for Designers.* New York: Prentice-Hall.

Yuikawa, K. (2001). Commercial production by VISCOTECS, digital printing onto fabric. *DPP 2001; International Conference on Digital Production Printing and Industrial Application*, 13 May 2001, Antwerp, Belgium, pp. 267–269.

Index

A Piece of Cloth, 339
A-POC *see* A Piece of Cloth
ABAQUS, 149, 155, 159
ABAQUS/Explicit, 159
active contour technique, 30, 32–4
active grid model, 27–9, 35–42
adaptive orientated orthogonal projective
    decomposition, 12
adaptive signal decomposition, 13
Adobe Illustrator, 248, 251
Adobe Photoshop, 248, 251
affine transformations, 84–6
    type I, 84–6
    type II, 84–6
ALE *see* Arbitrary Lagrangian Eulerian
ancient Greeks, 223–4
    ideal female figure proportions, 224
angle base flattening method, 76
angular energy spectrum, 25
animation, 308–10
    3D CAD system, 309–10
    skeletal-driven deformation, 308
    without skeletal definition, 308
anisotropy, 73, 74–5
ANN *see* artificial neural network
ANSYS, 149, 155
ANSYS-AUTODYN, 158
ANSYS/AUTODYN, 160
apparel
    design development using integrated
        digital processes, 326–46
        CAD and visualisation technologies,
        344–5
        conventional design, development and
            production processes, 327–32

future, 345–6
integrated processes in practice, 343–4
simultaneous design of textile and
    garment, 332–43
digital printing technology, 259–79
    colour technology and colour
        management, 266–72
    future trends, 277–9
    global developments, 263–6
    review, 260–3
    stages of computing, 272–7
apparel design, 298–9
three dimensional technologies,
    296–320
    animation, 308–10
    body surface development, 304–7
    generic vs individualised body models,
    310–13
    human body modelling, 299–304
    human body modelling applications,
    297–9
    virtual try-on technologies, 313–19
Apparel Design Framework, 331
Arbitrary Lagrangian Eulerian, 158
artificial neural network, 17, 19
Asian women, 225–8
    Chinese women presenting medals at the
        2008 Olympic Games, 227
ASTM D2255-02, 16
Athena Design System, 248
AUTODYN *see* ANSYS-AUTODYN
Autonomous Mental Development, 61

back-propagation, 54
backward difference formula, 181

ballistic impact modelling, 146–67
  computational aspects, 150–60
    commercially available finite element
      package, 159–60
    contact algorithm, 159
    Eulerian hydrocodes evaluation, 157
    Lagrangian hydrocodes evaluation, 156
    problem description, 156–8
    rezoning (re-meshing), 158
    schematic flow chart of hydrocode
      modelling process, 151
    spatial discretisation, 152–5
    time-integrating methods, 155
  finite element model of impactor–target
    interaction, 150
  modelled projectiles with different nose
    angles, 149
  numerical modelling of single and
    multiple layer fabric, 160–6
    finite element model simulation
      displacement profile, 163
    schematic response of projectile impact
      into a single fibre/yarn materials,
      163
    sequence of fabric deformation and
      damage, 162
    simulation of the five-layered fabric
      target perforated by the projectile,
      165
    weave architecture, RVE and quarter
      of RVE, 164
    'wedging-through' action of projectile
      into fabric during impact, 161
  Twaron plain woven fabric, 149
Bayesian networks, 19
bending springs, 182
Bezier curves, 71
bidirectional reflectance distribution
  function, 184–6
block, 216
body perception
  data gathering, 231–3
    hypotheses, 232–3
    questionnaire, 233
  Generation Y, 3D body scanning and
    virtual visualisation, 219–40
  ideal body image, 223–8
    Ancient Greeks, 223–4
    Asian women, 225–8
    proportion and ratio, 225
    Western culture, 225

  literature review, 221–31
    body cathexis, 223
    body image, self-concept and self-
      esteem, 221, 223
    body shape, 221
    3D body scanner, 229–30
    fit, 228
    sizing, 228–9
    smart card based on RFID, 230–1
    WMGPC manual body measurements,
      229
  research results, 233–8
    body cathexis, 234–5
    body shape, 235–6
    buying behaviour, 237–8
    fit satisfaction, 237
    participants' buying behaviour and
      preferences, 238
    sampling and demographics, 233–4
body shape, 221, 235–6
  body types, 222
  bust, 235–6
  bust, waist and hip measurement of
    participants, 235
  figure types, 222
  hip, 236
  ideal vs participants perception of body
    shape, 235
  mean body measurement, 236
  waist, 236
Browzwear International, 315

C-me, 315
CAD see computer-aided design
CAD software, 273–6
CAM see computer aided manufacturing
Canny's edge detection algorithm, 8
cathexis, 223, 234–5
CCD camera, 10
CEL see Coupled Eulerian Lagrangian
charged-coupled device, 24–5, 248–9
CIE colour space, 267–9
CIM see computer integrated manufacturing
CITDA see Computer Integrated Textile
  Design Association
cloth–object collision algorithm, 183
COLARIS, 264
  production rates, 266
  vs screen printing with 180 cm fabric
    width, 266
colour effects map, 123

colour pattern, 40–2
complete garment knitting, 336–7
computational attention-driven modelling,
    17
computer-aided design, 207, 245, 250, 259
    and visualisation technologies, 344–5
    case study, 289–93
        classroom management, 291–2
        curriculum design, 290–1
        teaching and learning effectiveness
            assessment, 292–3
        teaching and learning facilities, 290
    teaching approaches, 283–7
        comparison of four pedagogic
            approaches, 288
        constructivism, 283–5
        experiential learning, 286–7
        global-scope learning, 287
        problem-based learning, 285
    teaching in fashion and textile students,
        283–94
        areas for improvement in teaching,
            293–4
        challenges presented by teaching
            approaches, 287–9
computer aided manufacturing, 245
computer integrated manufacturing, 245
Computer Integrated Textile Design
    Association, 247, 252
computer technology
    benefits and limitations, 254–6
    future trends, 256–7
    main computer technologies in textile
        design, 248–54
        computer as a design editing tool, 252–3
        computer as a designing tool, 250–2
        computer as a presentation tool, 253–4
        computer as an information gathering
            tool, 249–50
    role in textile design, 247–8
    textile designer's perspective, 245–57
computer vision-based fabric defect analysis
    and measurement, 45–61
        fabric defect classification, 52–9
        fabric defect detection methods, 49–52
        fabric inspection for quality assurance,
            46–9
        fabric properties and colour
            measurement using image analysis,
            59–60
        future trends, 60–1

fabric defect classification, 52–9
    classification accuracy from two
        different cases, 58
    defect classification methods summary,
        58
    neural network-based fabric defect
        classification, 53–6
    neural network classification accuracy,
        five leather defects classifier, 55
    neural network classification accuracy,
        five stitching fabric defects classifier,
        56
    neural network structure with one
        hidden layer, 53
    segmentation flow diagram using neural
        network, 55
    support vector machine (SVM)-based
        fabric defect classification, 56–9
    SVM-based fabric defect classification,
        56–9
    two-class SVM decision boundary
        estimation, 57
fabric defect detection methods, 49–52
    best performing approach, plain and
        twill fabric inspection, 50
    patterned fabrics, 51–2
fabric inspection for quality assurance,
    46–9
    fabric defects, 46–8
    fabric inspection automation, 48–9
    patterned fabric defect image sample,
        48
    plain weave fabric defects image
        sample, 47
    quality assurance automation, 49
computer vision-based fabric measurement
    and defect analysis, 45–61
        fabric defect classification, 52–9
        fabric defect detection methods, 49–52
        fabric inspection for quality assurance,
            46–9
        fabric properties and colour
            measurement using image analysis,
            59–60
        future trends, 60–1
computerised grading method, 17
conceptual proposal, 206
constructivism, 283–5
contact algorithms, 159
content-based image retrieval, 19
continuous collision response, 192

continuous filament yarns, 97–8
continuous ink jet, 261
continuous multifilament yarn, 94
Coulomb friction laws, 180
Coupled Eulerian Lagrangian, 158
Courant condition, 188
cross-point classification algorithm, 31
Cyclops, 49

D'Alembert's principle, 176
Delaunay triangulation, 71
designers, 203
diagram of weave pattern, 135
digital-based technology
    AGM construction and weave pattern
        extraction methodology, 35–42
    AGM self-adjustment, 37–8
    colour configuration, 40–2
    cross-point classification, 39
    cross-point classification and result of
        correction, 40
    fabric grid model after adjustment, 38
    final fabric grid model, 41
    initial grid model, 37
    neighbour cross points possible types,
        39
    real fabric and stimulated images, 42
    template-based (warp) yarn detection
        demonstration, 36
    warp yarn detection result, 36
    yarn detection and AGM initialisation,
        35–7
    yarn interlacing classification, 38–40
background and literature review, 24–7
    China and Hong Kong, 27
    early Japanese studies, 24–5
    European and US developments, 25–6
fabric structure analysis, 23–42
    AGM construction and weave pattern
        extraction methodology, 35–42
    background and literature review,
        24–7
    dual-side images, fabric after alignment
        and merge image, 28
    dual-side imaging system, 28
    plain woven fabric 3D structure, 29
    systematic framework, dual-side
        scanning and AGM system, 29
    three basic weave patterns point maps,
        30
    weave pattern analysis, 29–34

weave pattern recognition digital
    system, 27–9
    yarn presentation demonstration, series
        linked grid elements usage, 32
weave pattern analysis, 29–34
    active contour and AGM self-
        adjustment, 32–4
digital printing technology, 345–6
    challenge of colour management, 269–72
        colour management solutions, 270–2
        colour specification, 269–70
        gamut mapping, 269
    colour technology and management,
        266–72
        CIE colour space, 268
        light emitters and light absorbers,
            266–9
    global developments, 263–6
        COLARIS production rates, 266
        COLARIS vs screen printing, 266
    review, 260–3
        components in inkjet image production,
            263
        effect of increasing flow rate, 261
    stages of computing, 272–7
        CAD software for creating designs,
            273–4
        Kaledo approach, 274
        prepare-for-print CAD software, 274–6
        RIP software for precision printing,
            276–7
    textiles and apparels, 259–79
        future trends, 277–9
digital signal processing, 26
digital technology
    and modelling for fabric structures,
        122–44
        experimental pattern analysis, 131–2
        fabric geometry structure models, 124–5
        fabric weave pattern, 126
        irregular patterns description and
            classification, 128–31
        methodology, 132–4
        regular patterns description and
            classification, 126–8
        results, 135–44
        weave pattern and fabric geometry
            surface appearance, 131
        woven fabric structure, 123–4
        yarn structure and appearance analysis,
            3–20

attention-driven modelling for abnormality detection, 18
attention-driven modelling for foreign matter detection, 18
digitised yarn feature analysis diagram, 4
future trends, 19–20
Gaussian functions calculated by AOP algorithms, 13
Gaussian functions reconstructed profile, 14
line sense of pixels, 15
measurement of yarn evenness, 4–5
optical system with a high-pass spatial filter experimental setup, 9
optical system with a low-pass spatial filter experimental setup, 5
original yarn image, 7
tracer fibre image captured by CCD camera, 11
yarn appearance grading, 16–18
yarn blend analysis, 14–16
yarn core detection, 8
yarn density profile, 12
yarn hairiness analysis, 5–9
yarn hairiness edge detection, 8
yarn hairiness estimation, 6
yarn snarl recognition, 11–14
yarn snarl sample, binary image, 12
yarn snarl sample, original image, 12
yarn snarls 2D projection schematic view, 12
yarn twist estimation, 10
yarn twist measurement, 9–11
drop on demand, 261
dual-side scanning technology, 27–9, 30
DYNA 3D, 159

e-fit Simulator, 317
edge springs, 182
EigenSkin, 83
eight-connectivity, 128
elastic repulsion, 183
elasticity theory, 176, 193
engineered design, 333
equation of state, 152
Euler equations, 152
Euler integration, 89
Euler integration method, 179
Eulerian coordinates, 156–7
expectation maximum (EM) criteria, 59

experiential learning, 286–7
extended free-form deformation, 72

fabric, 94
fabric defect classification, 52–9
classification accuracy from two different cases, 58
defect classification methods summary, 58
neural network-based fabric defect classification, 53–6
neural network classification accuracy, five leather defects classifier, 55
neural network classification accuracy, five stitching fabric defects classifier, 56
neural network structure with one hidden layer, 53
segmentation flow diagram using neural network, 55
SVM-based fabric defect classification, 56–9
two-class SVM decision boundary estimation, 57
fabric defect detection methods, 49–52
patterned fabrics, 51–2
fabric fractal scanning, 49
fabric geometry structure models, 124–5
fabric jam, 109–10
fabric structure analysis
digital-based technology, 23–42
AGM construction and weave pattern extraction methodology, 35–42
background and literature review, 24–7
weave pattern analysis, 29–34
weave pattern recognition digital system, 27–9
fabric structures
computation results
DOWP and human perception for PWEP, 138
DOWP and human perception for SWEP and JFEP, 139
DOWP and human perception for TWEP, 138
ISWP, RSWP with DOWP for PWEEP, 139
ISWP and RSWP with DOMP for SWEP and JFEP, 140
digital technology and modelling, 122–44
connectivity calculation, 130
fabric geometry structure models, 124–5
fabric structure and weave pattern, 126

fabric weave pattern, 126
irregular patterns description and
    classification, 128–31
methodology, 132–4
regular patterns description and
    classification, 126–8
taxonomy of fabric weave pattern, 126
transitions calculations, 129
weave pattern and fabric geometry
    surface appearance, 131
woven fabric structure, 123–4
experimental pattern analysis, 131–2
    fabric composites and manufacturing
        parameters, 131
    fabric ideal weave pattern schematic
        diagram, 132
results, 135–44
    calculation results for DOWP, ISWP
        and RSWP, 136, 137
    ISWP image comparison of nos 11 and
        12, 141
    ISWP image comparison of nos 4 and 5
        and nos 2 and 3, 141
    pattern complexity order, 142, 143
    real scanned fabric image, 143
fabric weave pattern, 126
Fabriscan, 49
false-twist texturing, 94
Fast Fourier Transform, 4, 26
Fast Walsh-Hadamard Transform, 4
feature points, 300
feedback, 206–7
fibre migration, 96–7
fibre packing factor, 96
fibres, 93, 94
fibrous yarn materials
    modelling and simulation, 93–120
        future development of textile
            modelling, 119–20
        weave modelling, 103–5
        woven fabrics, 105–19
        yarns, 94–102
finite difference methods, 175
finite difference scheme, 153
finite element methods, 175
finite element scheme, 153–4
four-connectivity, 128
Fourier analysis, 35
Fourier spectrum, 60
Fourier Transform, 25
Fourier Transform theory, 25

frame grabber, 24–5
fuzzy wavelets, 49

Gabor filters, 59
Gabor wavelet network, 50
garments
    advanced modelling techniques, 191–2
    computer graphics techniques for
        structure and appearance, 178–83
    constitutive model of cloth used in
        work of Selle et al., 182
    mesh model and three spring types in
        Provot's cloth model, 179
    considerations for real-time applications,
        188, 190–1
    future developments in simulating
        garment materials, 192–3
    model development, 174–8
        geometrical techniques, 174–5
        hybrid methods, 177–8
        physically based methods, 175–7
    modelling and simulation techniques,
        173–95
    rendering of garment appearance and
        model demonstration, 183–8, 189,
        190
        garment with long sleeves, 189
        model demonstration, 187–8, 189,
            190
        normal vectors used for computing
            three shade values for a thread,
            185
        rendering of garment appearance,
            183–7
        square cloth and a ball, 187
        square cloth and a spinning ball, 190
        square cloth and an object with sharp
            features, 187
        square cloth interacting with a
            deformable object, 188
        three layers of garments, 189
    sources of further information and advice,
        193–5
        computer animations and fashion
            shows, 194
        E-commerce, 194
        game engines, 194
        garment CAD/CAM, 194
        production of cloth simulation, 195
Gaussian function, 12, 57
Gaussian mixture model, 59

Generation Y
    3D body scanning for body perception
        and virtual visualisation, 219–40
        data gathering, 231–3
        future studies, 240
        limitations, 239–40
        literature review, 221–31
        recommendations, 239
        research results, 233–8
generic body models, 310–11
Gerber, 290
Gerber AccuMark PDS system, 317
global-scope learning, 287
golden image subtraction, 51

hash functions, 51–2
Hausdorff distance, 35
Havok Cloth, 195
HCI see human computer interaction
high-performance parallel computation
        techniques, 193
human body modelling, 299–304
human computer interaction
    definition, 204–5
        computer, 205
        interaction, 205
        user, 204
    future trends, 212–16
        challenges in apparel industry,
            212
    goals, 205–6
        framework development, 206
    methods for improving interaction,
            207–11
        conventional CAD garment design
            process, 207
        importance in textile manufacturing,
            207–8
    new features, 212–16
        garment database management, 216
        garment shape database, 216
        gesture-based garment shape editing,
            215
        gesture-based interaction, 213–15
        shape retrieval and dynamic
            association, 215–16
        shape retrieval and dynamic association
            of garment shapes, 215
        sketch interaction, 213
        sketching interface for garment design,
            214

    principles, 204–7
        basic, 206–7
    textile apparel industry, 201–17
    typical garment design methods, 208–11
        commercial 2D CAD system, 210
        2D/3D garment design, 211
        2D garment design, 208–10
        3D garment design, 211
        typical garment design methods, 209
hydrocode modelling, 150–2
    schematic flow chart, 151
hydrocodes, 149, 166

ideal simulation weave pattern, 135
Illustrator, 290
imaging, 299–300
    manual methods, 299–300
    photography manipulation, 300
    selecting best technique, 304
individualised body models, 311–13
    directly morphed and manually created
        avatars, 313
    examples, 312–13
infrared scanning, 302
integrated digital processes
    apparal design and development, 326–46
        CAD and visualisation technologies,
            344–5
        future, 345–6
        integrated processes in practice,
            343–4
    conventional design, development and
        production processes, 327–32
        approach and stages for textiles design,
            328–30
        textile design process, 329
    simultaneous design of textile and
        garment, 332–43
        challenges, 340–3
        cutting layout vs digitally engineered
            garment, 334
        engineered digitally printed marker,
            335
        engineered garment, 338
        garment cutting marker with
            traditionally engineered print, 335
        interface of textile design and apparel
            design domains, 332
        vs conventional processes, 337–40
inter-fibre gripping, 100
IQ-TEX system, 49

Jacquard figured effect pattern, 135
joint photographic expert group, 252
JPEG *see* joint photographic expert group

Kemp's racetrack fabric model, 107, 108
knitting, 336–7
Kuralon, 148

Lafortune model, 185–6
Lagrange's equation, 176
Lagrangian coordinates, 156
laser scanning, 302
least squares conformal maps (LSCM)
     method, 75
Lectra Systems, Inc., 316
linear interpolation, 71
localised binary patterns, 50
LS-DYNA 3D, 164
LS-PREPOST, 159
LSDYNA, 149
LSDYNA 3D, 159
LUSAS, 155, 160

macro-geometry, 123
Manual of Standard Fabric Defects in the
     Textile Industry, 48
MARC, 155
Markov random field models, 49
mesh parameterisation, 73
metamedium, 247
micro-geometry, 123
Microsoft Visual C++ 6.0, 27
milli-geometry, 123
minimum description length (MDL), 59
MIRACloth, 310
MIRACloth Software, 195
Modaris 3D Fit, 316
monofilament yarns, 94
MS-XR48, 264
MSC DYTRAN, 160
multiscale wavelet decomposition, 49
My Virtual Model Dressing Room and My
     Fit, 195

Nassenger VII, 264
NASTRAN, 155
neural network (NN), 53–6
No-Interval Coherently Phased Product
     Development Model for Apparel, 331
non-linearised Green–Lagrange tensor, 183
non-zero Gaussian curvature, 73

numerical modelling
     single and multiple layer fabric, 160–6
NVIDIA PhysX, 195

online garment-shopping system, 80–9
OpenGL mip-mapping technique, 186
optical grating method, 25
OptiTex, 307, 315–16, 319
OptiTex Runway 3D, 316

P ABAQUS/Explicit, 159
PAMCRASH, 160
panel scanner, 25
panels, 208
partial differential equations, 148
Particle-Swarm Optimisation - back
     propagation, 55–6
Peirce's model, 106–7
PET/rayon composite yarn, 15–16
photodiode, 5
photogrammetric methods, 302–3
photomultiplier tubes, 249
Photoshop, 290
PhysX SDK, 195
pin-jointed orthogonal approach, 160
plain weave effect pattern, 135
ply yarns, 95
point cloud, 301
point map diagram, 123
point map fabric model, 30
polygons, 305
Polynomial function, 57
Pose Space Deformation, 83
power spectrum, 133
pressure-sensitive surfacing, 307
PrimaVision, 290
principal component analysis, 55, 81
printed textiles, 345
problem-based learning, 285
prototyping, 206
PSO-BP *see* Particle-Swarm Optimisation -
     back propagation

quality assurance
     fabric inspection, 46–9
     automation, 48–9
     fabric defects, 46–8
quasi-Newton optimisation, 81

Raster Image Processor (RIP) software,
     276–7

ray tracing techniques, 183–4
real samples weave pattern, 135
repetitive unit cell, 164
representative volume element, 164
repulsion forces, 192
requirement analysis, 206
Revised Apparel Retail Product
    Development Model, 331
Revised Retail Product Development
    Model, 331
rezoning, 158
rigid impact zones, 192
row-by-row scanning algorithm, 7

satin weave effect pattern, 135
Shanahan and Hearle's lenticular fabric
    model, 108, 109
Shape by Example, 83
shape interpolation, 80
shape morphing see shape interpolation
Shirley Hairiness Tester, 5
sketch-based garment design, 70–3
    3D sketching and editing, 72
    generating 3D garment patch from 3D
        sketch, 72
    method 1: 2D sketches associated with 3D
        garment prototypes, 70–1
    method 2: fit and tight-fit design,
        parameterised 3D sketching, 71–3
    resulting garment surface from 3D
        sketching, 73
sketch-based interface and modelling,
    70
slippage factor, 98–9
sloper, 216
smooth skinning, 83
smoothed particle hydrodynamics,
    154–5
soft proofing, 253
spatial discretisation, 152–5
    finite difference scheme, 153
    finite element scheme, 153–4
    smoothed particle hydrodynamics,
        154–5
spectrum analysis techniques, 4
splitting cluster analysis, 17
spring-based repulsion force, 180
staple fibre yarns, 98–101
staple yarn, 94–5
'stiffness matrix,' 154
support vector machine, 19, 53

surface flattening
    virtual garments, 73–80
        conformal parameterisation with darts
            insertion, 75
        mesh parameterisation, 74
        pants model based on appropriate
            texture atlas, 80
        physically-based method, 79
        poor texture atlas, 79
        releasing bending configuration, 77
        spreading force, 78
        unfolding element with dihedral angle,
            78
surfacing 3D scan data, 306
surfacing techniques, 306–7
    pressure-sensitive surfacing, 307
    surfacing 3D scan data, 306
    texture mapping, 307

tag image file format, 252
tailor, 203
tan-hyperbolic function, 57
Taxonomy of Apparel Merchandising
    Systems, 331
TexGen, 113
textile composites, 93
textile design
    computer technology, 247–8
    main technologies in textile design,
        248–54
        computer as a design editing tool,
            252–3
        computer as a designing tool, 250–2
        computer as a presentation tool,
            253–4
        computer as an information gathering
            tool, 249–50
    three dimensional technologies, 296–320
        animation, 308–10
        body surface development, 304–7
        generic vs individualised body models,
            310–13
        human body modelling, 299–304
        human body modelling applications,
            297–9
        virtual try-on technologies, 313–19
textile designer
    computer technology, 245–57
        computers benefits and limitations,
            254–6
        future trends, 256

main computer technologies in textile
  design, 248–54
  role, 247–8
textiles
  digital printing technology, 259–79
    colour technology and colour
      management, 266–72
    future trends, 277–9
    global developments, 263–6
    review, 260–3
    stages of computing, 272–7
texture mapping, 307
TexWinCAD, 208
three-dimensional body scanning, 301–4
  3D point cloud of body surface, 301
  Generation Y body perception and virtual
    visualisation, 219–240
    data gathering, 231–3
    future studies, 240
    limitations, 239–40
    literature review, 221–31
    recommendations, 239
    research results, 233–8
  image processing and measurement
    extraction, 303–4
three-dimensional CAD system, 309–10
three-dimensional digitising, 299–300
three-dimensional garment design
  sketch-based garment design, 70–3
    3D sketching and editing, 72
    generating 3D garment patch from 3D
      sketch, 72
    method 1: 2D sketches associated with
      3D garment prototypes, 70–1
    method 2: fit and tight-fit design,
      parameterised 3D sketching, 71–3
    resulting garment surface from 3D
      sketching, 73
  techniques, 69–91
    automatic body measurement software
      system, 84
    body shape representing online
      shopper, 81
    bounding area attach to bone, 87
    challenges and future trends, 90
    conformal parameterisation with darts
      insertion, 75
    3D garment demonstration, online
      shopping store, 91
    drape style simulation, redressed
      garment, 89

dressing same items by various avatars
  of online shoppers, 90
fit/ease visualisation, 87
garment redressing pipeline, pose
  duplication, 85
information sources, 91
joint calculation anthropometries
  position, 84
mesh parameterisation illustration, 74
online garment-shopping system,
  problems and solutions, 80–9
penetration recovery, 88
poor texture atlas, scanned pants
  model, 79
rendering pants model, appropriate
  texture atlas, 80
sketch-based garment design, 70–3
spreading force, 78
surface flattening physically-based
  method, 79
surface flattening via releasing bending
  configuration, 77
type I affine transformation, 85
type II transformation, 86
unfolding element with dihedral angle,
  78
virtual garments surface flattening,
  73–80
three-dimensional technologies
  advanced surfacing techniques, 306–7
    pressure-sensitive surfacing, 307
    surfacing 3D scan data, 306
    texture mapping, 307
  animation, 308–10
    3D CAD system, 309–10
    skeletal-driven deformation, 308
    without skeletal definition, 308
  apparel and textile design, 296–320
  body scanning, 301–4
    3D point cloud of body surface, 301
    image processing and measurement
      extraction, 303–4
  body surface development, 304–7
    explicit methods, 305
    implicit methods, 305–6
  generic vs individualised body models,
    310–13
    generic body models, 310–11
  human body modelling, 299–304
    best imaging technique selection,
      304

human body modelling applications, 297–9
  apparel, 298–9
  imaging, 299–300
    manual methods, 299–300
    photography manipulation, 300
  individualised body models, 311–13
    directly morphed and manually created avatars, 313
    examples, 312–13
  virtual try-on example, 317–19
    draping simulation for tank top garment, 319
    fabric properties, 317
    fabric properties input, 318
    garment design and stitching, 317–18
    virtual stitching for a tank top garment, 318
  virtual try-on technologies, 313–19
    Browzwear, 315
    Lectra Systems, Inc., 316
    OptiTex, 315–16
    Tukatech, 317
three-dimensional woven fabrics, 103
TIFF *see* tag image file format
torsional buckling effect, 11
transverse wave, 163
triangulation, 302
Tukatech, 317
Twaron, 148
twill weave effect pattern, 135
twist factor, 96
three-dimensional Fourier Transform, 10

UniverWeave, 113
Uster tester, 4
Uster tester 3, 5
Uster tester 4, 5

V-Stitcher, 315
variance-length curve, 15
virtual try-on technologies, 313–19
  Browzwear, 315
  example, 317–19
    draping simulation for tank top garment, 319
    fabric properties, 317
    fabric properties input, 318
    garment design and stitching, 317–18
    virtual stitching for a tank top garment, 318

Lectra Systems, Inc., 316
OptiTex, 315–16
Tukatech, 317
virtual visualisation
  3D body scanning, and Generation Y body perception, 219–40
  data gathering, 231–3
  future studies, 240
  limitations, 239–40
  literature review, 221–31
  recommendations, 239
  research results, 233–8
visible microspectrophotometry, 60
Visual Studio Industry Partner, 25
visualisation technologies, 344–5

warp, 103
wavelet packet frame, 59
wavelet pre-processed golden image subtraction, 51
wavelet transform (WT), 51
Weave Engineer, 113
weave modelling, 103–5
  commonly used weaves for making fabrics, 104
  3D fabric weaves mathematical description, 105
  regular 2D fabric weaves mathematical description, 103–5
weave pattern
  digital system, 27–9
  theoretical background, 29–34
weaves, 103
WEB 2.0, 287
weft, 103
Window, Icon, Menu, Pointer (WIMP), 70
Windows XP operating system, 27
WiseTex, 113
wool/silk blended yarn, 16
woven fabric structure, 123–4
woven fabrics, 105–19
  finite element modelling, 113–19
    agreement between model and experiment, 116
    ballistic impact on fabrics, 115–19
    effect of fabric orientation on energy absorption, 118
    effects of varying yarn cross-sectional shape on fluid pressure, 115
    energy absorption vs impact energy, 117
    fabric in relation to planes, 114

filtration through fabrics, 113–15
modelled project velocity change due
    to impact, 117
stress distribution of 4-layer fabric
    assembly, 119
stress distribution on impacted fabric,
    116
geometric modelling, 105–13
    3D fabrics geometrical modelling, 113
Peirce's plain fabric model and its
    derivative models, 106–9
    derivative models, 107–9
    Kemp's racetrack fabric model, 108
    Peirce's model, 106–7
    Peirce's model for plain woven fabric,
        106
    Shanahan and Hearle's lenticular fabric
        model, 109
relationship between fabric parameters,
    109–13
    fabric jam, 109–10
    irregular cross-section expressed by
        polar coordinates, 112
    jammed fabric geometry, 110
    modular heights and yarn spacings,
        110–11
    yarn cross-section, 111–12
    yarn dimension and yarn linear density,
        112–13

Xenjet, 278
Xennia Technology, 278

yarn appearance analysis
    and yarn structure, digital technology,
        3–20
    future trends, 19–20
    yarn appearance grading, 16–18
    yarn blend analysis, 14–16
    yarn evenness measurement, 4–5

yarn hairiness analysis, 5–9
yarn snarl recognition, 11–14
yarn twist measurement, 9–11
yarn blend, 14–16
yarn core, 6
yarn evenness, 4–5
yarn hairiness, 5–9
yarn interlacing
    classification, 38–40
yarn snarl, 11–14
yarn structure
    and appearance analysis, digital
        technology, 3–20
    future trends, 19–20
    yarn appearance grading, 16–18
    yarn blend analysis, 14–16
    yarn evenness measurement, 4–5
    yarn hairiness analysis, 5–9
    yarn snarl recognition, 11–14
    yarn twist measurement, 9–11
yarn twist, 9–11
yarns, 93–4
    description of dimension, 101–2
        flattened yarn cross-sections, 101
    geometry, 95–7
        fibre migration, 96–7
        fibre packing factor, 96
        idealised helical yarn geometry, 95
        twist and twist factor, 95–6
        yarn migration, 97
    mechanics, 97–101
        continuous filament yarns, 97–8
        slip at fibre ends, 99
        staple fibre yarns, 98–101
    modelling and simulation, 94–102
    types, 94–5

Zweigle G565, 5
Zweigle G566, 5
Zweigle G580, 4

Printed in the United States
By Bookmasters